LONGMAN
CO-ORDINATED
GEOGRAPHY

LONGMAN
CO-ORDINATED
GEOGRAPHY

Series editor – Simon Ross

Longman

Longman Group UK Limited
*Longman House, Burnt Mill, Harlow, Essex, CM20 2JE, England
and Associated Companies throughout the World.*

First published 1990
ISBN 0 582 062993

*Set in 10/12 pt Palatino Linotron
Produced by Longman Group (F.E.) Ltd.
Printed in Hong Kong*

Cover photograph: Staithes, North Yorkshire; Images Colour Library

British Library Cataloguing in Publication Data
Longman coordinated geography.
   1. England. Secondary schools. Curriculum subjects:
Geography. G.C.S.E. examinations
I. Ross, Simon
910.76
ISBN 0-582-06299-3

Contents

LONGMAN CO-ORDINATED GEOGRAPHY

Series editor - Simon Ross

People and the Physical Environment

Peter Eyre

Acknowledgements

We are grateful to the following for permission to reproduce articles and other material;
The Daily Telegraph PLC for 'Volcanic Fumes kill 1,200 in Cameroon' by Arthur Max in *Daily Telegraph* 26.8.86 and 'Seven killed as resort is swept by avalanche' by Clare Hargreaves, Foreign Staff and Peter Hoffer in *Daily Telegraph* 14.3.88; Durrant Developments (Swanage Yacht Haven) Ltd for 'The Benefits of the Haven'; IPC Magazines Ltd for 'Skiing into trouble in Scotland' in *New Scientist* 14.1.88; Newspaper Publishing PLC for 'Landslips threaten Rio's hillside slums' by Richard House in *The Independent* 24.2.88 and 'Brazil warned of ecological 'catastrophe' by Isobel Hilton in *The Independent* 24.9.87; Robert McEnnis & Associates Ltd for 'No action over crumbling cliffs' in *Purbeck Mail* 17.12.87, 'Survey suggests that beach might not be affected' in *Purbeck Mail* 30.10.86 and a letter to the editor in *Purbeck Mail* 18.7.86; The Observer Ltd for 'Dying Swan' by Stephen Gardiner in *The Observer* 13.12.87 and 'Lesson for the South' by Geoffrey Lean in *The Observer* 22.11.87; C H Peacock (Westminster Press Ltd) for 'Flood chaos as River Colne's banks give way' by Marcus Duffield in *Watford Observer* 5.2.88; Southern Newspapers PLC for 'Fishermen slam marina scheme' in *Evening Echo*, Bournemouth, 9.12.86; Times Newspapers Ltd for 'California panics as earthquake shakes cities' by Foreign Staff in *The Times* 2.10.87, '100 vehicles crash in freezing M-way fog' by Staff Reporters in *The Times* 9.1.87 and 'Slow flow of warnings on river of death' by Gareth Huw Davies in *The Sunday Times* 16.11.86.

Inghams Travel; East Sussex County Council; Eyre & Gower: *Basic Processes in Physical Geography*, publ. Unwin & Hyman; Financial Times; *Geofile*, January 1988, No 102, publ. Mary Glasgow Publications Limited; Oxford and Cambridge Schools Examination Board.

We are grateful to the following for permission to reproduce photographs:
Aerofilms, pages 50, 55 *below*; Andrew Besley, page 56; John Cleare, Mountain Camera, page 22 *below left*; ESA/METEOSAT, page 65 (photo: Telegraph Colour Library); East Midlands Pictures, page 67; *Evening Echo, Southern Newspapers plc*, page 57; *Express Newspapers*; 17.10.1987, page 63; Friends of the Earth, page 74 *right*; *The Guardian*, 12.10.1987, page 62; Robert Harding Picture Library, page 2; Hulton-Deutsch Collection, page 6 *above*; International Photobank, page 70 (photo: Adrian Baker); Tony Morrison, page 77; Office Federal de Topographie, page 20 *below*; Oxfam, page 39; Rex Features (photo: M. Richards), page 6 *below*; Science Photo Library, pages 5 (photo: David Parker), 71 (photo: David Parker); Survival International, page 78 (photo: Sue Cunningham); *Watford Observer*, page 37; A.C Waltham, pages 26–27, 49 *below left*; West Air Photography, page 35; Map on page 41, Reproduced from the Ordnance Survey 1:25000 Watford & Rickmansworth, No. 1139, with the permission of the Controller of Her Majesty's Stationery Office, Crown copyright reserved; Map on page 16, Reproduced from Ordnance Survey 1:50000 Eastbourne & Hastings, No. 199, with the permission of the Controller of Her Majesty's Stationery Office, Crown copyright reserved. We are unable to trace the copyright holders of the following and would be grateful for any information that would enable us to do so, pages 28, 29.

Cover photo: Highlands-Glen More & Idairachradain, Images Colour Library

People and the Physical Environment

PM Eyre

Contents

To my parents

1 Volcanoes

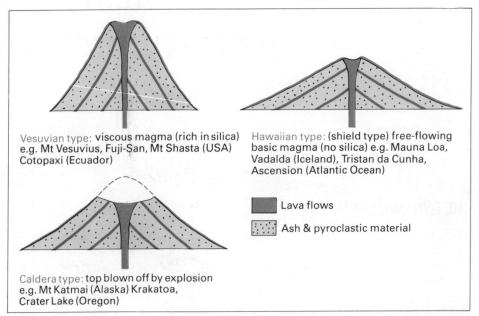

Figure 1.1 The eruption of Mt Lengai, Kenya

Vesuvian type: **viscous magma (rich in silica)** e.g. Mt Vesuvius, Fuji-San, Mt Shasta (USA) Cotopaxi (Ecuador)

Hawaiian type: **(shield type) free-flowing basic magma (no silica)** e.g. Mauna Loa, Vadalda (Iceland), Tristan da Cunha, Ascension (Atlantic Ocean)

Lava flows

Ash & pyroclastic material

Caldera type: **top blown off by explosion** e.g. Mt Katmai (Alaska) Krakatoa, Crater Lake (Oregon)

Figure 1.2 The major types of volcano

Volcanoes are the most explosive and destructive features on Earth. The amount of energy released in some eruptions makes even the largest man-made nuclear explosions look small. However, volcanoes do not usually kill as many people as earthquakes because there are often warning signs of a major eruption. Figure 1.1 probably fits your image of the typical volcanic eruption. Such eruptions are caused by hot molten rock (called **magma**) and gases pushing up through weaknesses in the Earth's crust. The force of the explosion pulverizes the solid rock around the opening in the summit (called the **vent**) to create **ash** and **pyroclastic** material. Pyroclastic material consists of broken rock fragments ranging in size from a few millimetres to several centimetres across. Most of the gas is steam. This condenses to give heavy rain which, together with the rock material, can produce destructive **mudflows**.

There are many types of volcano. The nature of the eruption, the kind of damage caused, and the shape of the cone all depend on the type of material produced. Figure 1.2 classifies the major types.

Most of the gases inside the magma escape when it reaches the surface. It is then known as **lava**. The shape of cone that is formed really

depends upon the type of lava that is produced. There are two major types.

1 **Acid** lava. This contains a lot of silica (SiO_2) and is **viscous**. This means it does not flow very far and builds up steep cones. These are called **Vesuvian Types** after the most famous example of Mt Vesuvius in Italy, or **composite** volcanoes.
2 **Basic** lava. This contains no silica and can flow very easily over long distances. It builds up shallow-sided cones. Mauna Loa in the Hawaiian Islands is one of the best examples of this type. The volcano is 250 km across at its base. Such cones are called **Hawaiian Types** or **shield volcanoes**.

In some volcanoes the magma is so viscous that it blocks the central feed-pipe which joins the vent to the molten material in the crust. The resulting explosion can be so violent that the top of the volcano is blown right off. This happened to Mount St Helens in the State of Washington, USA in 1980. There were no lava flows, but the ash, gas and mudflows that were emitted killed 61 people and destroyed 520 km² of valuable coniferous forest. Volcanoes that do 'blow their top' are called **calderas**. Crater Lake, in Oregan, USA is one of the most well-known examples of a caldera.

CASE STUDY

Cameroon 1986

Volcanic fumes kill 1,200 in Cameroon

By Arthur Max in Yaounde, Cameroon

GAS FUMES spewing from a volcanic lake in western Cameroon have killed at least 1,200 people. A further 300 are in hospital and thousands more are fleeing for their lives.

The figures were given last night by the stricken country's leader, President Paul Biya, after he had toured the area round Lake Nios near the town of Wum, about 200 miles north of the capital Yaounde. He declared the region a disaster area and appealed for international help.

Soldiers and rescue workers in gas masks were rushed to the area and the 10,000 population of Wum was being evacuated.

As the President ordered the immediate burial of the dead to prevent the spread of disease, some observers said that the eventual death toll could be more than 2,000.

At least three villages were enveloped by the gas – thought to be carbon dioxide – which was released by an explosion in a fissure beside the lake during Friday night. Many villagers died while they slept.

The scene of the disaster is some of Africa's most beautiful terrain, 150 miles north of the 13,353-ft Mount Cameroon in the Cameroon Highlands, a dramatic chain of volcanic peaks and valleys that reaches into Eastern Nigeria.

Villagers live in small clusters of straw and stick huts dotting the hillsides. They grow millet and cassava.

Drought has crippled crops in part of the region, and dwindling trees have made life harder. Higher up, herders raise livestock on some of the best cattle country on the continent.

The *Telegraph* 26 August 1986

Figure 1.3

1 a Draw a sketch of Mt Lengai in Figure 1.1 and describe its shape.
   b What type of volcano do you think it is? Justify your answer!

2 Read carefully the report on the Cameroon tragedy (Figure 1.3) and answer the following questions:
   a What was the main cause of death during the eruption?
   b What particular property of the gas made matters worse?
   c Why did the death toll rise after the eruption had ceased?
   d What other natural hazard has made matters worse for the people?
   e Why would Cameroon feel it necessary to call for international help?

3 Study Figure 1.4 showing the Shirane volcano.
   a What evidence is there around the crater to suggest that an eruption had occurred fairly recently?
   b How have the authorities actively encouraged tourism?
   c Do you think that tourists ought to be encouraged to climb Shirane? Give reasons for your answer.

The Cameroon tragedy raises the question of why people should live on the side of a volcano. Part of the answer lies in the fact that volcanoes can bring benefits. Soils developed on their surface are often very fertile, as in the case of Vesuvius in Italy and the volcanic island of Java. Villagers may lose their crops and houses but usually there is enough warning for people to leave the danger zone. The photographs in Figure 1.4 illustrate another way in which volcanoes can be exploited. Shirane gives out some very poisonous gases but that does not deter neither young nor old from climbing up it!

Figure 1.4 Left: the crater of Shirane, Shiga Heights National Park, Japan.
Right: tourists walking up to the crater of Shirane

2 Earthquakes

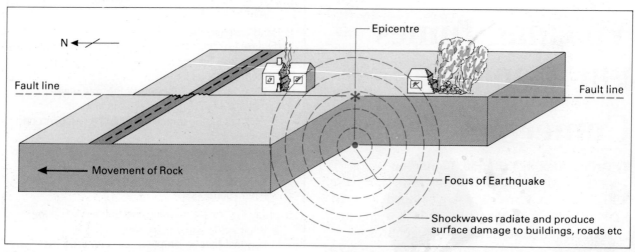

Figure 2.1 Diagrammatic representation of an earthquake on the San Andreas Fault, California

'I felt I was trying to stand on a bed of jelly when the main earthquake hit.'
'During the earthquake it was like walking on a rubber mattress.'

It is difficult to imagine what it is like to be in an earthquake unless you have had first-hand experience of one (or have stood on the vibrating plate in the Geological Museum!). The shaking and vibration of the ground are caused by shockwaves that radiate out from a sudden splitting of the rock in the Earth's crust. The splitting takes place at a point known as the **focus**. This lies on a line of weakness, or a **fault line**. The focus may be several kilometres below the surface. The point at ground level that is vertically above the focus is called the **epicentre**. (See Figure 2.1)

Earthquakes occur very frequently around the Earth, but they only hit the headlines when a severe one causes death, injuries or serious damage. The actual amount of damage done is one way of measuring earthquakes. This is known as the **Modified Mercalli Scale** and is shown in Figure 2.2. The values are shown by Roman numerals from I up to XII.

The second way of measuring them is by the magnitude of the earthquake. This is the amount of energy they produce and is measured by a **seismograph**, shown in Figure 2.3. The pen makes a tracing on a rotating drum as the earthquake shakes the ground on which the instrument

stands. These instruments are very sensitive: one in London could pick up the vibrations from just a mild tremor on the other side of the Earth.

The **Richter Scale** classifies earthquakes according to their magnitude. It ranges from the mildest − under 3.5 to the most severe at 8.9. The scale in numerals is shown on Figure 2.2 so that it can be directly compared with the Mercalli Scale.

Figure 2.3 A typical seismograph and its trace

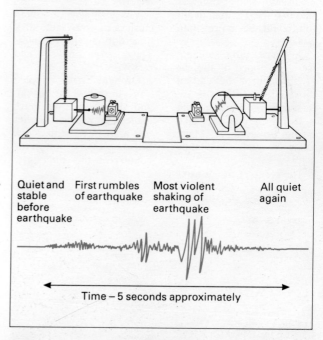

Figure 2.2 The Modified Mercalli Scale I to XII and the Richter Scale < 3.5 to 8.9

CASE STUDY

San Andreas Fault, California, USA

California lies on one of the most unstable parts of the Earth's crust. A major fault line runs the length of the State: the **San Andreas Fault**. The fault line itself can be seen in Figure 2.4. San Francisco lies right on the fault itself and was destroyed by the famous earthquake of 1906 (see Figure 2.5). (Although it was fire that caused most of the damage). Los Angeles lies less than 50 km from the fault. The last big earthquake here was in 1971 in which 64 people were killed. The newspaper report in Figure 2.6 describes a more recent earthquake on the 1st October 1987.

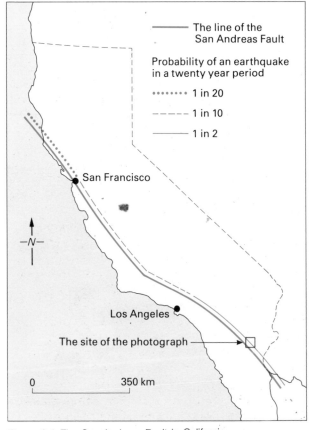

Figure 2.4 The San Andreas Fault in California

Figure 2.5 Damage in San Francisco 1906

ACTIVITIES

1 What are the 'epicentre' and 'focus' of an earthquake?

2 How long did the earthquake recorded on the seismograph in Figure 2.3 last?

3 Read the newspaper report in Figure 2.6 carefully and answer the following questions.

　a Describe in your own words the damage done directly by the earthquake to buildings and other structures.

　b A lot of damage was caused by fires: how were these caused?

　c What was the intensity of the earthquake on the Mercalli Scale?

　d Describe the likely effects if the earthquake had a magnitude of over 7.0 on the Richter Scale.

　e List what you think are the longer term effects of the earthquake to the area of southern California and its inhabitants.

California panics as earthquake shakes cities

By Our Foreign Staff

An earthquake rocked Los Angeles and towns in southern California yesterday, killing three people and injuring at least 100, 12 seriously.

The quake, which lasted about 20 seconds, started fires and landslides, destroying 20 buildings, damaging skyscrapers, and sending thousands of people racing into the streets.

The skyscrapers which dominate the heart of America's second biggest city were evacuated as hundreds of people were trapped in lifts, and glass from broken windows littered streets.

People at home ran into the streets and children who had gone to school early hid under their desks. Many cried with fright, teachers said.

A ring of 36 fires broke out around the city from ruptured gas mains, and several shops were destroyed. Landslides rumbled down the San Gabriel mountains and the Los Angeles canyons, where many people live.

Electricity was cut in many areas, blacking out 200,000 homes, and four men working in a hanging platform outside the 25th storey of a skyscraper were left swaying in the breeze. Judges, jurors and lawyers fled from concrete court buildings in the city centre, although prisoners were taken back to their cells.

One of the city's biggest freeways, the San Gabriel River Freeway, was closed, causing traffic jams, where it crosses the Santa Anna freeway. Concrete supports which hold up part of an overhead section of the road were badly damaged. King Juan Carlos of Spain and his wife, Queen Sofia, were in Los Angeles on a visit but were unharmed. They went ahead with their schedule.

"I felt I was trying to stand on a bed of jelly when the main earthquake hit," said Mr Bill Hernandez. "It was really frightening. It seemed to go on and on, but apparently only lasted for about 30 seconds. When we were told to leave the building we didn't need to be asked twice."

The earthquake, which registered 6.1 on the Richter scale, struck at 7.44 am local time. It was felt 150 miles away, and was followed by a dozen aftershocks, each measuring at least 3.0 on the Richter scale.

The California Institute of Technology placed the epicentre of the earthquake in the Montebello area of Los Angeles, 15 miles from the city centre.

The Times 2 October 1987

Figure 2.6

Figure 2.7 Damage in San Francisco 17 October 1989

3 Prediction and prevention

Earthquakes

'Half an hour before the earth moves all animals are seized with terror; horses whinny ... dogs bark, birds are terrified, rats and mice come out of holes.'

So wrote the natural historian Le Comte de Buffon in eighteenth century France. Mercalli (the man who invented the scale) noticed in Italy how restless animals became before an earthquake and how they tried to flee. Unfortunately animals are very unreliable. However, recent advances in earth science have enabled scientists to research into the problem of earthquake prediction.

One method is the use of **seismic gaps**. Where records of major earthquakes in a particular area go back a long way, an **'earthquake cycle'** is often seen to exist. The place where, according to the records, another earthquake is due is called the seismic gap. However such techniques do not necessarily produce accurate results and can even be socially and economically harmful. Residents and businesses that are in the seismic gap can experience falling property prices, lack of confidence and closure of industries. It may actually become more difficult to insure property once an earthquake hazard has been positively identified.

The seismic gap theory can only be used where earthquakes are frequent, as in parts of Japan and Mexico. (See Figure 3.1). Elsewhere only very careful monitoring of the crust can help. The build-up of seismic activity through small shock waves can be measured. The ground may start to bulge. There may be a change in the level of the water table or a sudden release of radon gas may also herald an earthquake. Continuous deformation of the ground *without* any tremors may also indicate trouble is on the way.

The attitudes that local inhabitants have towards the earthquake risk in their area is important. In a survey done in San Francisco in 1977, all the people interviewed put social and community problems before the earthquake hazard as disadvantages of living in the city. 4.2 per cent of those answering said they thought an earthquake would not occur at all! Over 40 per cent thought it would come 'in a few years' at the very latest.

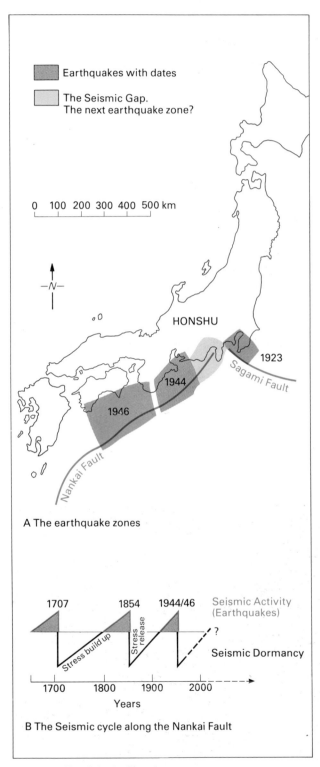

A The earthquake zones

B The Seismic cycle along the Nankai Fault

Figure 3.1 The Seismic Gap, Japan

Such attitudes will influence what people are prepared to do to minimise the effects when the earthquake does come. Table 3.1 shows what the residents of San Francisco are prepared to do.

Table 3.1 Preparations for an earthquake in San Francisco

	Percentage reply
Do nothing	37
Pray	37
Protect homes against looters	36
Take shelter in safe place	42
Run outside to open space	7
Take out insurance	7
Make structural changes to home	7
Evacuate the area	5

Local and national authorities have to be more positive. Figure 3.2 illustrates some of the precautions that are taken in earthquake prone urban areas like California. In some countries, however, planning controls and precautionary steps are not so vigorously applied, if at all. A rich state might regard the earthquake risk as too small to justify the expense (as in some eastern parts of America). In the case of many underdeveloped countries, the money and expertise are not available.

Volcanic eruptions

As with an earthquake, once a volcano has started to erupt, nothing at all can be done to stop it. However, volcanoes usually give more warning of an eruption and so safety measures like evacuations can take place. Scientists from all over the world had been keeping a close watch on Mount St Helen's in the USA for several weeks before it actually blew up in 1980. The main blast was preceded by small earthquakes and minor eruptions. Farmers, forestry workers, and tourists were evacuated from a large area around the mountain. Despite all the modern techniques available to them, the scientists were unable to say exactly when the eruption would occur or how big it would be. In the event, Mount St Helens erupted with the force of a 10 megatonne bomb and took even the most pessimistic scientists by surprise. The spread of the destructive gases and ash went beyond the area that had been evacuated. If a closely monitored mountain in the USA can spring nasty surprises, then there is little that can be done to reduce the volcanic hazard in less developed regions like Cameroon.

1 Buildings

Extra reinforcements in cross-members of tall buildings help prevent structure resonating with shock waves and provide strength with flexibility

Heavy furniture office equipment, computers etc. should be secured to prevent falling about when building sways

Gas mains and generators must be protected from damage to cut fire risk

2 Personal

Individuals should learn what to do to protect themselves both during and after a 'quake'

3 Public services

Well trained and prepared public rescue services including evacuation plans and warning systems

Figure 3.2 Earthquake Precautions

ACTIVITIES

1 Study Figure 3.1 carefully and answer the following questions using your atlas.
   a What large city was destroyed by the earthquake of 1923?
   b What cities are in danger from the next earthquake period?
   c Using the graph in B, try to predict when the next earthquake will occur.

2 Imagine you live in an earthquake zone. What advice would you consider giving to a new neighbour who had just moved in from a place that had no earthquake risk at all?

3 Explain why the attempts to predict earthquakes could do more harm than good.

4 Plate tectonics

By the early 1960s scientists knew a great deal about volcanoes, earthquakes, and mountain building. They knew *how* they were caused, but they were still puzzled as to *why* they occurred. Then a new theory began to take shape which helped scientists link all these things together and explain their occurrence. This theory is known as the **Theory of Plate Tectonics**. Its basic ideas are these:

a the crust of the Earth is made up of rigid **plates**;
b these plates move in relation to one another. They may move apart, or together, or slide alongside each other;
c the plates 'float' on the hot liquid **mantle** beneath and their movement is caused by rising hot currents within the mantle, known as **convection currents**;
d mountain chains, earthquakes, volcanoes and deep ocean trenches are caused by the processes that take place along the boundaries between the plates.

Figure 4.1 illustrates the structure of the Earth. The surface of the globe is called the **crust**. There are two types of crust: **oceanic crust**, which is only about 10 km thick; and **continental crust**, which forms the **continents**. The continents and oceans are shown on the world map in Figure 4.2.

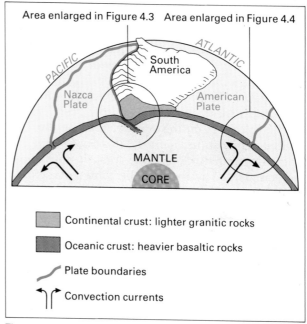

Figure 4.1 The structure of the Earth

Figure 4.2 The crustal plates

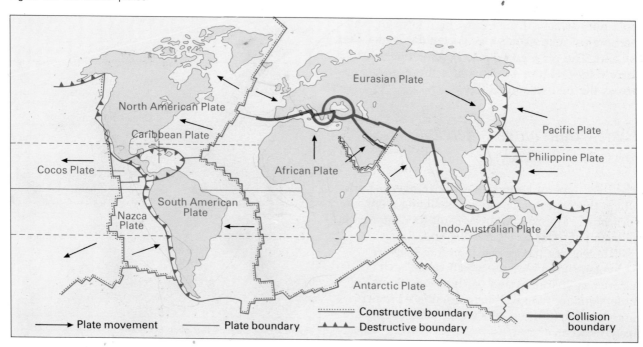

Plates that move together (**destructive boundary**)

Figure 4.3 shows what happens where plates move together. Here, the Nazca Plate is being forced below the South American Plate. The zone where the Nazca Plate pushes down into the mantle is called the **subduction zone**. Subduction is very erratic. Energy builds up in times of less movement to be released later, causing an earthquake. The oceanic plate also melts causing hot liquid rock (**magma**) to rise upwards. This causes volcanoes on the surface. Because the continental crust (in Figure 4.3 it is South America) is less dense than the oceanic crust, it rides above the boundary. The rocks of the continental crust become crumpled and folded to form **fold mountains**. Along the west coast of South America the fold mountains are the Andes.

Along some boundaries two continental plates collide, as shown in Figure 4.4. Here, the Indo-Australian Plate is colliding with the Eurasian Plate. The continents crumple to form the Himalayan Mountains.

Plates that move apart (**constructive boundary**)

Some plates move apart, mostly under the oceans. As they move away from each other the gap is filled by magma rising from below. This creates volcanoes which, if they grow high enough, will become volcanic islands e.g. Tristan da Cunha, and Iceland. The plate edges also become buckled to form ridges. Figure 4.5 illustrates these processes across the mid-Atlantic Ridge.

Plates that move sideways (**conservative boundary**)

The third type of boundary is the **tranverse fault** where the plates move sideways sliding alongside each other. The San Andreas Fault in California is an example (refer to Figures 2.4 and 2.5).

Plate tectonic theory can therefore help us understand the processes which cause such hazards as volcanoes and earthquakes. Such understanding may well help scientists predict more accurately when and where such events are likely to occur.

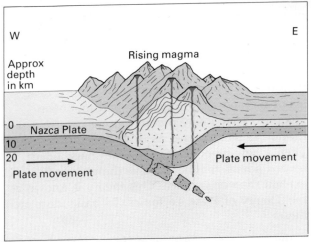

Figure 4.3 A destructive plate boundary

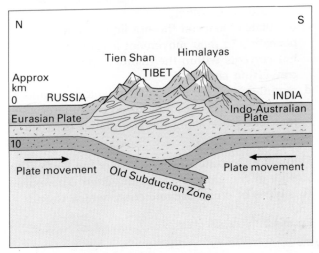

Figure 4.4 A collision boundary

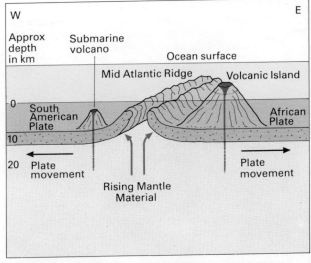

Figure 4.5 A constructive plate boundary

Table 4.1 15 significant earthquakes in 1986

	Latitude & Longitude	Place	Summary of effects
1	41° 55′ N 81° 16′ W	Ohio	Grade VI damage in Painesville, including Perry Nuclear Power Plant 17; people with minor injuries
2	27° 59′ N 139° 55′ E	Bonin Islands	No data
3	40° 37′ N 51° 56′ E	Caspian Sea	Minor damage at Baku
4	13° 41′ S 71° 59′ W	Peru	At least 16 people killed; 2000 houses destroyed; landslides near Cuzco.
5	37° 48′ N 121° 49′ W	Central California	6 people with minor injuries; grade VII damage in Fremont area.
6	13° 52′ S 166° 53′ E	Vanuatu Islands	No data
7	2° 39′ S 139° 31′ E	New Guinea	No data
8	32° 13′ N 76° 37′ E	Kashmir	6 people killed; 85 per cent of houses damaged
9	18° 40′ N 102° 47′ W	Mexico	Grade V damage in Mexico City area
10	37° 49′ N 37° 51′ E	Turkey	15 people killed, 100 injured; 4000 houses damaged
11	51° 42′ N 174° 58′ W	Aleutian Islands	Grade VI damage; 175 cm high tsunami waves.
12	41° 43′ N 43° 54′ E	Turkey/ USSR border	2 people killed, 1500 buildings destroyed.
13	10° 40′ N 62° 53′ W	Venezuelan coast	2 people killed, 45 injured; grade VII damaged
14	4° 45′ S 143° 54′ E	Papua New Guinea	Grade VII damage totalling $500 000 worth; submarine cables damaged. Widespread landslides.
15	34° 00′ N 116° 41′ W	California	29 injured; landslides; grade VII damage totalling $4.5 million, including Devers substation of Southern California Edison Co.

ACTIVITIES

1 Study Figure 4.2 and answer the following:
  a Which crustal plates do not have a continent?
  b With the help of your atlas, find and name as many mountain ranges as you can that are the result of
    i collision boundaries
    ii destructive boundaries.

2 Make a large, clear copy of Figure 4.3. Put on the following labels in their correct place on your diagram:
  SOUTH AMERICAN PLATE   THE ANDES
  VOLCANOES   RISING MAGMA   OCEAN
  TRENCH   SUBDUCTION ZONE   PACIFIC
  OCEAN

3 Add symbols to your diagram to show where you think the foci of earthquakes are likely to occur. Explain why you have chosen these locations for the earthquakes.

4 a Look through the first chapter on volcanoes and make a list of all the volcanoes that are mentioned by name.
  b With the help of your atlas, plot these volcanoes on a world outline map.
  c How well does the distribution pattern fit the plate boundaries in Figure 4.2?
  d Is any one of the plate boundary types more associated with volcanic activity than others? Try to explain your answer.

5 Test the following hypothesis:
  ALL MAJOR EARTHQUAKES OCCUR ALONG PLATE BOUNDARIES
  To do this, look at Table 4.1. This gives details of the 15 most severe earthquakes in 1986. Place a piece of tracing paper over the map in Figure 4.2.
  a Mark on this tracing paper the 15 earthquakes.
  b What conclusions do you come to?

5 Igneous rocks and metamorphic rocks

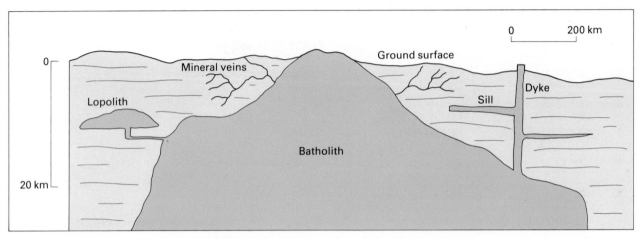

Figure 5.1 The occurrence of Intrusive Igneous Rocks

Igneous rocks

You have seen in the first chapter how lava from volcanoes solidifies to form solid rock. Rocks that form from the cooling and solidification of hot liquids are known as **igneous rocks**. Those that form from volcanic action *on* the earth's surface are called **extrusive** igneous rocks. Some are formed below the earth's surface, deep inside the crust. These are called **intrusive** igneous rocks (see Figure 5.1). Such intrusions are sometimes exposed at the surface when the rocks above them have been eroded away. This has happened at Dartmoor.

Dartmoor is made of **granite**, one of the most common types of igneous rock. It is very resistant to erosion and so protrudes as high ground. Cracks (or **joints**) occur roughly at right angles throughout the rock and these help to create the distinctive features on Dartmoor called **tors** (see Figure 5.2).

Igneous intrusions can also form valuable veins of **minerals** such as tin, copper, silver and gold. Until the latter part of the nineteenth century the hills of Wales and Scotland were famous for their metal mining and Cornwall was a major supplier of tin. However the veins are so difficult and expensive to exploit that they became uneconomic in the face of competition from cheaper sources in countries like Malaysia and South Africa.

Metamorphic rocks

All kinds of rock may be changed by heat and pressure into **metamorphic rocks**. These are usually hard and resistant to erosion. Granite is changed into **gneiss**, limestone into **marble** (a valuable decorative stone used by builders and sculptors) and clay into **slate**. (Limestone and clay are two sedimentary rocks that you will read about in the next chapter). Bethesda in Snowdonia has long been a source of high-grade slate for building purposes but as with so many natural stones it cannot easily compete with cheaper alternatives.

Figure 5.2 Dartmoor

In Cornwall, granite has been decomposed by hot gases within the crust to form **Kaolin**, or China Clay. This material has a wide range of uses from pharmaceuticals to paper. Unfortunately no-one has yet found a commercial use for the waste material — largely quartz — that is produced. This has been piled into huge conical heaps on which nothing will grow. (See the *Energy and Industry* book for more on the Kaolin industry.)

ACTIVITIES

1 a Study Figure 5.1 and write down your own definition of the following:
SILL DYKE LOPOLITH BATHOLITH

 b Why are these called intrusive igneous rocks?

2 Look carefully at the photograph of Dartmoor shown in Figure 5.2. Describe the scene either in words or by drawing a picture. Pay special attention to;

 a the shape and structure of the tor,

 b the landuse and vegetation.

3 a Draw an outline map of Devon and Cornwall. Show on this map (with the help of your atlas):
   i the areas of granite;
   ii the towns of Exeter, Plymouth, Okehampton, Bodmin, St Austell, Truro, Camborne and Penzance;
   iii the main roads.
     Do not forget to give your map a key, scale, and title.

 b Describe the relationships you can see between the geology of Devon and Cornwall and the towns and roads. How might you explain these relationships?

4 Find in your atlas a geological map of the whole of the British Isles.

 a On an outline map, shade on the distribution of igneous rocks and metamorphic rocks

 b On a second outline, shade on all areas that are higher than 400 m.

 c Describe, and try to account for, the relationship between the two maps.

6 Weathering and sedimentary rocks

All kinds of rocks that are exposed on the surface suffer from **weathering**. This means that they are gradually broken down physically and chemically by the agents of the weather. The two main types of **physical** weathering are:

1 **Frost-shattering**. Water in rock pores and cracks expands on freezing thus breaking up the rock.
2 **Exfoliation**. High daytime temperatures followed by low night-time temperatures cause the minerals in the rock to expand and contract. In this way the rock is weakened. This happens particularly in hot deserts.

In **chemical** weathering, acid in rainwater and soil-water dissolves and decomposes certain minerals in the rock. Granite is made up of mica, quartz and feldspar. Although it is generally a resistant rock, granite is affected very slowly by both physical and chemical weathering. The feldspar decomposes chemically to form **clay**. The mica and the quartz are physically broken into grains of **sand**.

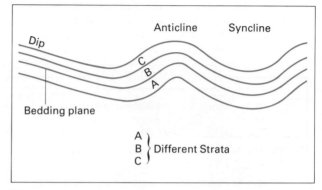

Figure 6.1 The folding of strata.

Both clay and sand are important examples of **sedimentary rocks**. These are rocks that are made from the debris formed from the weathering of rocks. Rivers and glaciers transport the sediments to lakes and the sea where they are deposited in layers called **beds**, or **strata**. Movements in the crust cause the strata to **tilt** or **dip**. They may be folded into **anticlines** and **synclines** as shown in Figure 6.1.

Figure 6.2 The structure of S.E. England

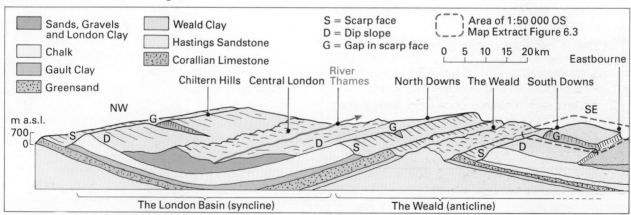

Table 6.1 (on p. 15) shows the main types of sedimentary rocks. Notice that some are harder than others. The harder ones will be more resistant to weathering and erosion. Rocks such as chalk and limestone stand up as lines of hills called **cuestas**. Cuestas have a steep side known as the **scarp face** and a gentle slope called the **dip slope**.

Clay, being soft, will become eroded to form **vales** between these cuestas. Figure 6.2 shows the structure and relief across the London Basin and the Weald of Sussex. The 1:50 000 OS map extract in Figure 6.3 covers part of this structure: the South Downs near Eastbourne.

Table 6.1 Types of sedimentary rock

Name	Description	Major uses
Sandstone	Grains of quartz cemented by calcium or silica. Often hard & resistant. Fairly permeable*	Building stone Grindstones Glassmaking
Clay	Soft and impermeable# Easily eroded.	Brick making Pottery Cement making
Limestone	Made from calcium carbonate, often from shells of minute sea creatures. Permeable* and fairly resistant but soluble in acidic water	Building stone Road stone Cement making
Chalk	A pure form of limestone	

* Permeable rocks allow water to pass through them
# Impermeable rocks do not allow water to pass through

Limestone scenery

Carboniferous Limestone, like chalk, is calcium carbonate. It is, however, much harder than chalk and has a well developed system of joints that run at right angles to each other. Rainwater percolates into these joints through **swallow holes**.
Because the rain is a weak acid it dissolves the rock and widens the joints. Underground **caverns** develop, sometimes to a very large size ... easily big enough to hold a four-bedroomed detached house! Sometimes cavern roofs will collapse resulting in subsidence on the surface which looks something like a **gorge**. Water carrying dissolved calcium bicarbonate drips from the roof of many caverns. The evaporation of the water causes calcium carbonate to be deposited. These deposits form **stalactites** which hang from the roofs of caverns. **Stalagmites** build up from the floor. The Karst region of Yugoslavia has some of the best-developed limestone features and has given its name to this kind of landscape: a **karst landscape** (see Figure 6.4).

Caving and pot-holing have become popular recreational activities in areas like the Mendip Hills in Somerset and the Peak District in Derbyshire. Both these areas have well-developed cave systems.

Much of England consists of gently dipping and folded strata of sedimentary rocks. Quite apart from the economic value of these rocks listed in Table 6.1, sedimentary rocks are also associated with very important fuel sources: coal and oil. North Sea oil is drilled from the rocks that lie under the sea. Many areas inland may well have oil deposits that are economically viable. A little oil is extracted from Eakring near Nottingham and Kimmeridge in Dorset. Oil companies are eager to drill in some environmentally beautiful areas like Ashdown Forest: quite a controversial issue. East Sussex Council has refused permission for BP to make test drillings in Ashdown Forest. (See the *Energy and Industry* book for more on coal and oil.)

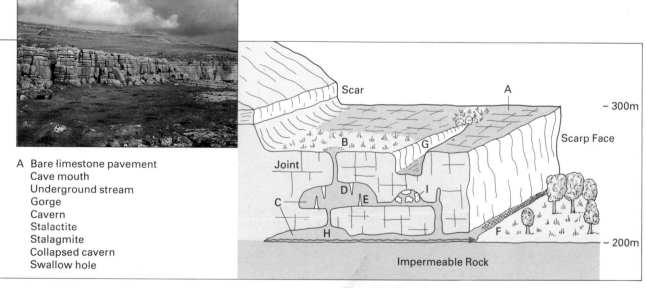

A Bare limestone pavement
Cave mouth
Underground stream
Gorge
Cavern
Stalactite
Stalagmite
Collapsed cavern
Swallow hole

Figure 6.4 Karst features

Figure 6.3

Figure 6.5 Geology of the Eastbourne Area

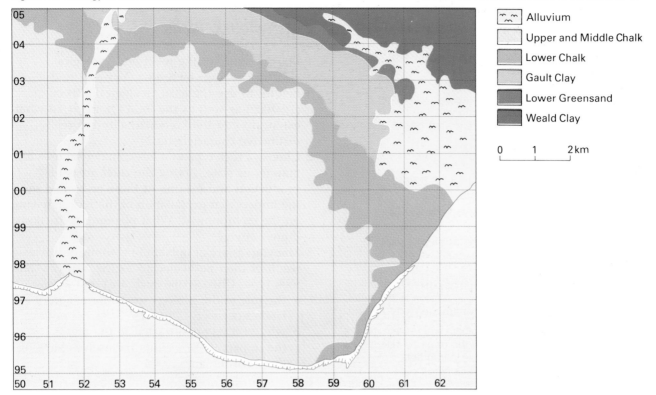

Alluvium
Upper and Middle Chalk
Lower Chalk
Gault Clay
Lower Greensand
Weald Clay

0   1   2km

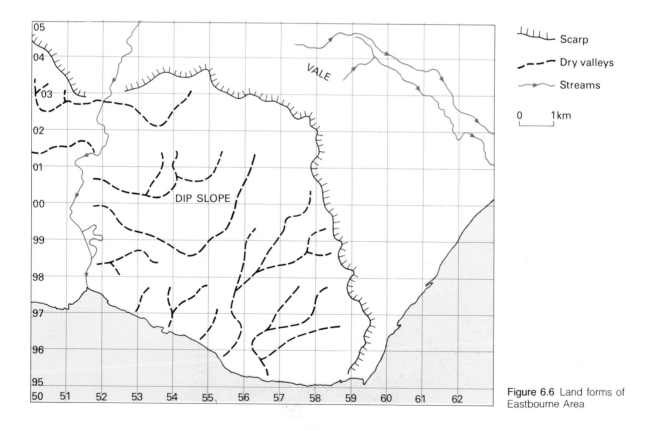

Scarp
Dry valleys
Streams

0   1km

VALE

DIP SLOPE

Figure 6.6 Land forms of
Eastbourne Area

1 Study carefully the 1:50 000 map extract of the Eastbourne area (Figure 6.3), the geology map (Figure 6.5), the landform map (Figure 6.6) and the photograph (Figure 6.7). Use these to answer the following questions.

   a If the photograph in Figure 6.7 was taken at 505038, in which direction was the camera pointing? At approximately what height was the photographer standing?

   b Make a sketch of Figure 6.7. Mark on your sketch the following:
   THE SCARP   THE CUCKMERE RIVER
   THE CLAY VALE   ALFRISTON

   c From what rock is the scarp made? (See Figure 6.5).

   d On what kind of rock are all the dry valleys found? Can you explain your answer? (Look again at Table 6.1).

   e With the aid of a sketch-map, describe the distribution of settlement in relation to the relief, drainage, and geology of the area.

2 Use Figures 6.1 and 6.2 to help you to write out the following paragraph. Fill in the gaps by selecting a word from the list below the paragraph.
The London Basin is an example of . . . .
The North Downs are made of . . . . and dip in a . . . . direction. The scarp of the North Downs faces . . . . The Weald is in the centre of . . . . and to the north and south of it is a vale of . . . .
Select from: CLAY, CHALK, AN ANTICLINE, SOUTH, NORTH, A SYNCLINE, SANDSTONE

3 Study Figure 6.4 which shows the main features of a karst landscape. Write out the list of features given and against each one put the correct letter from the diagram. Make a drawing of the view shown in the photograph. Label your drawing with as many karst features as you can. Title your drawing 'karst landscape above Malham, Yorkshire'.

Coursework ideas

1 Make a study of the relief and geological structure of your own area. You will find it helpful to use the local Geological Survey (1:50 000) as well as the local OS map. To what extent do you think your locality has really distinctive characteristics?

2 Choose a rock type that has commercial uses e.g. granite, or slate. Locate the chief areas of production and for one of these areas find out the impact that exploitation of the rock is having on the environment. What planning restrictions, if any, apply?

3 Many areas are of outstanding scenic beauty as a result of their particular geological structure and rock type. Choose one such area and assess:
   a the role of the geology in the formation of the landscape;
   b why visitors are attracted to the area;
   c the impact these visitors themselves have on the environment.

Figure 6.7 The South Downs

7 Ice in the mountains: glacial erosion

You have seen in the chapter on plate tectonics that the great mountain chains of the world like the Himalayas and the Alps were formed as the crust of the Earth became folded and crumpled along plate boundaries. Such mountain chains are called **fold mountains**. However, most mountain landscapes do not reflect the folds in the rocks underneath the surface because of **erosion** (or wearing away) by rivers and glaciers. Dramatic glaciated scenery attracts millions of tourists every year. Peaks and ridges are popular with climbers; and skiing on snow covered winter slopes has been a fast-growing sport in recent years. The tremendous development of tourism and recreation is now threatening the delicate environment of mountain regions.

The glacier as a system

A glacier is a very good example of an **open system**. A system is a set of things that work together to produce a particular result. An open system is one that has **inputs** and **outputs**. Figure 7.1 shows the inputs and outputs of a glacier system. Ice is lost by **ablation** which involves both melting and evaporation. This occurs particularly at the **snout**. If the glacier loses more ice by ablation than it gains from snowfall, then it will **retreat**. This usually happens when the climate gets warmer. Not only is the glacier more likely to melt, but there will be less snowfall and more rainfall. During the present century glaciers have generally been melting because of a gradual rise in the average temperature (about 0.5°C since 1900). If all the world's ice was to melt, the water produced would cause a rise in sea level of over 200 metres, flooding most coastal cities.

The development of glaciers and their work

Stage I
The changes that take place with glaciation are illustrated in Figure 7.5. High up on the mountainside it is much colder than lower down. This means it is more likely to snow than to rain. Snow patches will develop in hollows above a certain height (the **snowline**).

Stage II
As more and more snow accumulates it gets compressed into ice. This erodes and deepens the hollows to form **cirques**. Between the cirques, sharp ridges (known as **arêtes**) and **pyramidal peaks** are left. These are made more rugged by **frost-shattering**.

Stage III
There is now so much ice that it flows out of the cirques and into the river valley below. This valley glacier deepens and widens the valley into a **glaciated trough** or **'U'-shaped valley**. The frost-shattering on the rock slopes above the ice produces a lot of rock fragments called **moraine**. This moraine falls onto the surface of the glacier.

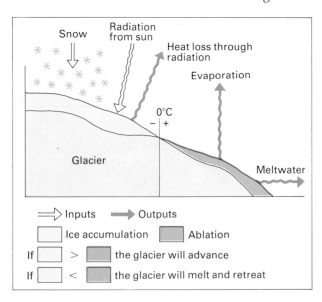

Figure 7.1 The glacier system

Much of it melts its way down to the base of the ice to form **ground moraine**. The ice itself is too soft to erode solid rock. The erosion is done by the ground moraine being carried under great pressure by the **sub-glacial melt-water**. Ice can freeze onto the rock face and pluck away particles of rock as it moves slowly downstream. The movement of glaciers is extremely slow. The Mer de Glace shown in Figures 7.2 and 7.3 travels at 3 millimetres an hour!

Stage IV
When the climate becomes warmer the ice melts away and the glacier **retreats**. The moraine is deposited in the valley floor where it may dam back the new river. This will cause a **ribbon lake** to form. Lakes may also form in some of the cirques. These lakes are called **tarns**.

Figure 7.3 The Mer de Glace

Figure 7.2 Swiss 1:50 000 map of the Chamonix Area

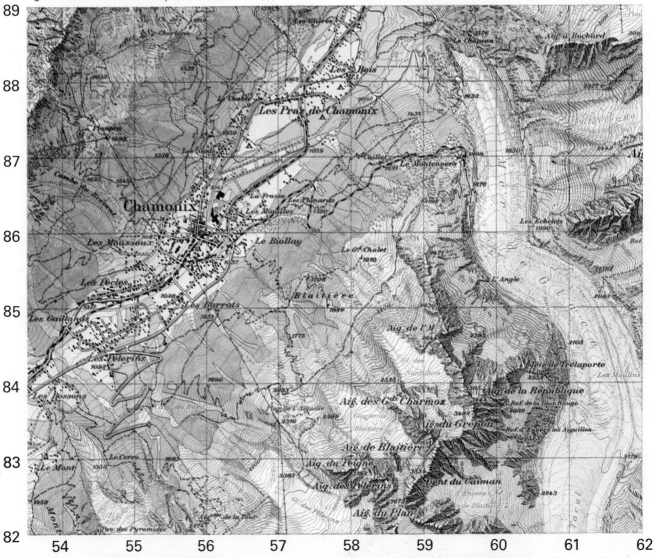

ACTIVITIES

1 A photograph of the Mer de Glace near Chamonix in France is shown in Figure 7.3. Make a sketch of this view and put on the following labels:
PYRAMIDAL PEAK ARETE CIRQUE CREVASSES LATERAL MORAINE
Use Figure 7.5 to help you.

2 What is the evidence in Figure 7.3 that the glacier is popular with tourists?

3 Study the photograph in Figure 7.4, which shows the site of Chamonix.

   a Describe the shape of the valley.

   b Do you think the valley has been glaciated? Why?

   c Describe the site of Chamonix.

4 Figure 7.2 is an extract from a 1:50 000 map of the Chamonix area. Study it carefully in order to answer the following questions.

   a Draw an accurate profile from 540890 to 580830. Use a vertical scale of 2 mm to 20 m. The contour interval on the map is 20 m. Mark on your profile the site of Chamonix.

   b The photograph in Figure 7.3 was taken at 596868. In what direction was the camera pointing?

   c How do tourists, who do not wish to walk or climb, reach the Mer de Glace from Chamonix? How far is their journey?

   d Glacier contours are shown in blue, and are at 20 m intervals. Calculate the average gradient of the glacier.

   e Find square 6184. Look carefully at the fine grey dots which show the moraine on the glacier surface. Draw a sketch map of this pattern. What can you learn from this about the movement of the glacier?

   f Locate the small glaciers in the north-east part of the map.
   i At what height above sea level is the snout of each one?
   ii These glaciers once joined the Mer de Glace. Why do you think they no longer do so?

Figure 7.4 The Site of Chamonix

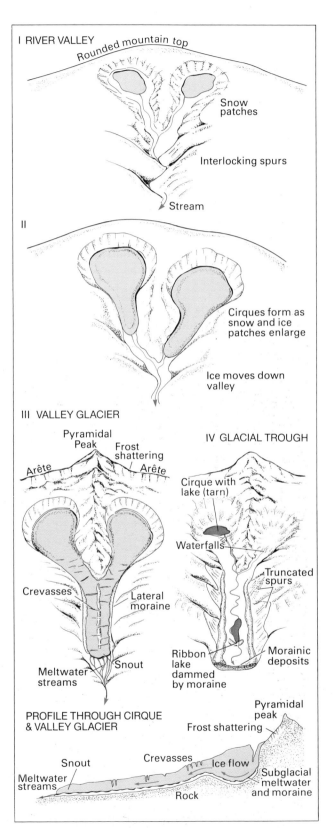

I RIVER VALLEY
Rounded mountain top
Snow patches
Interlocking spurs
Stream

II
Cirques form as snow and ice patches enlarge
Ice moves down valley

III VALLEY GLACIER
Pyramidal Peak
Frost shattering
Arête   Arête
Crevasses
Lateral moraine
Meltwater streams
Snout

IV GLACIAL TROUGH
Cirque with lake (tarn)
Waterfalls
Truncated spurs
Ribbon lake dammed by moraine
Morainic deposits

PROFILE THROUGH CIRQUE & VALLEY GLACIER
Pyramidal peak
Frost shattering
Snout
Crevasses
Ice flow
Meltwater streams
Rock
Subglacial meltwater and moraine

Figure 7.5 The evolution of a glaciated landscape

8 Avalanches

When heavy snow accumulates on a mountainside it is likely to collapse downhill as an **avalanche**. Scientists are still trying to unravel the causes of avalanches. The following factors are, however, thought to play an important part:

1 rising temperatures combined with heavy snowfalls in late winter and early spring;
2 less grazing on the grassy slopes in summer results in longer grass to which the snow is less likely to stick;
3 the development of the ski industry. Forests are cleared to make way for runs, hotels, car parks and roads.

There are two basic kinds of avalanche: the **powder** avalanche and the **slab** avalanche. An example of a powder avalanche is shown in Figure 8.1

At the Chamonix weather centre research is being done into the complexities of avalanches. The thickness, structure and temperature of the snow are fed into a computer which then predicts the level of the avalanche risk at a certain time and place. Warning signs are then placed in the skiing areas and other places where they could cause death and damage (see Figure 8.2). However, avalanches are very difficult to forecast accurately. They are often triggered off by controlled explosions before they can become a danger. Also, fence-like structures are often built across vulnerable slopes to protect people and property below (see Figure 8.3). Of course, an avalanche in an unpopulated area will harm no-one!

Figure 8.1 A powder avalanche

Figure 8.2 An avalanche warning sign at the top of a piste, Pra Loup, France

Figure 8.3 A structure for the prevention of avalanches, Pra Loup, France

CASE STUDY

Avalanche at St Anton, Austria 1988

Seven killed as resort is swept by avalanche

By Clare Hargreaves, Foreign Staff, and Peter Hoffer in Vienna

AN AVALANCHE smashed into hotels and chalets in the Austrian ski resort of St Anton yesterday morning, killing seven people in their beds and burying alive at least ten others.

Skiers have been banished from the slopes because of the hazardous weather conditions, and many British tourists returning home have been stranded as snow drifts and avalanches blocked roads and railway lines.

Among those cut off were ex-Queen Juliana of the Netherlands and her husband Prince Bernhard, who are on holiday in the nearby resort of Lech.

Police said five of the dead were Swedish holidaymakers staying in a guest house which bore the brunt of the avalanche. The other two were local people. No Britons were reported among the casualties.

One of the Austrian victims was Frau Aloisia Strolz, 75, a hotelier, who died when the avalanche smashed into her house on the eastern side of the town, completely destroying it.

Ten people were buried when the huge wall of snow smashed into a group of 15 hotels and guest houses 500 yards from the town centre, demolishing four buildings and crushing cars.

Four Swedes were rescued alive and flown to nearby hospitals, some with serious injuries.

At least 300 workers, including firemen, police, ski instructors, mountain rescue men and troops, fought against bitter winds to dig survivors and cars out of the snow.

Witnesses said the avalanche smashed through hotel windows into bedrooms where holiday-makers were asleep, sending many fleeing for safety in their nightclothes.

"It was terrible. I would say the snow was 10 to 15 metres (33 to 49 ft) high. It was really solid, not just a powder avalanche," said Andrea Waldner, hotel receptionist.

Blizzards have swept western Austria for the last 36 hours, triggering numerous avalanches and trapping some 45 000 tourists, mainly skiers.

Avalanches also killed three skiers in the Italian Alps yesterday: two West Germans near the Dolomite resort of Malles Venosta, and an Italian at Sondrio, near the Swiss border.

The *Daily Telegraph* 14 March 1988

Figure 8.4

ACTIVITIES

1 Study Figure 8.3 showing the precautions taken against an avalanche.
  a Make a sketch of the scene.
  b Label i the protective fence,
         ii the area from which an avalanche may come.

2 Figure 8.4 is a case study of the avalanche that hit St Anton in 1988. Read it carefully and then answer the following questions.
  a What kind of avalanche did it appear to be? What is the evidence for this?
  b Describe the weather conditions leading up to the disaster.
  c List the damage done to the town and the inconvenience caused to the people.
  d Do you think that this kind of publicity would influence people's choice of resort? Why?

3 Do you think that notices like that in Figure 8.2 are an adequate precaution? Discuss in class the advantages and disadvantages of completely closing ski runs that are in danger of severe avalanches.

9 Slopes and landslides

If you live in hilly countryside you will almost certainly have noticed that slopes are a very important part of the landscape. The steepness of the land has a very important influence over the way in which it can be used. In towns and cities slopes are not so obvious in everyday life (unless you have to cycle uphill to school!)

There are a number of factors that determine the steepness of a slope. You should notice the following points:

1 **Uplift**. This refers to the raising of the land surface during movements of the crust, e.g. in mountain-building and folding of strata. (See p. 14.) Generally speaking, the greater the uplift, the steeper the slope.

2 **Rock type**. Harder rocks will generally produce steeper slopes than softer rocks. For example, a chalk landscape is usually hillier than a clay one.
3 **Rainfall**. Rain drops break up soil particles and wash them down the hillside. This tends to make the slope less steep.
4 **Energy** input and output. As you've already seen, frost-shattering and exfoliation are a result of changes in temperature. Mechanical weathering like this will alter the slope angle. The build-up of loose material from mechanical weathering is called a **scree slope**.

Very often a slope becomes so steep that the rock material out of which it is formed is not strong enough to hold it up. Slope **failure** then takes place and a **landslide** will occur.

CASE STUDY

Rio de Janeiro, Brazil

Figure 9.1

Landslips threaten Rio's hillside slums

From Richard House
in Rio de Janeiro

AT THE Santa Genoveva clinic on a forested hillside above Rio yesterday, rescue workers wearing surgical masks and gloves were braving the rising stench of decomposing bodies to sift through the wreckage for 30 pensioners and mental patients who are believed to have died when a landslide struck their building during the weekend's devastating storms.

The debris, a wheelchair axle-deep in mud, a stretcher, and hastily thrown-aside bedclothes showed how suddenly the storm water struck last Friday, under-mining the building.

In all, 78 people have been killed around the city, and thousands more are homeless.

This is not the first time Rio has suffered in this way: in 1966 an identical storm claimed more than 200 lives. Since urban population growth has overstrained the city's infrastructure and urban planning, control has collapsed.

Long after the rain ceased the hillsides were streaked with cascades of dirty water, stripping the thin layer of soil off the rock face below like peel from an orange.

With the prospect of more heavy rain to come attention has now turned to the perilous conditions in the shanty towns or *favelas* on the steep hill sides above the city.

In the last **three** weeks the rain has killed 273 people in the Rio area, the majority of them black, jolting Brazilians into awareness that the *favelas* resemble nothing so much as South Africa's black squatter townships.

At least one-eighth of Rio's population lives in these unplanned and unsafe squatter communities, where the land is considered too precarious for apartment blocks.

As the *favelas* expand, they strip vegetation from the hillsides and block natural storm-drains. Experts say there are 60 areas in danger of collapse, threatening 50000 lives.

The *Independent*, 24 Feb. 1988

Mountainous and hilly regions are well known for **landslides** (see Figure 9.1). Rock and soil material is prone to slipping downhill under the pull of gravity. The steeper the slope, the more likely it is to be unstable. People make matters worse by clearing vegetation and erecting buildings and roads. Figure 9.2 shows the major factors, both natural and man-made, that contribute towards landslides. Notice how a naturally unstable slope can be made even more unstable by the actions of people.

People's interference with slopes does create major hazards, as in Rio de Janeiro. One area that suffers severely from landslides is the county of Los Angeles, California. The distribution of slope instability and some of the causes are illustrated in maps A–D in Figure 9.4. The built-up area of Los Angeles has covered virtually all the flat land of the coastal plain and rich suburban houses are creeping up the unstable sides of the Santa Monica Mountains and the San-Gabriel Mountains.

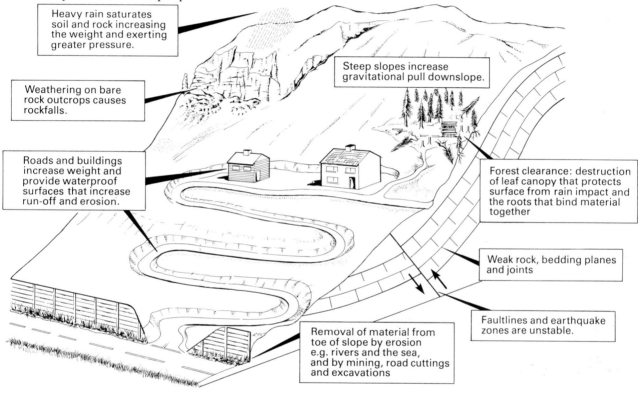

Heavy rain saturates soil and rock increasing the weight and exerting greater pressure.

Weathering on bare rock outcrops causes rockfalls.

Roads and buildings increase weight and provide waterproof surfaces that increase run-off and erosion.

Steep slopes increase gravitational pull downslope.

Forest clearance: destruction of leaf canopy that protects surface from rain impact and the roots that bind material together

Weak rock, bedding planes and joints

Faultlines and earthquake zones are unstable.

Removal of material from toe of slope by erosion e.g. rivers and the sea, and by mining, road cuttings and excavations

Figure 9.2 Factors which contribute to landslides

A typical house structure in the mountains of California is shown in Figure 9.3. Part of the slope is cut level. Downhill of this cut material is dumped to provide another level section; this is the fill. Because of the weak nature of the surface rocks, the house foundations will be on **piles** driven into firmer rock below. However, unless a reliable survey has been undertaken even a sound engineering job can result in disaster. The house in Figure 9.3 faces likely collapse from the ground sliding away beneath it.

Forest cleared

Road

Permeable sandstone

Shale

Cut

Fill

Clay

Slope cut away for another road and house

Figure 9.3 Typical house structure in Los Angeles

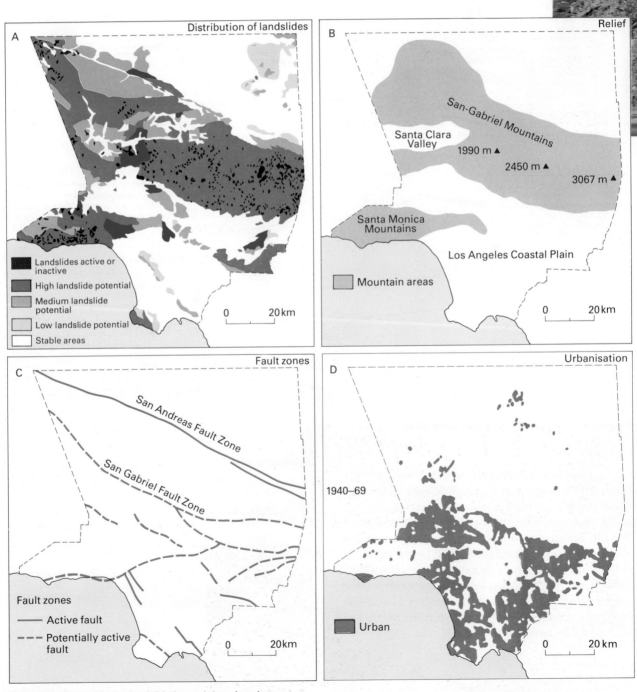

A Distribution of landslides

Landslides active or inactive

High landslide potential

Medium landslide potential

Low landslide potential

Stable areas

0   20km

B Relief

San-Gabriel Mountains

Santa Clara Valley

1990 m ▲

2450 m ▲

3067 m ▲

Santa Monica Mountains

Los Angeles Coastal Plain

Mountain areas

0   20km

C Fault zones

San Andreas Fault Zone

San Gabriel Fault Zone

Fault zones

Active fault

Potentially active fault

0   20km

D Urbanisation

1940–69

Urban

0   20 km

Figure 9.4 Factors in the landslide hazard, Los Angeles

Figure 9.5 Landslide damage in California

This has happened in Figure 9.5. Many people in Los Angeles County do realise the dangers, and much research is being done into both the physical processes that produce landslides as well as into the engineering problems created by development. However the problems are very complex and not everyone agrees on the possible solutions. The Los Angeles authorities attempted to control building in the Santa Monica Mountains. However, they were sued by a developer from New York on the grounds that such controls were not only against the city charter but discriminated unfairly against the Santa Monica community! The owners of very expensive houses do not want to see the price of their property decline as a result of environmental managers declaring their homes to be in a hazard zone.

ACTIVITIES

1 Refer back to the 1:50 000 OS map extract of Eastbourne Figure 6.3 and the geology map Figure 6.4. Choose two grid squares at random from the chalk area and two from the clay area.

   a Describe the slopes in each area.

   b Measure the average gradient in each pair of squares.

   c What conclusions can you come to from your answers to a and b?

2 Read the newspaper report in the case study of the Rio landslides in Figure 9.1 and answer the following questions:

   a How did the weather contribute to the disaster?

   b How does the spread of the favelas increase the risk of landslips?

   c Imagine that you live in part of the favelas. Give an explanation to a newspaper reporter of why you built your home where you did.

   d Other than physical injury, what other health risks are faced by the the people in the affected area?

3 Study carefully the maps A–D in Figure 9.4.

   a What areas have:
      i a high landslide potential, and
      ii a stable condition?

   b Using the evidence from maps B and C, explain the distribution of these two areas.

   c What is the evidence from map D that suggests landslides are a hazard?

   d Draw an outline of Los Angeles County. On this map shade the following areas using a suitable colour code:
      i housing with a high risk of collapse;
      ii housing with a low risk of collapse;
      iii housing with no risk of collapse.

4 From Figure 9.2 and Figure 9.3 explain why the house in Figure 9.3 is in danger of being destroyed by a landslide.

5 What has been done in Figure 9.6 to prevent slope failure? Use a sketch to illustrate your answer.

Figure 9.6 Stabilising a slope in Japan

10 Tourism in the mountains

Millions of people take their holidays in the mountains of Europe and North America both in the winter and summer. The air is clear and fresh, the scenery majestic and the resorts provide recreational facilities from skiing and climbing to tennis and golf. In Chamonix, France you can even have a go on an artificial louge run! Figure 10.1 illustrates the attractions of taking a holiday in Chamonix. However, as you have already

discovered, some activities, like skiing, create important environmental hazards. Many local people obviously benefit from the influx of tourists. They bring economic prosperity and jobs. On the other hand some people say that the traditional culture of a region is spoilt and devalued. Local crafts are commercialised and displays of local traditions like many of the dances degenerate into cheap tourist displays.

BEGINNERS ★★ INTERMEDIATE ★★★ ADVANCED ★★★

HAUTE SAVOIE 3,400 – 12,600 ft

Chamonix

The Chamonix Valley offers a vast and varied ski area almost without parallel.

Chamonix is a fascinating little town set amidst beautiful scenery at the foot of the majestic Mont Blanc (at 15,800 ft Europe's highest peak). The Chamonix area has been well known to British skiers and mountaineers for many years, and visitors come from all over the world to ski the famous Vallée Blanche which usually opens in early February. This run can only be undertaken with a guide. The skiing is divided into 5 main areas in or around Chamonix which are linked together by a very efficient ski bus, free to lift pass holders. The ski pass covers numerous resorts in the Chamonix Valley and in all there are 135 lifts and over 500 kms of pistes. An added attraction is the possibility of joining ski tours to such resorts as nearby Argentière, Megeve and Courmayeur in Italy. Apart from having something for every type of skier, Chamonix has an excellent range of sporting and leisure facilities, including a large sports centre with swimming pool, sauna and ice rink, cross country ski trails and beautiful cleared walks. There are a number of excellent restaurants as well as a host of bars and cafés. Film shows are held regularly in the evenings and Chamonix also has its own casino.

Lifts	Type	Rise ft	Lifts	Type	Rise ft
1 Plan de L'Aiguille	CC	4,200	9 L'Index	GL	1,600
2 Aiguille du Midi	CC	4,800	10 Lognan	CC	2,550
3 Vallée Blanche	GL	1,050	11 Grands Montets	CC	4,480
4 Planpraz	GL	3,000	12 Le Tour-Balme I, II	CL/DL	2,260
5 Brévent	CC	1,650	13 Chamillon I, II	GL	2,230
6 Bellvue	CC	2,650	14 Plan Joran	CL	2,263
7 Prarion	GL	2,750	15 La Pendant	CL	1,115
8 Flégère	CC	2,730	16 Plan Roujon	CL	787

Plus 1 more Gondola Lift and numerous Chair and Draglifts.
CC = Cable Car. DL = Draglift.
GL = Gondola Lift.

Mont Blanc
Les Houches
CHAMONIX
Les Praz
Argentière
Le Tour

YOUR RESORT

Marked piste: 500 kms.
Direction of slopes: All.
Number of pistes: 12 green, 17 blue, 20 red, 6 black.
Lifts: 6 cablecars, 6 gondolas, 14 chairlifts, 30 draglifts.
Ski school: Sun – Fri; morning lessons.
Ski bus: Included in the Mont Blanc ski pass.
Cross country trails: 25 kms.
CHILDREN
Children's ski school: From 4 – 12 years. 4 hours per day with midday meal included. Price: (1986/87) 150 FF per day, 750 FF for 6 days.
Kindergarten: Panda Club, daily, 8.30am – 5.30pm, includes ski school.
Price: (1986/87) 1 day 200 FF, 6 days 999 FF, including meals and ski lessons.
Age: 18 months – 14 years.
RESORT FACILITIES:
Variety of shops, banks and supermarkets, wide choice of restaurants and bars, several nightclubs, 4 cinemas, large sports centre with Olympic sized swimming pool, indoor tennis (2 courts), skating rinks, numerous mountain restaurants.
ACTIVITIES:
Swimming, sauna, skating, tobogganing, sleigh rides, tennis, hang gliding, cleared walks, excursions to Italy and Switzerland.

TRAVEL DETAILS

✈ BY AIR –

7 & 14 NIGHTS – DEPART SATURDAY

TO GENEVA FROM

GATWICK	GLASGOW
HEATHROW	MANCHESTER

Figure 10.1 Chamonix as a holiday resort

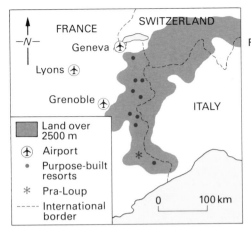

Figure 10.2 The distribution of new ski resorts in the French Alps

The boom in the skiing industry is causing a significant increase in avalanches, landslides, mudslides and floods. Trees, which bind the soil together and protect the surface from the worst of the elements, are cleared to provide pistes, access roads, and lifts. New hotels, shops, and car parks all provide water-proof surfaces that increase run-off and the risk of flooding. One commune in the French Alps had to spend nearly £6 million to protect itself from mudslides and floods that started to occur after the building of the new ski resort of Les Arcs in the 1960s.

Developers have built several purpose-built ski resorts in Haute-Provence and Haute-Savoie in France (see Figure 10.2). In many ways they are ideal for skiers. Their height (usually above 2000 m) makes the snowfall a lot more reliable, especially at each end of the season when lower resorts like Chamonix cannot guarantee a good cover. The number of hotels and the capacity of the ski-lifts have been phased together to prevent long queues and overcrowding. Several long, inter-connecting runs have been developed, and mountainside services like restaurants and first-aid posts have been carefully located.

Figure 10.3 The purpose-built ski resort of
Pra Loup, Haute-Provence

ACTIVITIES

1 From the brochure information in Figure 10.1 put the resort's holiday activities into two lists: one headed 'winter' the other headed 'summer' Put an asterisk by those activities you think are unique to mountain areas.

2 European journeys to Chamonix start from as far away as Glasgow. On an outline map of Europe use your atlas to mark on the following: the Alps (shade area above 1000 m), Chamonix, Geneva, Glasgow. Now draw a circle around Chamonix whose radius is the distance from Chamonix to Glasgow. What other important cities lie within this circle? You could mark some of these on your map. Give your map a suitable title.

3 From the Swiss map extract, Figure 7.2 on p. 20, what is the evidence that Chamonix could not easily develop its own airport?

4 Study the photograph in Figure 10.3 of Pra Loup and answer the following:
  a Describe the site of Pra Loup;
  b How can you tell from the photograph that it is a new resort?
  c Why should developers want to build a ski resort here?
  d Estimate the proportion of the forest below the tree-line that had to be cut down for ski-runs, lifts, and the village itself. (The tree-line is the highest level on the mountainside at which trees will grow to full height.)

Skiing into trouble in Scotland

THE skiing industry made an inauspicious start when the pistes opened up in the Cairngorms in the 1960s. Many ski operators made the mistakes familiar to the Alps, by bull-dozing mountain slopes of the arctic-alpine vegetation, above the timber line. They used heavy machinery such as "pisters" to move and pack the snow, and allowed skiing during periods of thin snow cover when the vegetation was most vulnerable to damage.

The first signs of trouble were visible soon after each skiing season. The natural vegetation was crushed, bare soil was exposed, and soil erosion set in. Because the skiing suffered, the ski operators took remedial action. Many snow fences were constructed to catch snow on the ski runs, and pylons and concrete for new ski lifts were flown in by helicopter to avoid damaging the ground. The operators built gravel tracks to channel mach-inery. Using fertilisers and commerical grass seeds, they attempted to restore damaged vegetation at the end of each skiing season.

Other problems are now surfacing e.g. changes in the vegetation near ski lifts. This was caused by machines, skiers and walkers. Species often found on heavily trampled footpaths were growing on reseeded pistes, such as creeping fescue and rye-grass. A survey in 1981 found more wide-spread serious damage to vegetation, decreases in vegetation cover and soil erosion.

The ski developments have led to other worries. On and near the ski grounds, the breeding success of red grouse has suffered due to lack of insect life in the new environment. The birds' breeding is also suffering from predation by crows, whose population is thriving on the scraps of sandwiches and other food left over from the tourists.

Many adult birds die when they hit the ski cable wires, particularly in Coire Cas where the large numbers of skiers flush out the birds.

The ski operations are also in con-flict with conservationists. A major application for a public road and several ski tows, in the northern corries west of Coire Cas, led to largescale objections from the Nature Conservancy Council (NCC), the Mountaineering Council of Scotland and several other bodies, culminating in the Lurcher's Gulley public inquiry in 1981.

The problem is a political hot potato. How much should the government and local authorities clamp down on the developers, in an area of chronic unemployment, which has traditionally had to rely on a short summer tourist season? Yet the booming ski industry is not so self-sufficient as it might appear. Many of the ski tows and other faclities have received large grants from public bodies in Britain and the EEC. The heavy costs of im-provements to roads and emergency policing are met by the ratepayers, and the high costs of skiing accidents and emergency rescue by the tax-payers. □

New Scientist 14 January 1988

Figure 10.4

ACTIVITIES

5 Study carefully the report on Scottish ski developments in Figure 10.4.
Prepare a report for the Lurcher's Gulley public inquiry for a representative from each of the following interest groups. Your report should be three or four paragraphs in length.
a Nature Conservancy Council
b Mountaineering Council of Scotland
c Local Authority
d Ski Club of Great Britain

6 Locate Planpra on the Swiss map extract on p. 20 (546873). A developer wishes to build a ski resort here. Outline the advantages and disadvantages of such a scheme. Imagine the new resort has been built. Prepare a page for a brochure for one of the leading holiday tour companies giving details of the resort and encouraging people to go there.

Coursework ideas

1 Choose a mountainous region, either in Britain or abroad. Make a study of the region you have chosen using the following topics to help you:
   a the role of glaciation in shaping the land;
   b the traditional activities of the people;
   c the impact of tourism;
   d industrial development.
   Include as many maps, diagrams and photographs as you can to illustrate the character of the region you've chosen.
2 Some mountainous areas in the world have not been developed very much, e.g. the Himalayas and the Andes. Select a country in such a region and find out why it has not developed as much as other mountainous countries. What advice would you give to such a country planning to develop its mountains for tourism?

3 Make a survey of slopes in your home area. This should be done along a number of different transects, i.e. lines between two points. The notes below show you how to do this. As you record the angle of slope, also write down the land use. Write up any conclusions you come to about any relationship between slope angle and the use of the land. If you live in an urban area you may like to test the following hypothesis: 'High class housing is more likely to occur on higher and steeper ground than low class housing.'

Notes on how to measure slopes

1 Slopes are measured with a clinometer. You can easily make a simple one like the one in Figure 10.5.

2 Two people of the same height stand 20 m apart along the line of the transect (see Figure 10.5). This is called the **measured length**. The uphill person looks towards the downhill person along the sighting pins. A third person records the slope at the point where the thread crosses the protractor markings. Work downhill along the maximum gradient. A number of measured lengths will be needed along one transect.

3 Record your results as shown in Figure 10.6

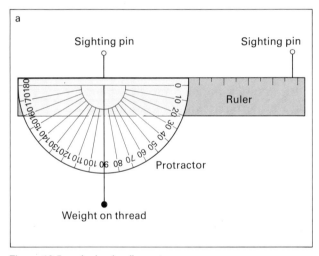

Figure 10.5 a A simple clinometer
   b Measuring slope along a measured length.

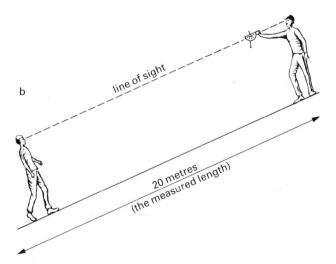

START	Post Office	FINISH	Duck & Feather Inn
	820963 (OS ref)		825964

Measured Length 20 metres	Angle x°	Uphill/ Downhill	Landuse
1	5	D	Detached & semi-detached housing
2	2	D	Shops
3	2	D	"
4	2	D	Terraced housing

Figure 10.6

11 The drainage basin

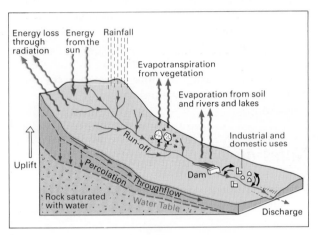

Figure 11.1 The drainage basin as an open system

Over much of lowland Britain and Europe the present day shape of the physical landscape is the result of the work of rivers. Your local river has probably created quite a distinctive valley. Some of these valleys are even detectable in a city landscape if you know what to look for. Over a wider area several neighbouring valleys will have combined to produce a recognisable **drainage basin**. This is the area that is drained by a river and its tributaries. The boundary of a drainage basin is called the **watershed**. Drainage basins are a good example of an open system with inputs and outputs, as shown in Figure 11.1.

You can see that an important input is rainfall. This provides the water that flows in each river. The amount of water in the river is known as the **run-off**, or the **discharge**. However, a number of things can happen to the water besides entering the rivers:

1 it can be **evaporated** into the atmosphere;
2 it can **percolate** into the soil and rock;
3 it can be absorbed by plants;
4 it can be used by people in homes, industries and farms.

Figure 11.2 shows two contrasting river basins: the Exe in Devon and the Colne in Hertfordshire. There are a number of important differences between them:

1 the Exe Basin is on mainly impermeable rock whilst the Colne is on mainly permeable rock;
2 the rainfall into the Exe Basin is much higher than that in the Colne Basin;
3 the gradient of the Exe Basin is steeper than that of the Colne Basin.

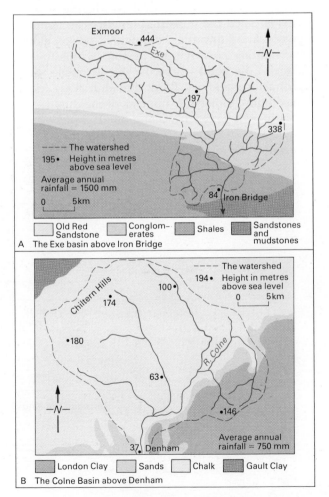

Figure 11.2 The drainage basins of the Exe and the Colne

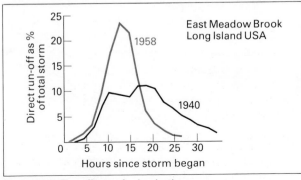

Figure 11.3 The effects of urbanisation

Table 11.1 Rainfall and Run-off for the Rivers Exe and Colne

River Colne	Oct	Nov	Dec	Jan	Feb	Mar	Apr	May	Jun	Jul	Aug	Sep
Run-off (mm)	20	16	15	15	14	14	13	12	12	12	15	15
Rainfall (mm)	30	40	60	65	30	50	60	50	70	120	65	125
River Exe												
Run off (mm)	50	100	145	215	35	70	50	35	30	75	55	80
Rainfall (mm)	140	145	180	205	15	120	70	100	105	180	110	145

The activities of people have a considerable impact on the drainage basin. In cultivated regions the landuse type influences the amount of run-off significantly. Ploughed land increases run-off because rain drops compact the soil particles making an impermeable layer. You have already discovered that tree-felling greatly increases run-off. It is this that is contributing to the frequent flooding of the lower Ganges in Bangladesh.

A major reason for increased run-off in many rivers is the building of towns and cities. Vegetated areas are replaced with water-proof surfaces of tarmac and concrete. These direct water straight into the rivers through the drains making rivers rise rapidly (see Figure 11.3).

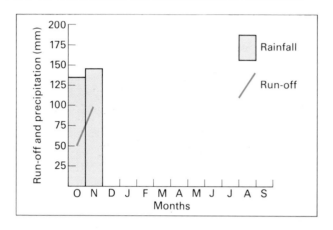

Figure 11.4 The construction of rainfall and run-off graphs.

ACTIVITIES

1 From the maps in Figure 11.2 work out the drainage density of each river basin. To do this, carry out the following steps for each basin:

   a measure the total length of rivers in kilometres;

   b make a tracing of the watershed and place it over a piece of graph paper;

   c count the total number of squares on the graph paper that are enclosed by the watershed;

   d from the scale bar on the map find out the area of each square of the graph paper in km^2. Now multiply this area by the total number of squares within the watershed. This figure is the area of the basin in km^2;

   e divide the area of the basin into the total length of the rivers. This is the drainage density of the basin in kilometres per square kilometre.

2 Compare your drainage density results with each of the following from Figure 11.2:

   a the average rainfall for each basin;

   b the rock types in each basin;

   c the average gradient of each basin.

   What conclusions can you draw from your observations?

3 Table 11.1 shows the rainfall and run-off for the River Colne and the River Exe in a typical hydrological year. Study this and then answer the questions that follow.

   a From the table find out the following for each basin:

   i the value and month of the maximum run-off and rainfall

   ii the value and month of the minimum run-off and rainfall

   iii the average monthly run-off and rainfall.

   b Draw a graph to show the relationship between rainfall and run-off for each basin. Figure 11.4 shows you what to do.

   c Describe the relationship between run-off and rainfall for the Colne and the Exe during the year.

   d Describe and attempt to explain the differences between the Exe and the Colne in their run-off characteristics.

4 Study carefully Figure 11.3 which shows the run-off characteristics for East Meadow Brook, USA before and after urbanisation in two similar storms. Urbanisation took place between 1940 and 1958.

   a What was the peak run-off value in 1940 and 1958?

   b How long after the start of each storm did it take for these peaks to occur in each of the years shown?

   How can you explain the differences between the run-off characteristics in 1958 and 1940?

12 The river and its valley

As you have already discovered, a good deal of the water that enters a drainage basin eventually finds its way into individual rivers. The amount of water flowing in a river at any one time is known as the **discharge** of the river. It is measured in cubic metres per second. The level of the discharge is very important in helping to determine the kind of work that the river can do. As the discharge increases so does the speed or **velocity** of the river.

Basically a river will do three things:

1 **transport** material [boulders, pebbles, silt] called the **load**;
2 **erode** its bed and banks. Particles carried by the river will also be eroded;
3 **deposit** its load.

As the discharge and velocity increase, the river is able to transport bigger and bigger particles and carry out more erosion. As the level of the river drops so does its velocity. This means it can only transport the smaller particles. The heavier material is deposited and erosion will get less.

The discharge and the velocity give the river **energy** to do work. When the discharge and velocity are high (e.g. when the river is in flood) the river can carry very large boulders which hit the bed and banks of the river's channel as they are swept along. This causes erosion of the channel. When the water level is low, there is much less energy. The larger material will be deposited and only the smaller particles can be carried. There will be very little erosion. The load carried by the river not only erodes the bed and banks, but the particles also erode each other. They will collide as they are carried along and so

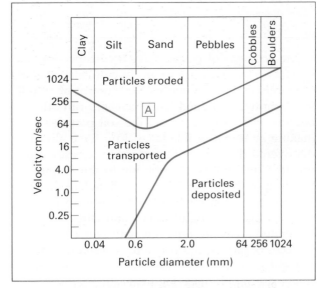

Figure 12.1 Relationship between velocity and work of a river

become smaller in size. This process is called **attrition**. Figure 12.1 shows the important relationships between river velocity and the work the river does.

You will notice that when particle size drops below about 0.8 mm in diameter (at the point marked A on the graph in Figure 12.1) the velocity of the river has to increase in order for them to be eroded. This is because they are so small they tend to stick to one another and to the channel bed.

Figure 12.2 is a **long profile** of the River Exe in Devon. Notice that the gradient is steepest higher up near the source and gradually flattens out towards the mouth. This is because downward erosion is a dominant process in the upland section, whilst deposition is the dominant process in the lowland section.

Figure 12.2 The long profile of the River Exe from source to Iron Bridge

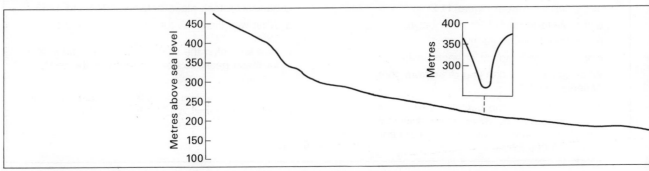

Figure 12.3 The upper reaches of the River Exe

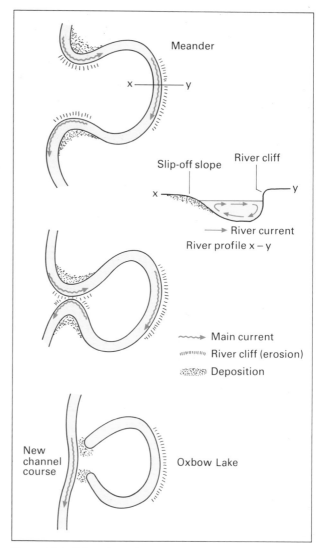

Figure 12.4 Oxbow formation

The photograph in Figure 12.3 is of the upper reaches of the River Exe. Here it is high above sea-level and it is this height that gives it **potential energy** to erode downwards into its bed. Erosion is carried out in the following ways:

1 by **corrasion** – boulders and pebbles hit the bed and banks;
2 by **hydraulic action** – the force of the water dislodges material;
3 by **corrosion** – bed and bank material is dissolved by acids in the water;
4 by **pressure-release** – when the water level drops, the pressure on the banks is reduced causing the bank to crumble.

Interlocking spurs are formed on the inside of each bend. Erosion of the valley sides, including soil creep and landslips, helps the valley to widen and the sides to slope away from the river. In this way the valley becomes wider and less steep further downstream.

In lowland stretches of a river, downward erosion is far less important. Although the discharge is greater, a greater proportion of energy is used to transport material. Less energy is used in erosion. Bends enlarge into **meanders** which wander across a wide **flood plain**. The current of the river flows around the outside of each meander causing the undercutting of the outside bank.

Deposition takes place on the inside of the meander where the current is slack. This causes the shape of the meander to change, with interesting results shown in Figure 12.4. Meanders also move their position. This is called **meander migration**.

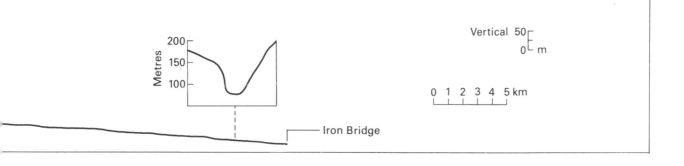

Figure 12.5 The river flood plain: a the Cuckmere Valley

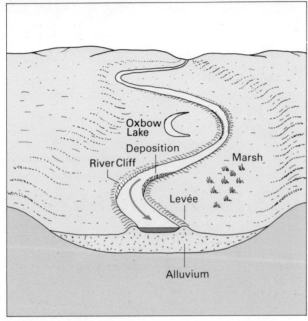

b Flood plain features

Meander 'migration' and frequent flooding both help to create quite distinct valley landforms. These are illustrated in Figure 12.5. A lot of fine-grained material is deposited in times of flood, especially near the banks where it builds up to form embankments or **levées**. The deposited material is called **alluvium**. It is particularly fertile but is unsuitable for ploughing. When artificially drained and protected from further floods farmers do use the flood plain for pasture.

ACTIVITIES

1 Study very carefully the diagram in Figure 12.1 and then answer the following.

  a Find out the lowest river velocity that is needed to carry:
    i silt particles 0.6 mm in diameter;
    ii sand particles 2.0 mm in diameter;
    iii boulders 1000 mm in diameter.

  b Why do you think it takes a faster velocity to carry boulders than to carry sand?

  c Describe the kinds of particle that a river can carry if it flows at
    i 4.0 cm/sec
    ii 50 cm/sec

  d What is the minimum velocity at which any attrition will take place?

2 Look again at Figure 12.3. From this photograph:

  a draw a picture to show the interlocking spurs;

  b draw a sketch contour map at a scale of about 1:25 000. Assume the river is at 200 m above sea level and the top of the valley sides are at 350 m. Use a contour interval of 20 m.

3 Study Figures 12.4 and 12.5. Then turn back to the map extract Figure 6.3 on p. 16 (the OS map extract of the Eastbourne area). Answer the following questions.

  a Draw a sketch map of the course of the river Cuckmere and its valley south of Exceat Bridge (514994). Label in the correct places:
  MEANDER   OX-BOW LAKE
  ARTIFICIAL DRAINAGE CHANNEL.

  b Draw an accurate cross-section from 500980 to 530980. Use a vertical scale of 2 mm to 20 m.

  c Measure the width of the flood plain from your section.

  d Levées run along the banks of the Cuckmere. Why do you think they are not shown on the OS map?

  e From the photograph in Figure 12.5 describe how people are making use of the river Cuckmere and its valley.

  f What is the evidence from the OS map that:
    i the meanders below Exceat Bridge are unlikely to change their present shape whilst people are occupying the valley;
    ii the river is depositing material beyond its mouth. Draw a sketch map to illustrate your answer.

13 Flooding

Flooding will occur when the the amount of water flowing in a river (the discharge) is too great to be contained by the channel of the river. The water will then overflow onto the surrounding land, or the flood plain. In the countryside flooding may drown livestock, ruin crops and cut off villages from the outside world. In urban areas there may be many people at risk from drowning and the damage done to domestic and industrial property can be very costly indeed (see Figure 13.1). It is clear that if floods can be predicted then measures can be taken to protect lives and property (see Figure 13.4).

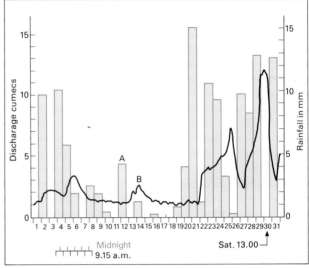

Figure 13.2 Discharge and rainfall for the Colne, January 1988

Figure 13.1 Flooding in Watford, January 1988

Flood chaos as River Colne's banks give way

Marcus Duffield

Flash floods hit Watford's Water Lane at the weekend, bringing chaos and concern in their wake.

Thames Water Authority issued red alerts after the wettest January in 50 years brought the threat of serious flooding.

The River Colne burst its banks in the early hours of Saturday, and motorists, students, pedestrians and animals in Water Lane have all suffered as a result.

The police were aware of the rising water level at 12.30am and 30 minutes later the Colne burst its banks.

By daylight the water level had stopped rising and Water Lane was passable. Police positioned warning signs for drivers and the road was still flooded at 5pm on Saturday evening.

"The road was passable all day although we were concerned during the morning. We left motorists to find alternative routes if they wanted to" said a police spokesman.

Watford College car park was under three feet of water and some students were sent home because B block was flooded and two of the fire escapes were closed. The college library was also shut.

Engineers told the college that the recent remedial work carried out on the River Colne —diverted to make way for the new Tesco superstore—should have lessened flooding risks.

The Tesco site was also threatened. "The rainfall has been exceptional over the last few weeks but the work done to the river means that it should be able to cope" said a spokesman for the superstore.

Passers-by concerned about six donkeys and two ponies in the flooded field next to Water Lane, were relieved to see them moved after the heavy rainfall of recent weeks.

The field was almost completely under water and Mrs Brenda Hughes of Shepherds Lane said "It was upsetting for the children to see them in that state".

Owner, Mr Fred Eames, of Church Hill, Bedmond said "Those animals are my livelihood and if I do not look after them I will lose out".

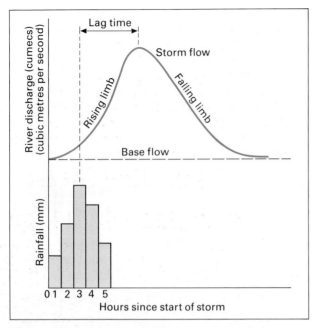

Figure 13.3 The flood hydrograph

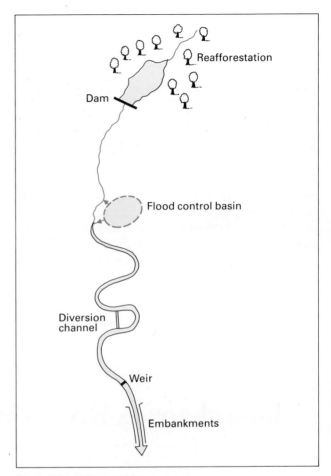

Figure 13.4 Methods of flood control

Floods have to be understood before they can be predicted. The **flood hydrograph** is a way of describing a storm and the river's response to it. A flood hydrograph is shown in Figure 13.3. The **lag time** is a very important part of the flood hydrograph as it tells the authorities the time available between the peak rainfall and when the river may burst its banks (the peak storm flow). Usually this is no more than a few hours, but is, perhaps, long enough to evacuate people from the areas most at risk. For effective flood protection, hydrologists study the records of past floods. From these they can work out the probability of floods of different magnitude happening.

Very severe floods may only occur once in perhaps one hundred years. A flood of this magnitude is called a **one hundred year flood.** Most British towns are protected against such a flood. Protection against a one hundred year flood is obviously more expensive than protection against a fifty year flood or a twenty year flood. A one thousand year flood would involve very costly protection measures. However it is important to realise that one thousand year floods do not occur at regular intervals of 1000 years. It could occur two years running and then not for 2000 years. It could work out that the cost of protection from a very infrequent but severe flood might be greater than the flood damage itself. Much depends upon what facilities there are to be protected.

There are a number of personal and public adjustments that can be made to the risk of flooding.

1 *Accepting the loss* This is the most common adjustment, particularly for poor individuals in the developed world and for whole communities in the developing world.
2 *Relief Funds* Funds from central governments, charities and individuals are given to help flood victims and to compensate for damage to life and property. Studies done in North America, however, show that the granting of public relief does nothing to discourage the settlement and use of flood-prone areas.
3 *Emergency action* This includes the evacuation of areas in immediate danger of flooding. Some of the largest rivers in the world have mass evacuation programmes, e.g. the Lower Mekong, the Mississippi and the Indus. Also included here is flood fighting such as the last minute use of sandbags around buildings.

4 *Structural changes* Buildings can be made water-proof or put on stilts.

5 *Regulation of landuse* Not using flood plains at all would reduce the flood hazard considerably, but this is not always possible or economically sensible. Functions that can withstand flooding without costly consequences, like recreation facilities, or functions that can afford the 'natural tax' of flood losses e.g. some industrial uses, can be located on the flood plain.

6 *Insurance* Although it is possible to insure against flood damage, premiums may be high because only occupants of flood-prone areas are going to take out such policies.

7 *Flood Control* Most flood-prone areas in the developed world have a variety of control methods ranging from highly expensive dams to the least expensive diversion channels and embankments. Figure 13.4 illustrates some of the techniques that may be employed.

CASE STUDY

The Bangladesh floods 1987

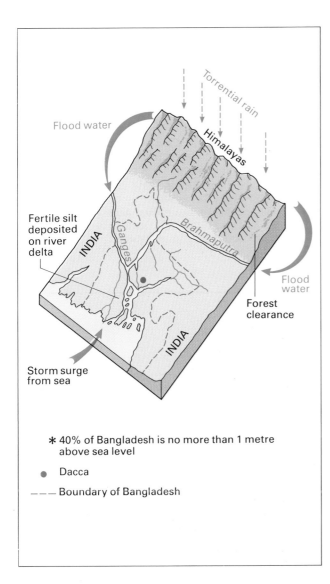

* 40% of Bangladesh is no more than 1 metre above sea level

● Dacca

--- Boundary of Bangladesh

24 million hit by Bangladesh deluge

Figure 13.5

During September 1987 devastating floods hit Bangladesh (see Figure 13.5). Large areas of the country lay under six feet of water. The main effects were:

1 24 million people made homeless;
2 4.3 million acres of crops ruined;
3 over 700 people killed;
4 1800 miles of road and more than 1200 bridges destroyed;
5 the spread of dysentery and diarrhoea.

The causes of the flood are outlined in Figure 13.6.

Figure 13.6 The causes of the Bangladesh Flood

ACTIVITIES

1 Study the local newspaper report of the floods in Watford (Figure 13.1) and the graph showing the discharge and rainfall for the River Colne for January 1988 (Figure 13.2).

   a What was the peak discharge of the river and on what day did it occur?

   b How much rain had fallen in the previous 24 hours?

   c What was the lag-time between the peak rainfall at 'A' on the graph and the peak discharge at 'B'? How might you account for the much shorter lag-time before the flood of the 30th?

   d How much flooding was there at Watford College?

   e What do you consider were the most serious effects of the flooding?

2 The 1:25 000 OS map Figure 14.1 on page 41 covers the Colne flood plain in Watford. From this map find the answers to the following questions.

   a Give the six-figure grid reference for the college. What is the height of the college above sea level?

   b What kinds of landuse appear to occupy the flood plain downstream of the college?

   c Describe the distribution of housing in relation to the flood plain upstream of the college.

   d How are the railway lines and the M1 that cross the flood plain protected from flooding?

3 The flooding in Watford on the 30th of January 1988 was not very serious: probably a one year flood. A one hundred year flood could reach the edge of the CBD. Imagine a one hundred year flood is imminent. Write a 'Flood Emergency' leaflet that would be distributed to

   a householders, and

   b shopkeepers.

In the leaflet outline the steps that ought to be taken to protect lives and property. Also, design a warning poster to be displayed in public places.

4 With the help of your atlas draw a map of the area that lies from 20° N to 30° N and from 85° E to 95° E. Mark on your map:

   a the Ganges and the Brahmaputra rivers;

   b the Himalayas;

   c the cities of Dacca, Barisal, Rangpur and Calcutta;

   d the border of Bangladesh;

   e a key, scale, and title.

5 Study all the information given about the Bangladesh floods of 1987 and answer the following questions:

   a What two major rivers flow through Bangladesh?

   b What particular physical feature do they create near the coast?

   c Why does this particular feature encourage the people to settle in an area that so easily floods?

   d Explain why the rivers became swollen and overflowed their banks.

   e How did the sea contribute to the flooding?

   f The immediate effects of the floods have been listed. Which one of these would particularly hamper relief supplies and workers reaching the needy? What do you think are the longer term effects of the flooding?

   g Do you think the citizens of the developed world, some of whom themselves may be in danger of flooding, should make contributions to relief funds for flood victims in countries like Bangladesh? Justify your answer.

14 Rivers and recreation

CASE STUDY

Watford's Colne Valley Linear Park Plan

Figure 14.1 The Colne Valley in Watford (O.S. 1:25 000 map extract)

The 1:25 000 OS map extract in Figure 14.1 covers some of the area of Watford's Colne Valley Linear Park. It runs from the Borough boundary in the north to beyond the south west corner of the map extract. The width of the park is limited by the Borough boundary to the east. Just over 75 000 people live in the Borough of Watford and the amount of recreation land available to them is limited within the Borough itself. You have already seen that flooding is a hazard in the flood plain. In a fairly dense built-up area there is little to encourage wildlife. The Linear Park attempts to deal with these problems. Its main aims consist of the following points (see Figure 14.2).

1 To improve the recreational and amenity value of the Colne Valley. Providing for organised recreations like soccer and cricket is difficult within the park area because of the narrow width of the valley floor. However, there is scope for the more informal activities like walking, cycling, having picnics and so on. Unsightly areas like factories and the M1 motorway are being screened with the planting of shrubs and trees. The amenity value of the river itself can be improved with the clearing out of rubbish that has been tipped into it.
2 To improve the flow of the river. Clearing banks of dense vegetation and dredging the bed will help to increase the rate of flow of the river and help reduce the flood risk.
3 To improve the value of the river to wildlife. There are a number of wildlife habitats in the valley that are largely free from human interference. The railway embankments and the disused marl pit in the north of the park area are two important habitats. Public access will remain limited in these places. The tree planting programme elsewhere will help to enhance the wildlife communities and reinforce the valley as a routeway for the movement of birds, animals and insects. Two sites of particular interest are the water meadows in the area around 103952 and at 116954 where swans nest each year.

a 112953

Figure 14.2 Sites in Watford's Linear Park

Obviously if the general public are going to be able to enjoy the benefits of the valley they must be able to get to it and be able to walk along its length. The 1:25 000 map does show where people have right of way (green pecked lines). There are however significant sectors where there are no rights of way because of problems of land ownership and physical obstacles like the railway lines and the river channels themselves. Unfortunately the length of the valley is now to be the proposed route of a link road to the M1 motorway. Should this road be built it could well undermine many of the proposals in the Linear Park Plan.

ACTIVITIES

1 Each of the four photographs labelled a to d in Figure 14.2 shows the scene at different places along the Linear Park. The six-figure grid reference for the 1:25 000 OS map given below each picture shows where the photograph was taken. Locate the site of each photograph.
   a In cases a and b describe which aspects of the park plan need to be implemented and why.
   b In cases c and d explain what aspects of the plan have already been implemented.

2 The stretch of the valley between 117966 and 105951 has been given priority. Why do you think this has been necessary?

3 Imagine that the disused marl pit at 122984 is scheduled for use as
   a a bird sanctuary, OR
   b a motor-bike scrambling course, OR
   c a tip for domestic refuse (to be landscaped when full).

Write out or discuss within your class the likely points of view on the matter of the following people: the secretary of the local motor cycle club,
an official from the council's refuse department.
a local resident.

b 108952

d 117970

c 116962

15 Water supply

'There is no life without water. It is a treasure indispensable to all human activity'
[European Water Charter 1968]

You probably use over 100 litres of water everyday. The average household water use is shown in Figure 15.1.

The amount of water used in industry is on a much larger scale. The manufacture of one tonne of steel needs about 250 m³ (250 000 litres) of water.

Water for use in homes and industry comes from three basic sources:

1 from underground water bearing rocks (known as **aquifers**) through wells and boreholes;
2 directly from rivers;
3 from reservoirs created by building dams across upland streams.

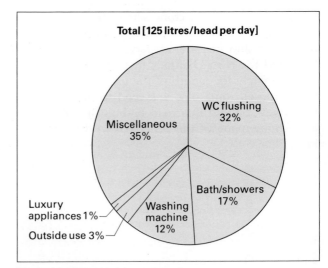

Figure 15.1 Household water usage

CASE STUDY

Watford's water supply

Watford is in the supply area of the Colne Valley Water Company. An average of about 30 megalitres of water are supplied each day to the 75 000 people living in the Borough. The sources of this water are shown in Figure 15.2 and Figure 15.3. There are three **boreholes** at The Grove: these supply about 70 per cent of the total. Hillfield Park reservoir is used for emergency supplies.

After being pumped up from the chalk **aquifer** (a water bearing rock), the water must be treated to make it fit for human consumption. This is done at Clay Lane and The Grove. Treatment includes

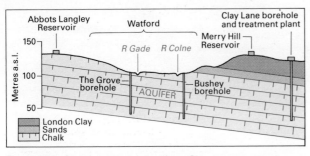

Figure 15.3 Sketch cross section from Clay Lane to Abbots Langley

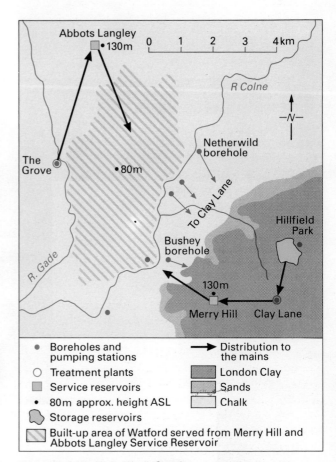

Figure 15.2 Watford's Water Supply

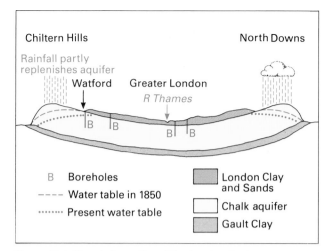

Figure 15.4 The chalk aquifer of the London Basin

ACTIVITIES

1 Study the information in Figures 15.2 and 15.3 which show Watford's water sources.

a From what rock does the water come?

b How is it extracted from the rock?

c How far away are the two service reservoirs from the centre of Watford? What advantage does the site of each service reservoir have for supplying water to Watford?

d Where do you think Hillfield Park Reservoir gets its water from?

e Approximately what is the depth of the boreholes shown in Figure 15.3?

2 Hillfield Park is an emergency reservoir. Under what circumstances do you think its water might be used?

3 What is an aquifer? Why do you think that chalk makes such a good aquifer?

4 Study the map in Figure 15.5 alongside a rainfall map of Britain (in your atlas).

a In general water is to be moved from the west and north to the south and east. Why should this be so?

b Where are the new reservoirs planned? How is water to be taken from these to where it is needed?

c The River Dee could be used for 'estuarial storage'. What do you think this means?

5 Make a diary of your daily water consumption. How does your use of water compare with the figures in Figure 15.1.?

sterilization with chlorine, softening (removal of some of the calcium carbonate), and the addition of fluoride. From the treatment works the water is pumped into **service reservoirs** and from there it enters the mains.

Watford is only one of the many communities that get their water from the chalk aquifer that is found in the **syncline** of the London Basin (see Figure 6.2 p. 14). Rainwater percolates into the chalk of the Chiltern Hills and the North Downs. The water is stored in the rock pores and fissures. The upper level of the water is called the **water table**. The water table level is normally kept up by the amount of rain that falls, but since so many boreholes and wells have been sunk to extract water, the level has fallen about 50 m over the last 140 years.

Boreholes are able to tap the water from the chalk as it moves through the pores and fissures of the rock. Figure 15.4 is a cross-section through the London Basin. London itself gets about 15 per cent of its supply from wells sunk into the chalk. 71 per cent comes from the River Thames and therefore consumes water that originates well beyond its own boundaries.

For most of lowland England demand for water outstrips local supply, whereas in the highlands of the west and north there is a water surplus. Reservoirs in central Wales like Vyrnwy have supplied Birmingham since the last century. There are now plans for a **national water grid** similar to the electricity grid to distribute water from the surplus to the deficit areas. The outline of this is shown in Figure 15.5. However the original scheme has been shelved and, with the privatisation of the water supply industry, its future is very uncertain.

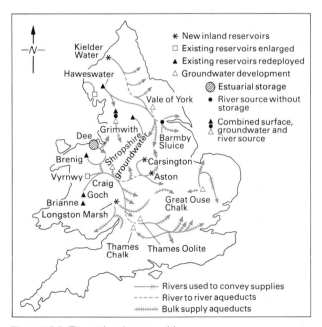

Figure 15.5 The national water grid

16 River and groundwater pollution

CASE STUDY

The Rhine pollution disaster 1986

The case of the Rhine diasaster described in the newspaper report in Figure 16.1 may be an extreme one, but the fact remains that many rivers throughout Europe and the rest of the world are severely polluted. The sources of **river pollution** are essentially the same as for marine pollution (see Figure 21.1 on p. 59).

River quality in England and Wales

Table 16.1 shows the changes in river water quality in England and Wales between 1980 and 1985. There have been notable successes in cleaning up rivers, especially the Thames which is now so clean that salmon have been seen in the river upstream as far as London.

Table 16.1 Water quality in England and Wales

Fresh water, rivers and canals				
Class	1980		1985	
	km	%	km	%
Good (1A)	13 830	34	13 470	33
Good (1B)	14 220	35	13 990	34
Fair (2)	8 670	21	9 730	24
Poor (3)	3 260	8	3 560	9
Bad (4)	640	2	650	2
TOTAL	40 630	100	41 390	100

Unfortunately there have also been examples of notable deterioration.

1 Rivers in Devon and Cornwall have a significant increase in pollution from cattle slurry.
2 The Mersey Basin has 31 per cent of the UK's polluted waterways. Between 1983 and 1986 the length of river polluted by farming and industry went up by 90 km.
3 Between 1974 and 1981 the amount spent on sewage treatment fell by 50 per cent.

Slow flow of warnings on river of death

The greatest ecological disaster to hit the River Rhine in modern times had a slow almost casual beginning. Yesterday, anti-pollution experts in many parts of Europe were still trying to come to grips with the scale of the catastrophe and the lessons it presents for the future.

Early on Saturday, November 1, firemen began fighting a blaze at the Sandoz chemicals factory near Basel, Switzerland. There was no immediate air of impending crisis.

But the first and only line of defence against environmental accident was breached within minutes. A basin to collect water was built to hold only 50 cubic metres. During the blaze firemen sprayed at least 25,000 cubic metres of water on to the building. Every minute more than 25 cubic metres of water polluted with chemicals from exploding drums poured into the Rhine and began its unstoppable 740-mile progress to the sea.

For almost two days none of the governments of countries along the Rhine knew the true nature of the cocktail of the 34 chemicals which were passing along Europe's greatest waterway in a slick 50 miles long. By the time the Swiss authorities put out their first urgent telex to river monitoring stations downstream, 7.30pm on November 2, the disaster was beyond human intervention.

The chemicals were so lethal to river life that parts of the Rhine may never recover, says Walter Wallmann, West Germany's environment minister.

Even on November 3 the Swiss could not tell the countries downstream exactly what had been stored in the stricken plant or even the amount of chemicals involved.

Crucially, they did not know that the 1,246 tonnes of agricultural chemicals flushed into the river included 12 tonnes of organic compounds containing mercury, one of the most harmful non-radioactive substances which can be introduced into the environment. Mercury becomes more concentrated as it passes through the food chain to accumulate in highly lethal doses in animals such as otters. And it lingers—to devastating long term effect—in river silt.

By the time the Swiss authorities sent full details of the Sandoz chemicals to France on Tuesday November 4, the stretch of Rhine from Mannheim to Basel was ecologically dead—from micro-organisms to fish and river birds. Near Strasbourg, two sheep were reported to have died after drinking from the river.

Riverside communities in West Germany were put on high alert. The governments ordered children, animals and dogs to be kept away from the river. Villages such as Unkel, 15 miles from Bonn, which would normally receive water pumped from the Rhine, were sent emergency supplies. This weekend they were still without mains water.

By midnight on November 8, the toxic waters had begun to reach the small Dutch border town of Lobith. But water engineers, with the few days of warning, had initiated emergency procedures. They used centuries-old flood defences to ensure that the pollution did not spread through the Netherlands network of inland waterways.

Using sluice gates and pumping stations, engineers were able to direct the main flow of the Rhine along the most direct of four possible outlets, straight into the sea and by-passing the canals.

By last Wednesday most of the poisonous slick had reached the North Sea near Rotterdam, but the crisis has not passed. The consequences are likely to be felt throughout Europe and the states bordering the North Sea for a long time.

Environmentalists believe the Rhine's ecological clock has been put back more than 20 years to the time when the river was known as the longest open sewer in the world.

Neelie Smit-Kroes, the Netherlands' minister responsible for water resources, has warned that the full extent of the damage to river life and agriculture will take up to a year to assess. The Dutch are particularly concerned that their Zeeland oyster beds may have been ruined.

Hank Kersten, of Greenpeace in the Netherlands, says it could take 30 years to restore to life the upper section of the Rhine.

After an accident at a chemical plant in Seveso, Italy in 1976, caused a leak of dioxin, the EEC introduced laws laying down special safety rules for chemical storage and manufacture. Stanley Clinton Davis, the EEC commissioner responsible for environmental safety, told The Sunday Times that the Rhine at Basel might not have been poisoned if Switzerland had applied similar legislation.

Sunday Times 16 November 1986

Figure 16.1

The consequences for the general public are potentially very serious. Wastes from industry are very poisonous. **Sewage** is broken down by bacteria in the water but these bacteria absorb oxygen. The consequent fall in oxygen levels gradually kills other aquatic life. When most of the dissolved oxygen has been removed other chemical changes take place which produce hydrogen sulphide. This has the smell of rotten eggs and is associated with stagnant water.

Nitrates: the threat to drinking water

Since the Second World War farmers have been applying more and more fertilizers to the soil in order to increase crop yields. The **nitrates** from these fertilizers have percolated into the soil and the rock below. Since all our drinking water supplies come from **groundwater** and rivers, such contamination is now regarded as a serious problem. Some sources of groundwater contain well over the EEC limit for nitrates. As it can take up to 40 years for the nitrates to reach the water table, the problem is a long term one. Nitrates have been proved to cause 'blue-baby' syndrome in infants, and there *may* be a link with stomach cancer in adults.

ACTIVITIES

1 Read carefully the newspaper article (Figure 16.1) before answering the following questions.

 a Where did the accident occur?

 b Describe in your own words what happened at the Sandoz chemical plant.

 c Why was it difficult for the countries downstream to take immediate effective action?

 d With the aid of your atlas, draw a sketch map to show the course of the River Rhine. Mark on your map the effects of the pollution at the places downstream of Basel.

2 a From Table 16.1 draw pie charts to show the percentage of each class of water for 1980 and 1984. Figure 16.2 shows you how to start. Remember that in a pie chart every 1 per cent occupies 3.6° of the circle.

 b Are there any significant differences between the values for 1980 and 1985?

3 a Study carefully Figure 16.3. With the aid of your atlas, list the counties that are most affected by nitrate pollution.

 b Why do you think it is these counties that are most affected?

Figure 16.2 The construction of a pie chart

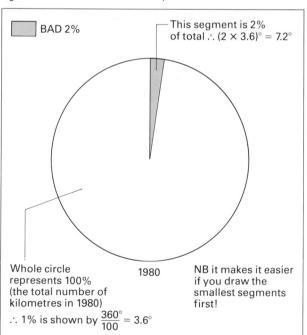

BAD 2%

This segment is 2% of total ∴ (2 × 3.6)° = 7.2°

1980

Whole circle represents 100% (the total number of kilometres in 1980)

∴ 1% is shown by $\frac{360°}{100} = 3.6°$

NB it makes it easier if you draw the smallest segments first!

Figure 16.3 Distribution of serious nitrate pollution in England and Wales.

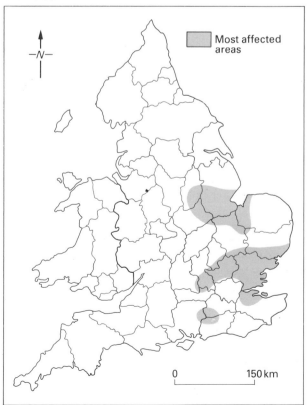

Most affected areas

−N−

0   150 km

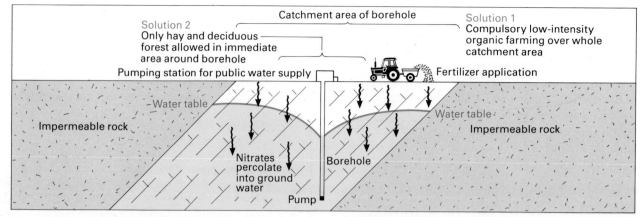

Figure 16.4 Nitrate pollution and possible solutions

ACTIVITIES

4 Two possible solutions to the nitrate problem are given in Figure 16.4. Imagine you were a successful arable farmer who farmed the whole of the catchment area. Write down your likely reactions to each suggested solution. What additional information might you need before finally supporting one solution or the other?

Coursework ideas

1 Calculate the discharge of your local river. To do this, carry out the following steps. See also Figure 16.5.

a Collect together the following equipment:
  i a tape measure;
  ii a metre rule;
  iii a stopwatch;
  iv an orange;
  v two poles.

b Find a straight stretch of the river and along the bank measure off a distance of 20 m. Mark the start and finish of this section with the two poles.

c At the first pole, measure the width of the river in metres.

d At every metre across the width of the river measure the depth of the water.

e Place the orange into the river at the first pole and time the number of seconds it takes to float down to the second pole, 20 m downstream. Do this five times at different places across the width of the stream.

The discharge of the river can now be calculated as follows.

f On graph paper plot the width and depth measurements found in steps c and d. This will give you the cross-section of the river.

g By counting the total number of squares covered by the cross-section you can work out the area of the cross-section (A).

h From the recordings made in c with the orange, find out how far the orange floated in one second in each of the five places. Find the average of the five readings. This is the velocity (V) of the river.

i The area of the cross-section is now multiplied by the velocity (A×V). Because of frictional drag on bed and banks, this figure must be multiplied by 0.8. The answer gives you the discharge of the river in **cumecs** (cubic metres per second).

2 Find out if your town is liable to flooding. When was the last serious flood? (Your local library or water company may be able to help you). What steps have been taken in your area to protect people and property from floods? Illustrate these measures on a large scale map.

3 Where does your local area obtain its water?

4 Do a geographical research project along the whole course of a river from source to mouth. Find out how the valley changes shape. What uses have people made of the valley and how has the nature of it and the river been modified?

Figure 16.5 How to measure the discharge of a river

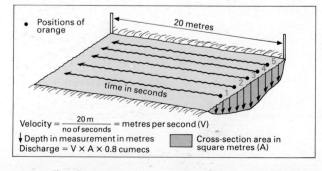

Velocity = $\dfrac{20\ m}{\text{no of seconds}}$ = metres per second (V)

↓ Depth in measurement in metres

Cross-section area in square metres (A)

Discharge = V × A × 0.8 cumecs

17 The power of the sea: coastal erosion

The photograph in Figure 17.1 shows waves attacking the promenade at Eastbourne during a south-easterly gale. Eastbourne can be seen on the OS map extract on Figure 6.3 p. 16. The whole length of coastline between about 603973 and 620992 has been artificially strengthened with a sea wall. However, as waves can exert pressures of up to 30 tonnes per square metre as well as hurl pebbles and boulders onto the shore, these sea defences are often damaged. Eastbourne has to spend about £200 000 a year on the maintenance of its promenade.

At the western end of the promenade the chalk cliffs begin. You can see these in Figure 17.2. Because the chalk rock is fairly hard, erosion of the cliff by the waves is slow and the cliff face is steep. Notice the groynes on the beach. These are built to trap material and protect the bottom of the cliff from erosion. Take care to read the notice in the foreground. This part of the beach is very accessible.

Some cliffs erode much more easily than the chalk near Eastbourne because they are made of much softer material. At Skipsea in Humberside (Figure 17.3) **boulder clay** (deposited by the ice sheets during the last ice age) is very weak. The people who built the houses on the cliff top did not seem to notice this! Easily eroded boulder clay cliffs are found along much of the east coast of England.

Figure 17.1 Wave attack on Eastbourne's promenade

Figure 17.2 The chalk cliffs at Eastbourne

Figure 17.3 Skipsea, Humberside

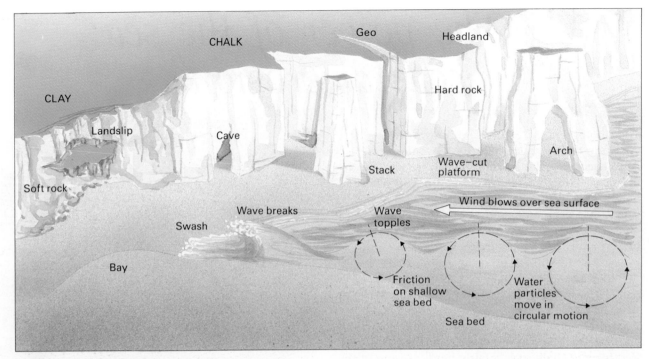

Figure 17.4 The major features of cliff erosion.

Waves are caused by the wind blowing over the surface of the sea. The distance over which the wind blows is called the **fetch**. The longer the fetch, the larger the waves. Wave size also increases with wind strength. The larger the waves, the more power they have and so the more destructive they can be. The base of the cliff is eroded by the waves to form a **wave-cut platform**. Landslips and weathering contribute to the erosion of the cliff above wave level. Weaknesses in the rock like joints and faults are eroded more quickly to produce features like **caves, geos, arches** and **stacks**. These are illustrated in Figure 17.4.

Figure 17.5 Cliff features at Elretat, France

ACTIVITIES

1 How do waves erode the shore?

2 Study Figure 17.1 and answer the following questions:

   a is the profile of the promenade wall concave or convex?

   b how does this profile help to protect the promenade? Draw a diagram to show how the wall shape affects breaking waves.

3 Study the OS map extract of Eastbourne.

   a The seawall runs from 603973 to one kilometre North East of the pier at 617988. What is its total length in kilometres?

   b Find the height of the cliffs shown in Figure 17.2.

   c What is the height of the cliff at Beachy Head? How can you tell from the contour pattern that this cliff is vertical?

   d The prevailing wind along this section of coast is south-westerly. Explain why a south-westerly wind would not have much effect on the shore at Eastbourne (hint: think about the fetch). With the help of your atlas find out which direction of wind would create the most damaging waves at Eastbourne.

   e What map evidence is there to the south-west of East Dean (55 97) that the area is popular with walkers and holidaymakers?

4 Draw a sketch of the scene in Figure 17.5. Identify and label as many cliff features as you can using Figure 17.4 to help you.

18 Protecting the shore

Figure 18.1 Groynes protecting the beach at Eastbourne

Seaside resorts like Eastbourne want to protect their beaches for the enjoyment of holidaymakers. Cliff erosion in many places needs to be checked to protect buildings, roads and other structures near the cliff top. Where the coast is low-lying, defence may be needed against flooding at high tide and stormy periods. Seawalls can be effective but are very costly to maintain and build. It is much cheaper and often more effective to use the work of the sea itself to protect the coast.

Notice how in Figure 18.1 a series of groynes along the coast at Eastbourne has caused the beach material to build up behind each groyne. The sea transports a great deal of material along the coast in a process known as **longshore drift**. How this works is illustrated in Figure 18.2.

Unfortunately, if one section of the coast is collecting material like this, another section further along is being deprived of sand or shingle. This results in the lack of a beach and consequently more erosion as beach material forms a buffer against sea attack.

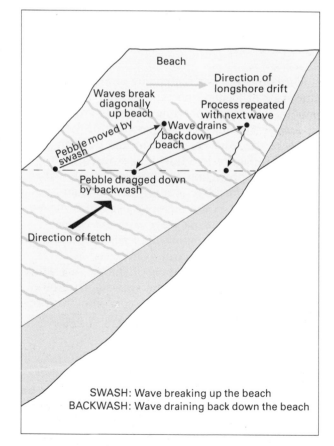

Figure 18.2 The process of longshore drift

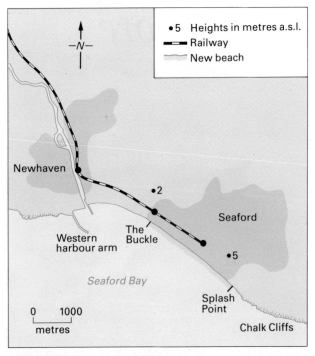

Figure 18.3 The position of Seaford

CASE STUDY

Seaford

This situation was happening at Seaford, 15 km due west of Eastbourne. The seawall at Seaford was in real danger of being seriously eroded. This in turn would have caused flooding of the lower levels of the town. Notice from the position of Seaford in Figure 18.3 how it is exposed to the full force of the prevailing south-westerly gales. In addition, Seaford Bay is much deeper than anywhere else on the Sussex coast which means larger, more damaging waves can get closer inshore. But, most important of all, the building of the Western Harbour Arm to the west at Newhaven cut off the supply of shingle that would otherwise have protected the seawall. To protect Seaford the following plan was put into effect by the Sussex Division of Southern Water. (Figure 18.4)

1 A groyne was built at the eastern end of the wall (Splash Point).
2 A dredger pumped shingle from the sea bed to the shore where it was spread out by bulldozers.
3 The dredger was moved to new positions westwards as far as The Buckle.

This was not only the cheapest method but also thought to be the most environmentally attractive way of protecting Seaford. It has proved successful. The coastal defences stood up to the severe storm of 16 October 1987.

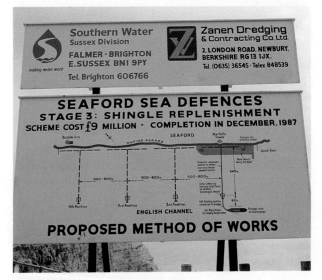

Figure 18.4 Defending Seaford

Figure 18.5 The new beach protecting Seaford.

No action over crumbling cliffs

Residents at the northern end of Swanage Bay were shocked to learn this week that Purbeck District Council proposes to take no action over the crumbling cliffs near their homes — after hearing that it would cost nearly £300 000 to make the cliffs safe.

For years, cliff falls at the north end of the beach, from Ocean Bay onwards, have caused parts of the gardens of clifftop houses and even garden sheds to tumble down to the beach and sea below.

Now a shock report shows that to halt the coastal erosion — which is caused not by the sea but by rain and surface water seeping through the cliffs above — would cost an estimated £297 000.

Of three specific problem sites identified, the consultants say 'there is the option of doing nothing and allowing slips and falls to continue as they have in the past.

'We do not see any potential catastrophic failure if this was done although a large part of the cliff above Ballard Beach cabins could cause some considerable damage.'

Such a fall would also bite into the gardens of 26 Burlington Road and jeopardise the stability of the cliff at 28 Burlington Road.

Action

They continue: 'The slippages north of the area are likely to continue, on occasions blocking the promenade and eventually requiring work to be done to prevent further loss of back gardens.'

The council agreed that no further action be taken on the proposals submitted by the consultants for the time being.

But it is going to have the cliff monitored and measures taken for the safety of the public as in the past.

Purbeck Mail 17 December 1987

Figure 18.6

ACTIVITIES

1 In what direction do you think longshore drift is at Eastbourne? Refer to the OS map extract (p. 16) and to Figure 18.1.

2 Explain in your own words how longshore drift works. Use diagrams to help your explanation.

3 With the help of Figure 18.4 draw a diagram to show how Seaford has been protected.

4 Study the map in Figure 18.3 and explain why:
   a the Western Harbour Arm deprived Seaford of its beach material;
   b Seaford was in danger of flooding.

5 The completed beach at Seaford is shown in Figure 18.5. Which beach do you think is more attractive for holidaymakers: the one at Seaford without groynes, or the one at Eastbourne with groynes? Why?

6 The local Water Authority met the costs of the Seaford Scheme. After reading the newspaper article in Figure 18.6 discuss who should be responsible for coastal defences:
   a the Local Authority;
   b the local residents directly affected;
   c central government.

19 Coastal deposition

Figure 19.1 Sand dunes at Perranporth and the St. Piran Oratory

Every year thousands of holidaymakers enjoy the large expanse of **sand dunes** on the north Cornish coast at Perranporth (see Figure 19.1). The extent of the sand can be seen on the map in Figure 19.2. The dune area is known as the Penhale Sands. However, despite their attractions the dunes do present real problems for coastal management. Thousands of pairs of feet wear away the **marram grass** cover which binds the sand together. Without this grass the dunes become eroded and the sand blows inland creating a hazard to nearby buildings and roads. A small chapel (the St Piran Oratory) was finally buried by blown sand. Further north, on the Camel Estuary, the famous church of St Enedoc had been inundated by sand until it was dug out in 1863 and protected by a screen of evergreen trees. In an effort to prevent more dune erosion and drifting sand, the local authority is taking the action shown in Figure 19.3.

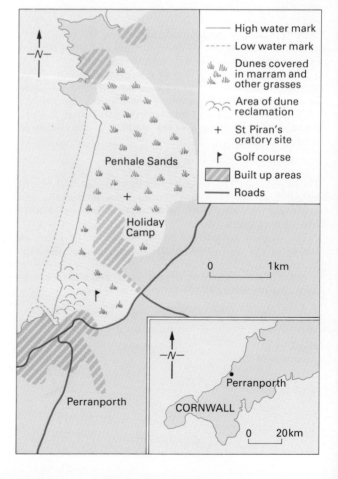

Figure 19.2 The location of the Penhale Sands

Figure 19.3 Dune conservation measures at Perranporth

CASE STUDY

Orford Ness, Suffolk

Like Holderness, the Norfolk coast has soft clay cliffs that are rapidly eroded. The material from the cliffs is being transported southwards by longshore drift. It is then gradually being deposited as a long **spit** (called Orford Ness) that extends along the coast south from the little town of Aldeburgh. At one time Aldeburgh was on the mouth of the river Alde but the river has been diverted a long way south by the growth of the spit. Behind the spit **saltmarshes** have developed with important bird and plant life. Orford itself was once a flourishing port in the twelfth century. It went into decline as the spit cut it off from the sea. It is now a centre for sailing on the river.

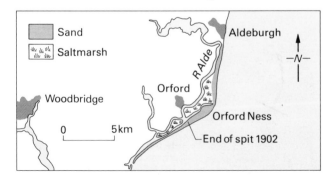

Figure 19.4 Orford Ness

Figure 19.5 Orford Ness from the air

20 Coastal development and conflict

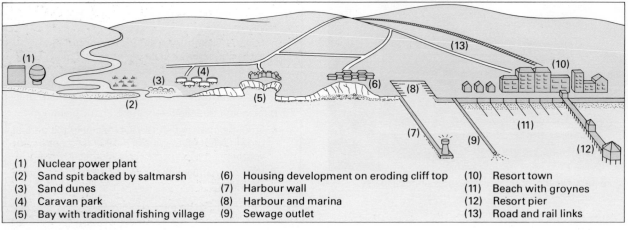

(1) Nuclear power plant
(2) Sand spit backed by saltmarsh
(3) Sand dunes
(4) Caravan park
(5) Bay with traditional fishing village
(6) Housing development on eroding cliff top
(7) Harbour wall
(8) Harbour and marina
(9) Sewage outlet
(10) Resort town
(11) Beach with groynes
(12) Resort pier
(13) Road and rail links

Figure 20.1 Types of coastal development

The coastline is coming under increasing pressure from urban and industrial developments and the demands of tourism. Some of these are shown in Figure 20.1.

Often these developments cause bitter arguments between the local residents and conservationists on the one hand and the developers on the other. Conservationists succeeded in protecting Holm Bush, Dorset as shown in Figure 20.2. A significant part of the South Downs coast near Eastbourne has been protected from urban development by purchases of land by the National Trust and Eastbourne Borough Council. The information in Figure 20.5 examines the case of the Yacht Haven at Swanage, Dorset.

Figure 20.2 An example of conservation

Holm Bush, 29 acres of meadowland west of Lyme Regis, Dorset, which has been bought by the National Trust. The meadows adjoin the coastal footpaths, with a nature reserve further inland, and are renowned for their cowslips, orchids and gorse.

The land was bought for £38,000 with the help of the residents of Lyme Regis, who set up the Lyme Regis Society, and raised £26,000 in six months; but for their efforts this stretch of coastline would have been threatened by developers.

The purchase is part of the National Trust's Enterprise Neptune Campaign, which aims to save 900 miles of coastline in the British Isles; so far, 490 miles are now protected.
The *Independent* 28 March 1988.

1 Study Figure 20.2 and answer the following:

   a Locate the stretch of coast near Lyme Regis in your atlas.
   How far is it from Swanage?
   From London?
   How long would it take for you to get there from your home if travelling by car or coach assuming an average speed of 64 km (40 miles) per hour?

   b What features of Holm Bush make it desirable for protection?

   c How much of Britain's coastline has been purchased by the National Trust? What is this as a proportion of the total that is aimed for?

2 Study the map of Swanage and the photograph in Figures 20.3 and 20.4.

   a Describe the site of Swanage and the proposed yacht haven.

   b What are the physical advantages of the site for the development of a yacht haven?

3 Read carefully all the newspaper reports on the yacht haven development that was proposed for Swanage, in Figure 20.5. Your teacher will divide the class into four groups; each group represents one of the following:

   a the Swanage Development Action Group;

   b the Swanage Coastal Protection Association;

   c the Development Company;

   d the Purbeck District Council.

   Each group is to prepare a report that will be read by a representative to a public enquiry (the rest of the class). Questions can be asked by anyone at the enquiry after each report has been given. Afterwards the class can vote on the scheme.

Figure 20.3

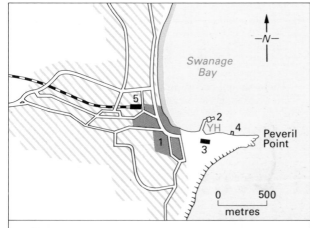

1 Town centre
2 Pier
3 Hotel Grosvenor site
4 Lifeboat house
5 Railway station
YH Yacht Haven site
Sandy beach
Cliffs

Figure 20.4 The site of Swanage

Figure 20.5

Dying Swan

ARCHITECTURE

Swanage threat

STEPHEN GARDINER

IT WOULD be a good thing if Prince Charles sometimes turned his attention to vulnerable spots beyond cities and architecturally famous buildings.

Take Swanage. Here is a still largely tranquil and atmospheric seaside resort at present under threat from a highly damaging proposal to build a vast boat-park/ car-park (for around 300 boats and 300 cars) and a mass of 'holiday homes.' These will, if permitted in the coming week, overwhelm the promontory of Peveril Point, utterly ruin the town's very special architectural character and lead to the developer, Durrant Developments, taking control of a large slice of the bay.

Swanage makes an excellent study in character and scale, its architecture having evolved from the geographical form of the bay, orientation and ideal conditions for a small fishing industry: hence the modest scale, but hence too the distinctive pervasive character of large roofs sheltering balconies with lookouts over the sea. But to recapitulate: when I first took up this story, a plan to develop the old Grosvenor hotel (now being demolished) and surrounding slopes along Peveril Point surfaced last year as a hard proposition.

The boat-park/car-park, which brought immediate, and stiff, opposition from local people, was planned to extend from Swanage Pier to the Wellington Clock Tower, and to cover a vast piece of beach much enjoyed by holiday makers. In addition, a 15ft concrete causeway was to be built right out into the bay from the Clock Tower and the Victorian pier transformed into a wall that would block all views of this part of the coastline.

Then there are the 'holiday homes', 100 of them planned for the steep slope immediately above the car-park and promoted as a 'fishing village'. Divided into stages, the first was given planning permission by the Purbeck District Council and building has begun.

13 December 1987

Fishermen slam marina scheme

Evening Echo 9 December 1986

The town's fishermen's association is preparing to fight the Bill, being promoted in Parliament in February by Christchurch-based developer David Durrant.

Key point in their opposition is a clause which gives complete control of the mooring area of the bay to the owner of the yacht haven.

Another clause bans commercial boats from the planned yacht haven except at the discretion of the owner of the haven.

Fishermen's association vice-chairman Peter Haine, said: "At stake here is a third of a million pound export industry for shellfish as well as the local market fishery.

"If this Bill is passed by Parliament, Swanage will cease to be a fishing harbour," he added.

Earlier this year, Durrant Developments planned a much larger harbour, enclosing a quay for up to 60 fishing boats.

Mr. Durrant said: "My scheme met with hostility. They felt I was interfering with their freedom. They went away in August to discuss it, and I heard no more from them.

"We tried to accommodate the fishermen from the word go. We wanted them in the harbour because they would add life and activity all the year round," he added.

Survey suggests that beach might not be affected

Purbeck Mail 30 October 1986

The survey which caused a row at Monday's meeting was prepared as a study of existing data supplemented by a site appraisal.

The study suggests that the main beach might not be affected by the harbour development.

And it recommends that further reports should be prepared on the 'wave climate' in the bay.

The conclusions are:

'In the main part of the bay the beaches are stable with no evidence of any significant longshore drift.

'Further south between Peveril Point and the Pier there is some evidence of an east to west drift although not large, or necessary to preserve healthy beaches further north.

'The main problem areas with the present beaches are between the jetty and the Mowlem, caused by the shape of the reclamation and the vertical seawall, east of the slipway where there is no permanent beach; and the west side of the solid part of the pier. This is not a suitable shape to retain a beach.

'Part of the town beaches and the existing harbour area are suffering some pollution due probably to the sewer outfall at Peveril Point and the east to west drift.

'The effect of the harbour will be to intercept the east to west drift from Peveril Point, although this will not affect the beaches further west adversely.

'Increased wave action near the foot of the proposed new breakwater could occur, although this could be countered fairly easily.

Minor changes in the beach to the north might occur but could be controlled by adjustment of the existing groynes.

It was unlikely that any significant changes in the weak tidal currents would occur in the bay following construction of the proposed new harbour.

There should be no problem retaining a sailing club beach to the west of the new harbour.'

Swanage divided over yacht haven project

The Benefits of the Haven

- It will generate trade for businesses in the town.
- It will generate jobs for the people of Swanage both during the construction stage, and when it is in operation.
- When fully restored, the Pier will be a splendid promenade once again.
- The outfall from the town's sewage system will be improved significantly, so reducing the risk of inshore pollution, at nil cost to the authorities and the ratepayers.
- It will provide moorings for 250 boats.
- Moorings at half the price of those at Poole and Lymington will be guaranteed to the nine full-time fishing vessels of Swanage. The fishermen will be able to land their catches on the quay of the breakwater.
- The necessary relocation of any moorings in or around the area of the haven will be undertaken by us if required by owners.
- The new bunkering facilities at the haven will be available to any visiting vessels not carrying cargo.
- Vessels will also be able to take refuge in the haven in emergencies, when there is extreme weather.
- There will be significant improvements in the facilities for other boat owners resulting from extensive works to be carried out to the town boat park.
- Swanage Sailing Club has been offered new premises at Marine Villas, together with a new slipway for their members.
- Swanage's Diving School will also have new premises.
- The beaches of Swanage will be as fine as they have always been.
- The public slipway to the east of the haven will be improved and will particularly benefit those with moorings in the area.
- The lifeboat will have its own special berth in the yacht haven and bunkering facilities for its exclusive use. There will be space for an additional fast inshore rescue boat.
- A new RNLI station will be built and the existing lifeboat house will be kept, possibly for a relief boat.

Durrant Development company

Figure 20.5 (continued)

21 Marine pollution

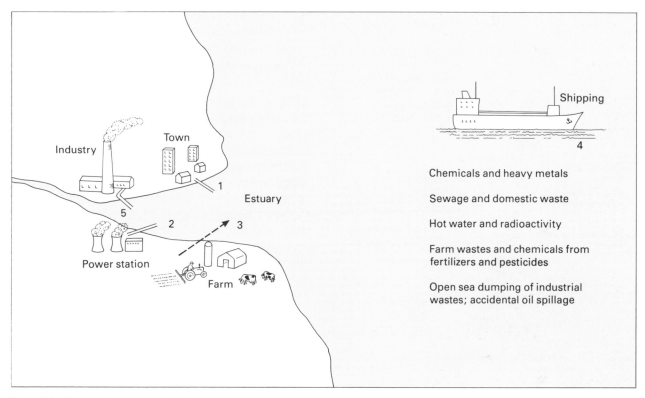

Figure 21.1 The main sources of marine pollution

Evidence is growing that coasts, estuaries and the open sea are dangerously polluted. Figure 21.1 shows the sources of the main kinds of pollution.

A clean beach?

The beach shown in Figure 21.2 may look clean enough, but it is not one of the EC's 'Eurobeaches'. These are beaches that comply with the minimum standards of cleanliness laid down by the EC. The best standards of all are given 'Blue Flag' status and are shown in Figure 21.3. The most common forms of beach pollution are sewage, oil, industrial waste, litter and dog faeces. Much of the first three of these is dumped out to sea but gets washed ashore by tides and currents. In 1984 over 9 million tonnes of sewage sludge was dumped around the shores of Britain, often through outlet pipes that are far too short to allow safe dispersal out to sea.

Figure 21.2 A clean beach?

Figure 21.3 The Blue Flag Beaches of Britain, 1987

CASE STUDY

Polluting the North Sea

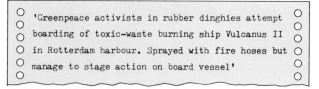

'Greenpeace activists in rubber dinghies attempt boarding of toxic-waste burning ship Vulcanus II in Rotterdam harbour. Sprayed with fire hoses but manage to stage action on board vessel'

Most of the countries around the North Sea have for a very long time been depositing wastes of all kinds into the North Sea. The particular characteristics of the North Sea encourage these wastes to accumulate, especially in the eastern margin of the southern half. Several large polluted rivers enter the sea here, the circulation of water in the sea (shown in Figure 21.4) ensures an easterly drift, and the water is shallow.

A table of North Sea heavy metal pollutants is shown below.

Table 21.1 North Sea pollutants (heavy metals)

Pollutant	Quantity		Source	
Iron	331 000 tonnes	⎫		
Manganese	25 000	"		
Lead	14 000	"	River Rhine/Meuse	51%
Copper	10 000	"	River Elbe	20%
Chromium	6 000	"	River Weser	10%
Zinc	4 500	"	River Schelde	10%
Cadmium	1 120	"	River Humber	5%
Mercury	1 000	"	River Thames	4%

The effects of heavy metals on the body can be particularly horrible, causing severe damage to the nervous system. Pesticides used on farms get washed into rivers and so to the sea. These are toxic to fish and build up in the food chain. They stay in the ecosystem for a very long time. The Wadden Sea (between the Dutch mainland and the Frisian Islands) is the most polluted area of the North Sea. This is a very important spawning ground for one-third of the fish in the North Sea. Between 1967 and 1983 the porpoise population died out and the seal population was cut by one third.

The Dutch Government now has strict pollution controls over the Wadden Sea. There are very many EEC reports and directives on the North Sea but a united approach is still required. In November 1987 the Bremen Declaration member countries accepted the need to 'take timely preventive measures to maintain the quality of the North Sea'.

ACTIVITIES

1 Make a copy of Figure 21.1. In a key write the correct label against each of the numbers 1–5.

2 Study Figure 21.3 showing the distribution of 'Blue Flag' beaches in Britain.
   a Where are most of the Blue Flag beaches?
   b Suggest reasons why there are so few in
      i NW Scotland, and
      ii the mouth of the Thames.

3 Copy the map of the North Sea from Figure 21.4. On your map:
   a label the rivers listed in Table 21.1.
   b show, by using a suitable method, the percentage contribution of each river to heavy metal pollution in the North Sea
   c with the help of your atlas mark on the major industrial and urban areas that are likely to be contributing towards the pollution of the North Sea.

4 Find the Wadden Sea on Figure 21.4. Why do you think this is the most polluted area? Why does this have serious implications for all the countries around the North Sea?

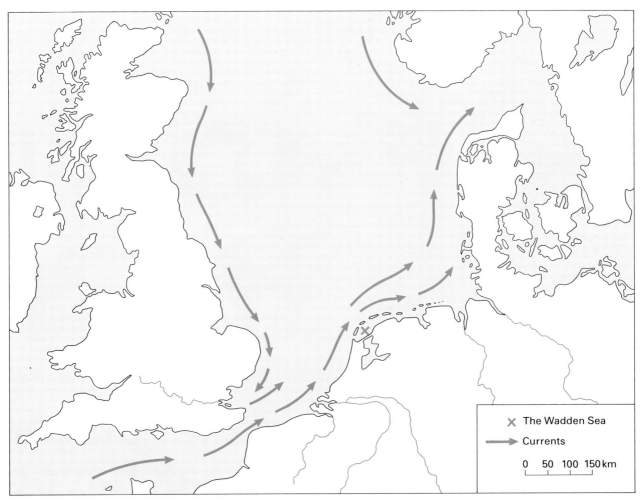

Figure 21.4 Circulation of water in the North Sea

Coursework ideas

1 Find out as much as you can about the National Trust's Enterprise Campaign. Choose one stretch of the coast that has been bought by the National Trust and find out what features make it worth preserving. These may include the plant and animal life, the geology, scenery, or some particular connections with famous people or events in the past.

2 Look again at Figure 20.1 on p. 56 Choose one (or more) of the types of coastal development shown from any part of the coast you know well. For the area you've chosen:

a find out if there are any conservation issues involved like damage to wildlife habitats or particular ecosystems like saltmarshes;

b try to discover through local newspapers or other sources the arguments for and against development;

c show by means of maps, diagrams and sketches the likely changes that are to take place between now and the end of the century.

If a section of the coast is within easy reach of your home or school you should be able to use your own personal observations and data collection.

3 Adopt a stretch of coastline (preferably one you can get to personally) and study how each of the following factors have influenced its main characteristics:

a geology (rock type and structure);

b physical processes that are operating (wave action, erosion, deposition);

c human activities.

Record your observations using maps, sketches and diagrams.

22 Depressions

Praying for sunshine

CORNWALL'S tourist industry will be praying for the good weather to continue for since the sun came out there has been a perceptible increase in the number of casual visitors to the county in an otherwise "patchy" peak season period.

State of emergency after crops ruined in hottest known weather

Lightning kills cricket player as freak storms wreak havoc

Hundreds die during heat wave in Greece

'The outlook is continuing unsettled'

The weather is probably the most popular subject of day to day conversation. This is perhaps because it never ceases to spring surprises, especially in Britain where it is so variable and unpredictable. It can directly influence our daily activities. When 'good' it can boost the profits of industries like brewing; when 'bad' it can cause loss of life.

Our weather is so changeable because Britain lies in the path of weather systems known as **depressions**. These are areas of low atmospheric pressure (usually between 960 and 1000 millibars). They form over the North Atlantic where cold air from the Arctic meets warm air from the Tropics. The air spirals upwards in an anticlockwise direction and as it does so it cools down. This cooling causes the water vapour in it to condense, giving clouds and rain. The rain (or snow in winter) is particularly heavy and prolonged along the **warm front** and the **cold front**. These fronts are where the cold and warm air directly meet each other. The whole system moves west to east across, or close to, the British Isles.

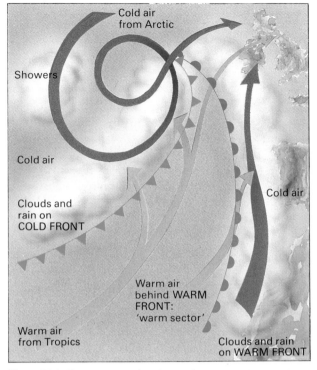

Figure 22.1 The structure of a depression approaching Britain

Figure 22.3 Trees blown down at Orpington, Kent on 16 October 1987

A typical depression is shown in Figures 22.1 and 22.2. The numbered lines on the weather map (Figure 22.2) are **isobars**. Isobars join places with the same pressure. The closer these are together the stronger the wind is. Unfortunately for the weathermen, depressions can form very rapidly – in a matter of hours – and move very fast and change direction suddenly. Each weather station on the map is shown by a circle with various symbols indicating the wind speed, direction and weather conditions.

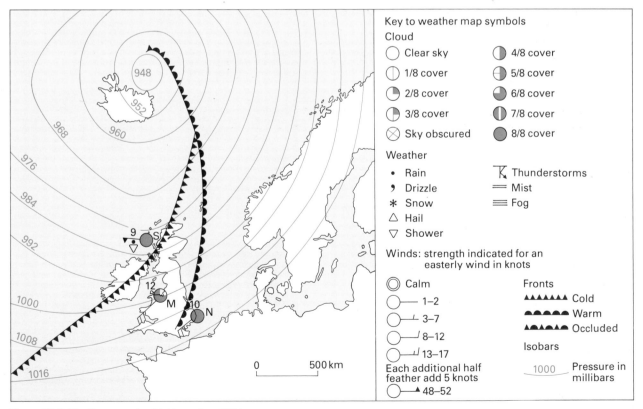

Figure 22.2 Weather map for 24 November 1984, noon

CASE STUDY

'Hurricane Ethelred', Southern Britain 1987

At about 6 pm on October 15 1987 an unexceptional depression off the coast of Brittany began to deepen very rapidly. As it moved north eastwards towards Britain its central pressure fell to 959 mb producing exceptionally strong winds. Over a ten minute spell Dover recorded an average wind speed of 122 kmh while a single gust over the London Weather Centre reached 152 kmh. The worst storm since 1703 had hit the south-east of England. Notice from Figure 22.4 that the storm did not take the path predicted. The weathermen also underestimated the strength of the wind. The storm was dubbed 'Hurricane Ethelred' because we were not ready for it!

A Weathermap at 03.00 hrs 16th October

B The storm track

C The change in pressure at Reading

Figure 22.4 Recording the storm of October 16th 1987

Lesson for the South

by GEOFFREY LEAN

THE WIND that blew in from the sea in the early hours of 16 October, killing 19 people, may change more than the face of south-east England. For the very power of the storm has brought a new awe of the forces of nature to a community that had come to forget them.

In a few short hours, 15 million trees were blown down. Giants that had stood for hundreds of years were plucked effortlessly from the ground, whole woods flattened by a single gust, priceless collections of rare trees built up over centuries destroyed literally overnight. Great ships were tossed on to the shore, 3,000 miles of telephone lines brought down, and hundreds of thousands of homes deprived of electricity in the worst power failures since the Second World War. The region came to a virtual standstill. Schools and businesses were closed, and the Stock Exchange shut—contributing to the severity of the great shares crash of the following Monday.

It was the greatest storm to hit the south-east since records began, possibly the worst since the gale of November 1703 when more than 8,000 people died. It was chronicled by Daniel Defoe, and his account reads uncannily like last month's newspapers, reporting 'whole Parks ruin'd, fine Walks defac'd and Orchards laid flat.'

In 1703, no one pretended that man had 'conquered nature'—and even present-day inhabitants of hardier parts of the country reacted with some bewilderment to the outcry. For though the wind reached record speeds for south/east England, it was gentle compared to the gales that fairly regularly sweep Scotland (in 1981, for example, 140 mph winds battered the Cairngorms). Natural disasters get far less attention when they strike far from the centre of power and from the homes of Government ministers, civil servants—and national newspaper journalists.

The unique feature of October's storm was that it happened to hit one of the most densely populated and prosperous areas of Western Europe normally insulated from rough winds, whether economic or metereological. Although this was almost certainly a natural freak, scientists agree that man-made climatic changes—caused by burning fossil fuels and the use of the gases that destroy the ozone layer—threaten to cause increasing disruption of this sort over the next decades.

The appalling devastation—and real tragedies—of the storm, may yet have some point if they lead the decision-makers who live in the south-east to a more healthy respect for the forces of nature. If they do, this ill wind may yet blow us all some good.

Observer Weekend 22 November 1987

Figure 22.5

Figure 22.6 A photograph from the Meteosat satellite

Figure 22.7 The Beaufort Wind Scale

Beaufort force	Type of wind	Effects to look for	Speed in kph
0	Calm	Smoke rises vertically	0
1	Light air	Smoke drifts	1-5
2	Light breeze	Wind felt on face, leaves rustle, weather vanes move	6-11
3	Gentle breeze	Leaves and small twigs move, flags extended	12-20
4	Moderate breeze	Dust and loose paper blow about, small branches move	21-30
5	Fresh breeze	Small trees sway, wavelets form on water	31-40
6	Strong breeze	Large branches sway, umbrellas used with difficulty, telegraph wires whistle	41-50
7	Moderate gale	Whole trees sway, hard to walk into the wind	51-60
8	Gale	Twigs break off trees, very hard to walk into wind	61-74
9	Strong gale	Chimney pots and slates blow off	75-87
10	Storm	Trees uprooted, serious damage to buildings	88-100
11	Violent storm	Rarely occurs inland, causes widespread damage	101-115
12	Hurricane	Disastrous results	115+

ACTIVITIES

1 Look back to the newspaper headlines at the beginning of this chapter. For each one describe the particular weather condition and discuss the ways in which people may be affected. (You may find it helpful to do this in pairs or small groups)

2 Study carefully Figure 22.2 and answer the following questions:

a Where is the depression centred?

b What is the air pressure value in the centre?

c Describe the position of the warm and the cold front.

d What are the weather conditions at Norwich, Manchester, Stornoway? (Labelled N,M,S.)

e Give reasons for the weather conditions you have described (use Figure 22.1 to help you)

3 If the depression moves east to northern Norway in the next 24 hours, what changes do you think will happen to the weather in London?

4 Figure 22.6 is a photograph taken by the Meteosat satellite.

a There is a depression shown on the photograph: try to locate it.

b How can you tell it is a depression?

5 Study the photograph in Figure 22.3.

a If the camera was pointing south-east, what was the direction of the wind?

b What evidence can you find from Figure 22.4 to back up your answer?

c Describe the types of damage and inconvenience shown in the photograph.

6 Read carefully the newspaper article in Figure 22.5

a List the effects of the storm mentioned in the article that are not shown in Figure 22.3.

b In what area of Britain would this storm not be unusual?

c Why was the storm considered to be such a serious hazard for the south-east?

d Consider what the likely effects may have been if the storm had arrived six hours later.

7 Answer the following questions from the information in Figure 22.4.

a What counties of England were most affected. Can you say why?

b Between what times were the strongest winds experienced?

c If the storm had actually taken the predicted path, what countries would have been affected? Illustrate your answer with a sketch map.

8 Figure 22.7 shows the **Beaufort Scale** which classifies winds according to their speed and strength.

a Where on the Beaufort Scale do the winds recorded in the October storm appear?

b Where on the scale do the winds on the more normal depression in Figure 22.2 appear?

c Try to estimate the Beaufort strength of the wind on the day you are doing these exercises. You could record wind strength for several days and record the results in the form of graphs.

23 Anticyclones

'The present settled spell is likely to last for the next few days'

Settled weather is usually associated with **anticyclones**. These are areas of high pressure (over 1016 mb). The air in them gradually sinks and moves outwards in a clockwise direction as shown in Figure 23.1. As the air sinks it becomes warmer and drier. This usually means sunshine. In summer the weather can be very warm but in winter the short hours of sunshine cannot compensate for the loss of heat during the night and so the weather is often very cold and frosty. **Fog** is a problem with anticyclones. Cloudless nights and light winds mean that the earth loses a lot of heat through outward radiation. Under these conditions condensation may occur, forming mist and fog.

Anticyclones can block the passage of rain-bearing depressions causing long dry spells. Between May 1975 and August 1976 frequent high pressure over the country caused the total rainfall over England and Wales to be only about 50 per cent of the average.

Figure 23.1 The movement of air in an anticylone

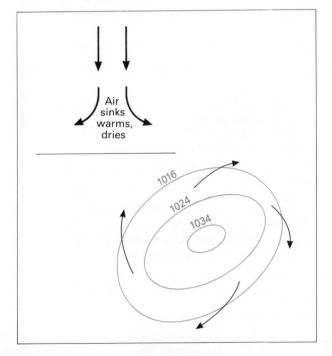

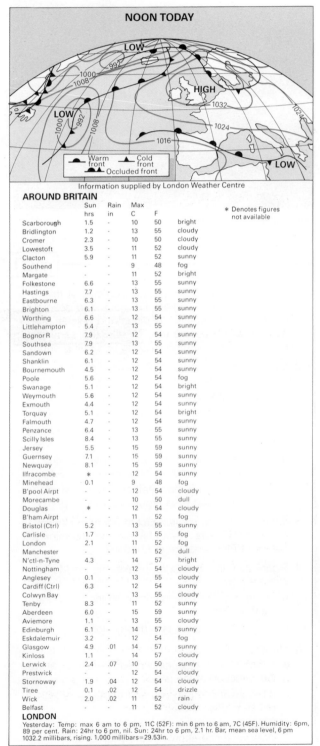

Information supplied by London Weather Centre

AROUND BRITAIN

	Sun hrs	Rain in	Max C	F	
Scarborough	1.5	-	10	50	bright
Bridlington	1.2	-	13	55	cloudy
Cromer	2.3	-	10	50	cloudy
Lowestoft	3.5	-	11	52	cloudy
Clacton	5.9	-	11	52	sunny
Southend	-	-	9	48	fog
Margate	-	-	11	52	bright
Folkestone	6.6	-	13	55	sunny
Hastings	7.7	-	13	55	sunny
Eastbourne	6.3	-	13	55	sunny
Brighton	6.1	-	13	55	sunny
Worthing	6.6	-	13	54	sunny
Littlehampton	5.4	-	13	55	sunny
Bognor R	7.9	-	12	54	sunny
Southsea	7.9	-	13	55	sunny
Sandown	6.2	-	12	54	sunny
Shanklin	6.1	-	12	54	sunny
Bournemouth	4.5	-	12	54	sunny
Poole	5.6	-	12	54	fog
Swanage	5.1	-	12	54	bright
Weymouth	5.6	-	12	54	sunny
Exmouth	4.4	-	12	54	sunny
Torquay	5.1	-	12	54	bright
Falmouth	4.7	-	12	54	sunny
Penzance	6.4	-	13	55	sunny
Scilly Isles	8.4	-	13	55	sunny
Jersey	5.5	-	15	59	sunny
Guernsey	7.1	-	15	59	sunny
Newquay	8.1	-	15	59	sunny
Ilfracombe	*	-	12	54	sunny
Minehead	0.1	-	9	48	fog
B'pool Airpt	-	-	12	54	cloudy
Morecambe	-	-	10	50	dull
Douglas	*	-	12	54	cloudy
B'ham Airpt	-	-	11	52	fog
Bristol (Ctrl)	5.2	-	13	55	sunny
Carlisle	1.7	-	13	55	fog
London	2.1	-	11	52	fog
Manchester	-	-	11	52	dull
N'ctl-n-Tyne	4.3	-	14	57	bright
Nottingham	-	-	12	54	cloudy
Anglesey	0.1	-	13	55	cloudy
Cardiff (Ctrl)	6.3	-	12	54	sunny
Colwyn Bay	-	-	13	55	cloudy
Tenby	8.3	-	11	52	sunny
Aberdeen	6.0	-	15	59	sunny
Aviemore	1.1	-	13	55	cloudy
Edinburgh	6.1	-	14	57	sunny
Eskdalemuir	3.2	-	12	54	fog
Glasgow	4.9	.01	14	57	sunny
Kinloss	1.1	-	14	57	cloudy
Lerwick	2.4	.07	10	50	sunny
Prestwick	-	-	12	54	cloudy
Stornoway	1.9	.04	12	54	cloudy
Tiree	0.1	.02	12	54	drizzle
Wick	2.0	.02	11	52	rain
Belfast	-	-	11	52	cloudy

* Denotes figures not available

LONDON

Yesterday: Temp: max 6 am to 6 pm, 11C (52F): min 6 pm to 6 am, 7C (45F). Humidity: 6pm, 89 per cent. Rain: 24hr to 6 pm, nil. Sun: 24hr to 6 pm, 2.1 hr. Bar, mean sea level, 6 pm 1032.2 millibars, rising. 1,000 millibars=29.53in.

Figure 23.2 Weather chart and reports for noon November 3rd 1987

100 vehicles crash in freezing M-way fog

By Staff Reporters

Freezing conditions and fog brought accidents involving more than 100 vehicles on the M1 and its junction with the M6 yesterday, including a coach crash. A total of 19 people were injured, several seriously.

About 50 vehicles were involved in the first pile-up on the M1, and two more accidents occurred on the M6 near the junction—the first involving 15 vehicles and the second about 30.

Thirteen people were injured in the M1 coach crash (above) but it is believed none was seriously hurt. The crashes also left a car on fire and resulted in a chemical spillage.

The accidents caused the closure of a 10-mile section of the M1 southbound carriageway on the Leicestershire-Northampton border for three hours.

Last night two people taken to Leicester Royal Infirmary were said to be seriously injured. One was the driver of a coach, the other a lorry driver.

Police said that fog warning lights had not been switched on before the accidents.

In North Wales a traffic patrolman, Police Constable Michael Evans of Porthmadog, Gywnedd, died when his patrol car skidded on ice.

A baby was killed last night when he went through the window of his family's BMW car as it collided with another car on the A303 Wincanton by-pass in Somerset.

Martin Webber, aged seven months, of Gillingham, Dorset, was dead on arrival at hospital. His mother, Lynne, aged 24, who was driving, was treated for severe cuts.

In Maidenhead, Berkshire, a burst water main closed the A330 with a covering of ice two inches deep and a mile long.

Freezing fog at Heathrow last night forced the cancellation of 61 flights—25 arrivals and 36 departures. Another 50 arriving flights were diverted to other airports in Britain and Europe. More than 70 flights were delayed by up to five hours.

Twenty-four airliners were diverted from Gatwick to fog-free destinations.

Frost and freezing fog are expected in the east this morning, with rain, sleet, and snow in the west later.

The Times 9 January 1988.

Figure 23.3 The fog hazard

ACTIVITIES

1 Study Figure 23.2 which shows the weather chart for November 3rd 1987 and the weather reports around Britain.

  a Where is the anticyclone situated?

  b What was the atmospheric pressure in London at 6 pm? Is this what you would expect?

  c In what direction was the wind over
    i SE England;
    ii NW Scotland? Try to explain the difference. (N.B Wind direction is always given as the direction from which the wind has come)

  d Describe the weather conditions that were reported around Britain. Give reasons for the weather you have described.

  e On the basis of the information given in Figure 23.2 write a weather forecast for the following:

    i BBC Radio 1;
    ii BBC TV;
    iii the motorway police;
    iv a local farmer;
    v a cross Channel ferry.

2 Read the newspaper article in Figure 23.3.

  a What parts of the country were affected by fog on this occasion?

  b Why is fog such a dangerous hazard on motorways?

  c What other forms of transport are affected? What solution has been adopted?

3 Look again at the Meteosat photograph Figure 22.6 on p. 65. What parts of Europe seem to be under the influence of high pressure? How can you tell this from the satellite photograph?

24 Local climates

Climatic conditions can vary a great deal over very short distances. You may have noticed how it never seems as frosty in the town centre as it does in the park on a winter's morning, or how mist always seems to form first in low-lying fields on an Autumn evening. These two observations can be explained by two important local climates: the **urban heat island** and the **frost hollow**. These are illustrated in Figure 24.1. Both occur to their maximum extent on calm, clear nights.

The heat island occurs because:

1 buildings radiate heat, especially if not insulated well;
2 bricks, tarmac and concrete absorb more of the sun's energy than plants;
3 there is much less vegetation losing water by **evapotranspiration** into the atmosphere. Evapotranspiration uses heat from the air and so vegetation makes it cooler.

Frost-hollows are found on clear, calm nights when:

1 outward radiation of heat from the ground makes the air in contact with the ground cold,
2 this cold air, being heavier than warm air, sinks downhill to the valley bottom.

Figure 24.1 Two examples of local climate

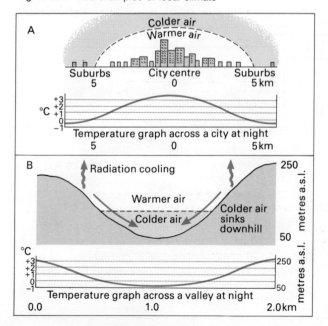

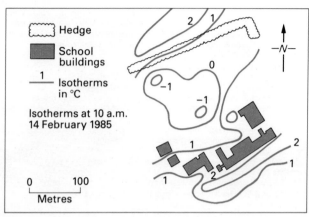

Figure 24.2 The heat island of a school

ACTIVITIES

1 Study Figure 24.1 A
   a Which part of the city is the warmest, and which part is the coldest?
   b What is the maximum temperature difference in °C between these two places?
   c How might the heat island effect be reduced,
      i by changing weather conditions,
      ii by human activity?

2 Study Figure 24.1 B
   a What is the temperature difference between the top of the valley and the bottom?
   b Why does the cold air sink down into the valley?
   c How might the frost-hollow affect a farmer or gardener?

3 Figure 24.2 was compiled by a group of GCSE pupils in the grounds of their school.
   a At what time of the day and year was the survey done?
   b Make a copy of the map. On your copy shade in blue the areas that were below freezing at the time of the survey. Shade in red the areas that were above freezing. Try to explain the pattern of temperature produced.
   c How might you account for the fact that:
      i the front of the school is warmer than the back, and
      ii the temperature rises to just over 2°C beyond the northern end of the school field? (The pupils have left off some information here!)
   d What important implications does information like that shown on Figure 24.2 have for the design and running of buildings such as schools?

25 Holidays in the sun

Many people will have unhappy memories of their annual holiday in Britain being spoilt by cold, rainy weather. However, down on the Mediterranean coast the sun is almost guaranteed to shine for up to 12 hours each day in August. This compares very favourably with just over 7 hours for the south coast of England! It is no wonder that in 1980 nearly 5 million holidaymakers went to Spain for their holidays. Resorts like Benidorm, on the Costa Blanca, Torremolinos on the Costa del Sol and Blanes on the Costa Brava have grown rapidly over the past 20 years from small fishing ports to large resorts with multi-storey modern hotels and apartments.

Smooth sandy beaches are an important attraction, but the main local resource is sunshine. While Britain and north-west Europe remain in the path of the depressions (and occasional anticyclones) the Mediterranean is for the most part under the influence of high pressure from May to October. As you have already seen, high pressure results in very little cloud and low rainfall. The rain that does occur in summer comes in the form of short, sharp showers and storms. It is also much warmer than Britain since it is further south and the sun is higher in the sky.

Tables 25.1 and 25.2 summarise the climates of Palma de Mallorca and London. The figures for temperature (°C) and rainfall (mm) are the averages for each month, calculated over a long period of time, usually more than 30 years. Generalised conditions of the atmosphere over a long period of time are known as a **climate**. Weather is the specific condition of the atmosphere at one particular time.

The climate of the Mediterranean area has been very much the same for thousands of years, but it has only been since the 1960s that the coasts of Spain and other countries have become so popular. Amongst the most important reasons for the growth of the holiday business are the development of inexpensive jet flights and the 'package holiday' industry. The general rise in wages and higher living standards have also boosted the holiday industry. Earlier this century the Mediterranean coast was very much the preserve of the wealthy few.

Many resorts are now reaching saturation point in the more popular locations. Long stretches of coastline have been developed for tourism and the natural environment has suffered as a result. Pollution is a severe problem in many places. For these reasons tour operators are looking further afield for new resorts e.g. north Africa, the Canary Islands and Turkey. The nature of the resort is also changing, from high-rise hotels with restricted activities like sun-bathing and swimming, to holiday 'villages' with a greater range of more organised leisure pursuits (see *Energy and Industry* for more on tourism).

Table 25.1 Climatic figures for Palma de Mallorca, Spain

	Jan	Feb	Mar	Apr	May	Jun	Jul	Aug	Sep	Oct	Nov	Dec
°C	10	11	12	15	17	21	24	25	23	18	14	11
mm	39	34	51	32	29	17	3	25	55	77	47	40

Table 25.2 Climatic figures for London

	Jan	Feb	Mar	Apr	May	Jun	Jul	Aug	Sep	Oct	Nov	Dec
°C	4	5	7	9	12	16	18	17	15	11	8	5
mm	54	40	37	37	46	45	57	59	49	57	64	48

CASE STUDY

Tourism in Turkey

Milta Village

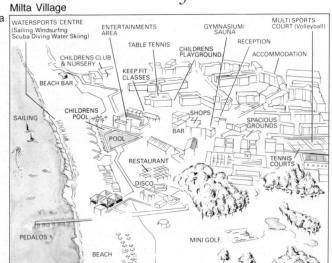

1 Study Tables 25.1 and 25.2 and then answer the following questions.

  a Draw climate graphs for London and Mallorca.

  b Describe in your own words the differences between the climates of London and Palma de Mallorca.

  c Why do you think the figures in Tables 25.1 and 25.2 are inadequate in giving a full picture of the climate of each place?

2 Why might someone living in southern Spain prefer to go to London for an annual holiday?

3 Study all the information in Figure 25.1 and answer the following questions:

  a from the picture a list those activities that depend on sunshine and those that are all-weather;

  b From the data in b describe how the cost of staying at the Milta village is linked to the time of year and the likely weather conditions;

  c From photograph d describe the site of Bodrum (on the south-west coast of Turkey). What do you think are this town's attractions for a tourist from Britain?

  d If a tour company were to build a holiday village on the outskirts of Bodrum, how do you think the following people are likely to react and why?
    i The chamber of commerce
    ii Local farmers
    iii Local fishermen

4 Map c in Figure 25.1 shows the distribution of other holiday villages run by the same company as Milta.

  a Find out which of these are in Common Market countries. Why might a tourist consider this important after 1992?

  b Using suitable climatic data in your atlas consider which places a holiday-maker might choose for:
    i the least likelihood of summer rain;
    ii the highest August temperatures;
    iii the mildest January.

5 Find out the areas around the world that have a climate similar to that of the Mediterranean. Use the climate data in your atlas to help you.

Figure 25.1 Aspects of tourism in Turkey

b

Prices per person in £s	MILTA		FIRST CHILD SHARING REDUCT-IONS %
VIEWDATA CODE	CMT		
Board Arrangement	Full Board		
No. of Nights	7	14	
1 May - 7 May	444	674	40%
8 May - 19 May	451	681	30%
20 May - 1 Jun	471	719	15%
2 Jun - 9 Jun	463	713	20%
10 Jun - 16 Jun	479	725	20%
17 Jun - 30 Jun	492	749	20%
1 Jul - 14 Jul	518	784	20%
15 Jul - 21 Jul	548	798	15%
22 Jul - 14 Aug	568	839	15%
15 Aug - 21 Aug	554	815	15%
22 Aug - 31 Aug	544	794	15%
1 Sep - 11 Sep	519	769	20%
12 Sep - 30 Sep	507	753	30%
1 Oct - 24 Oct	477	699	30%
REDUCTION per night for 3rd adult sharing	£3.00		

(Departures on or between)

Average Maximum Temperature °F	LONDON	TURKEY	Ave Max.hrs of Sun
April	56	70	8
May	62	72	9
June	68	77	9
July	71	82	
Aug	71	82	12
Sept	66	75	11
Oct	58	66	8

26 Atmospheric pollution

The warm sunshine of California does have its drawbacks, at least for the citizens of Los Angeles. Here the exhaust fumes from thousands of motor vehicles react with strong sunlight to produce a dangerous chemical **smog**, shown in Figure 26.1. Every morning local radio stations tell their listeners the likely level of pollution for that day. When levels get too high, people with respiratory problems are well advised to stay indoors.

Vehicles are not the only source of atmospheric pollution. Figure 26.2 shows the major sources of air pollution in Britain.

These pollutants are dispersed by the wind within the atmosphere. Places downwind of a source are particularly at risk. **Lichens** (a form of algae) are very sensitive to dirty air and soon die in very polluted areas. Figure 26.3 shows the distribution of lichen in Britain. This map is therefore a very good indicator of the distribution of polluted air.

Figure 26.1 Atmospheric Pollution, Los Angeles

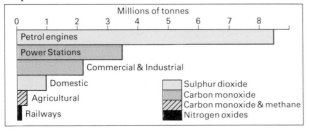

Figure 26.2 Major sources of air pollution in Britain

As well as being dangerous to health, atmospheric pollution seems to be causing changes in the climate of the planet. The increase in the amount of **carbon dioxide** in the atmosphere may be resulting in a general warming of the air. The gas is transparent to the sun's rays but absorbs the radiation from the earth. It is estimated that within the next 50 years or so the average global temperature could go up by about 2°C. Changes will not be uniform. Much of western Europe, for example, may become warmer and drier whilst northern Russia may get warmer and wetter. It is predicted that the eastern Mediterranean could even become colder. Perhaps the most serious implication of the **greenhouse effect** is the gradual melting of the Polar ice caps. This would raise the sea-level by about 50 metres and so flood many coastal areas.

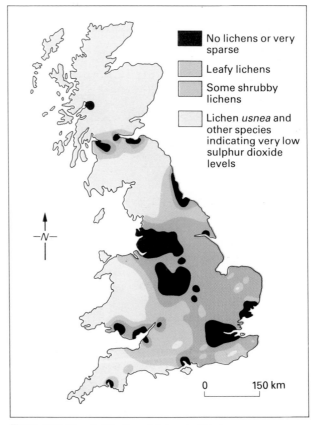

Figure 26.3 The distribution of lichen in Britain

Acid rain

The burning of fossil fuels (oil, gas, coal) releases sulphur compounds − particularly **sulphur dioxide** − which is absorbed by cloud droplets. The resulting rainfall is therefore acidic. Some measurements show that rain can be as acidic as vinegar or lemon juice. This acid accumulates in rivers and lakes and kills the fish. Trees and other plants absorb the **acid rain** and become diseased or even die. Igneous rocks break down chemically under the influence of percolating acidic water. Minerals like aluminium are released which poison the vegetation. Dissolved sulphur compounds can travel 1000 to 2000 km in 3 to 5 days, which makes the problem truly international.

1 In West Germany more than 33 per cent of trees have been damaged, which includes 75 per cent of all Fir trees
2 In Sweden, 18 000 of the 90 000 lakes are badly affected, 4 000 are totally devoid of fish and most plants.
3 In the UK 67 per cent of conifers have some damage and 29 per cent are severely damaged.

11 lakes and streams in the Lake District are seriously affected.
4 The exteriors of many buildings are becoming severely weathered.

The map in Figure 26.4 shows the weight of sulphur deposition in Europe and the percentage of this in each country that comes from other nations.

The EC is attempting to cut the levels of sulphur dioxide and nitrogen oxides from all sources. It calls for a cut of 60 per cent on 1980 levels. However, environmental groups say that the position is so bad that nothing short of a 90 per cent cut will do. Britain has not agreed with the EC Directive. The Central Electricity Generating Board is not entirely convinced by the arguments linking acid rain with generating stations. The British Government also says that Britain is a special case because British coal has a higher proportion of sulphur in it than other European coal. The cost to the consumer of cleaning up electricity generating stations could be a 10 per cent increase on a person's electricity bill. However, three of the largest coal-fired stations are being fitted with cleaning equipment.

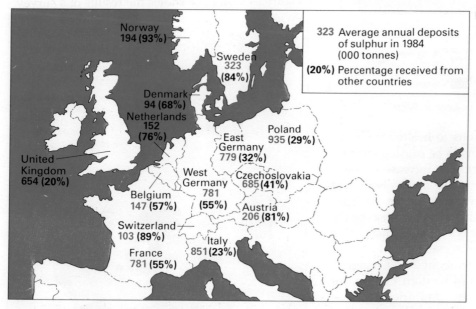

Figure 26.4 Sulphur deposits in Europe 1984

The ozone problem

High in the atmosphere, between 25 and 50 km above the ground there is a layer of **ozone** (O_3). This protects all life on Earth from the harmful ultra-violet rays from the sun. It is estimated that a

1 per cent reduction in the ozone layer lets in an increase of ultra-violet radiation to cause the number of cases of skin cancer to go up by 5 per cent. In 1986 scientists discovered a 'hole' in the ozone layer above Antarctica. It has been known since the 1970s that gases used in aerosol sprays,

refrigerators and in plastic cartons are damaging to the ozone layer. These gases are known as **chlorofluorocarbons** (CFCs). In 1987 the United Nations sponsored a treaty to control the production and use of these gases. The EEC and the USA agreed to cut their production by 50 per cent. However, since it takes 30 to 40 years for the CFCs to penetrate the atmosphere, the future picture is still very uncertain. Furthermore, industrial growth in many parts of the Third World could mean increases in the emission of CFCs.

ACTIVITIES

1 Study Figure 26.3.

 a Name the four largest areas in Britain where there are no lichens or their distribution is only sparse.

 b How might you account for the distribution of these areas?

 c Why are lichens a good indicator of pollution levels?

2 Make two large copies of the map shown in Figure 26.4 but do not write in the statistics.

 a On the first outline use a suitable shading method to show the distribution of sulphur deposits in the countries of Europe.

 b Describe the pattern produced.

 c On the second map outline draw proportional columns to show the amounts of sulphur deposits in each nation that comes from other countries. Which countries seem to 'import' most sulphur. Can you explain why?

3 Why is it difficult for individual nations to pass atmospheric pollution laws that are really effective?

4 Describe ways in which an individual person can help to cut down the amount of atmospheric pollution.

Coursework ideas

1 Table 26.1 is an extract of the weather recordings made by a boy at his own weather station in Hertfordshire. The figures are for the first seven days of September 1986.

 a If you have access to weather instruments at school or at home, keep a similar set of readings for at least a month in each season of the year.

 b Present your data in the form of graphs.

 c Collect the weather maps from one of the daily newspapers day by day. How do the weather conditions you record fit in with the distribution and occurrence of depressions, anticyclones, and fronts?

2 If you do not have access to weather instruments you can still keep a visual record of the weather each day and find out what the temperatures and rainfall have been for the previous day from the daily newspaper. Find out from any of the following how the weather may have affected activities:

 a a local farmer;

 b an outdoor sports centre;

 c the local bus depot;

 d the local authority highways department;

 e the parks department.

You may be able to think of other organisations or individuals to ask.

3 Figure 24.2 on p. 68 shows the results of a survey done by a group of GCSE pupils to find out the effects that the school buildings may have had on temperature distribution in the school grounds after a cold night.

 a With the help of your teacher organise a group to do a similar survey around your own school. You will need one dry bulb thermometer for each member of the group.

 b Try to explain the reasons for any patterns you observe.

 c How do your results differ under different weather conditions the previous night? e.g. if it was cloudy, or windy, or calm and clear.

 d What conclusions might you come to about the design and heating of *your* school buildings?

Table 26.1

Year 1986
Month September

Day	Hour	TEMPERATURE Min.	Max.	Dry Bulb	Wet Bulb	Humidity %	RAINFALL Daily	Monthly	Barometer	Wind	Cloud
1	6·00	11·5°C	13°C	14·5°C	14°C	94%	·5 mm	·5 mm	1016 mb	Force 0	8 oktas
2	6·00	10·5°C	20°C	19°C	16·5°C	76%	·0 mm	·0 mm	1010 mb fall	Force 8	4 oktas
3	6·00	10°C	14°C	13·5°C	11°C	71%	·3 mm	·3 mm	1015 mb rise	Force 3	7 oktas
4	6·00	4·5°C	18°C	16·5°C	14°C	72%	·0 mm	·0 mm	1020 mb rise	Force 0	1 okta
5	6·00	5°C	17·5°C	16°C	13·5°C	74%	·0 mm	·0 mm	1016 mb fall	Force 2	4 oktas
6	6·00	8°C	16·5°C	14°C	13·5°C	90%	·0 mm	·0 mm	1016 mb "	Force 0	5 oktas
7	6·00	6°C	18°C	16·5°C	13°C	64%	·0 mm	·0 mm	1010 mb "	Force 2	1 okta

27 Forest and grassland

The natural vegetation of the Earth over much of its surface is some kind of forest or woodland. The major exceptions are the hot deserts, the cold areas around the Poles and the high mountains. In these forests there are a wide variety of tree species. At ground level shrubs and herbs flourish. These plant communities take their nourishment from the soil below and are watered from the rain above. When leaves die and fall to the ground, they return the **nutrients** to the soil. Bird, animal and insect life are supported by the vegetation and of course many feed off each other. Such a community of living things within a particular environment is known as an **ecosystem**. Regional and local variations are produced in different environments from the **tropical rain forests** of Brazil to the **deciduous broadleaved forest** of southern England.

Figure 27.1

Many of these ecosystems have been affected by human activity. Figure 27.2 is a photograph of an ecosystem: the broadleaved deciduous forest. The trees include oak, ash and beech. They lose their leaves in winter. Shrubs and herbs near ground level include ferns, wood sorrel and blue-bells. Most of Britain and Europe was once covered in such forest but much of it has been cleared by man for agriculture, industry and settlement. Even the natural-looking forest in Figure 27.2 has been interfered with by man. In this particular case the wood in the past was managed as a source of timber. Today it is a recreation area in which diseased or dead trees are removed, bridleways kept clear and car parks and picnic sites are provided. Figure 27.3 illustrates the relationships within a forest ecosystem.

Figure 27.2 The broadleaved deciduous forest

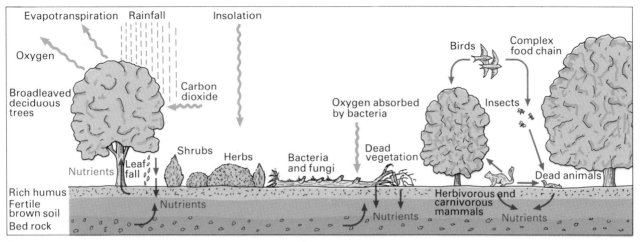

Figure 27.3 Some features of the broadleaved deciduous forest ecosystem

CASE STUDY

Seven Sisters country park, East Sussex

The Seven Sisters Country Park (see Figure 27.4) is an area of chalk grassland on the South Downs between the Cuckmere valley and East Dean. The land was purchased by the East Sussex Council.

The main aims of the Park are:

1 to conserve the scenic beauty and wildlife;
2 to provide opportunities for people to enjoy and appreciate its qualities.

Figure 27.4 The Seven Sisters Country Park

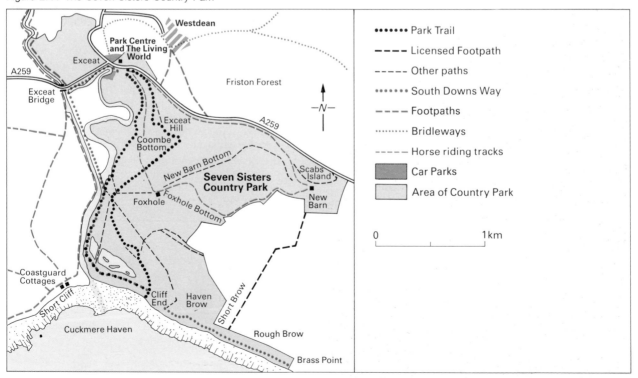

Figure 27.5 a and b
On the park trail

The area of grassland is a fine example of a **chalk downland** ecosystem. This ecosystem, unlike the forest ecosystem, is very much influenced by the activities of people (see Figure 27.5). Left undisturbed it would eventually revert to forest (see Figure 27.6). The ecosystem depends on sheep **grazing**. This prevents the development of shrubs and trees and allows a wide variety of grasses and herbs to flourish. Many flowering species are found, e.g. cowslip, buttercups, milkworts, thyme, scabious and felwort. These in turn are an important source of nectar for numerous butterflies like the chalkhill blue and insects like the wart-biter grasshopper. Some of the plants and insects are very rare like the orchid. The sheep fertilise the grassland with their droppings. Any use of chemical fertilisers would mean death for many of the more sensitive plant species.

The downland ecosystem is destroyed by ploughing. This is banned within the Seven Sisters Park area (only grazing licences are issued). Elsewhere on the South Downs (defined as an **Environmentally Sensitive Area:** see *Agriculture and Rural Issues*) farmers are encouraged not to plough up by giving them an annual payment of £60 per acre. Unfortunately this is less than the current price for crops like rape and barley.

About 250 000 visitors come to the Park each year. The environment is preserved by restricting the car parking facilities. No motor vehicles are allowed beyond the car parks and activities like horse riding and camping are only allowed in certain areas. At Exceat there is a Park Centre where displays, exhibitions, books and leaflets encourage people to learn about the local environment and its care.

ACTIVITIES

1 Study Figure 27.3 and answer the following questions.
   a List the inputs and the outputs of the system.
   b Describe the movement of nutrients through the ecosystem. Why do you think that what you have just described can be called the **nutrient cycle**?
   c Why are the trees known as **broadleaved deciduous** trees?
   d What is the **food chain**?
   e Draw another ecosystem diagram to illustrate the changes that would take place in the ecosystem shown in Figure 27.2 if the woodland was to be used as a Country Park.

2 For the following activities refer to the 1:50 000 OS map extract Figure 6.3 on p. 16 as well as Figure 27.4 and 27.5.
   a Work out the size of the area of the Seven Sisters Country Park in km²
   b Many of the visitors to the Park are spending holidays in Eastbourne. How many kilometres do they have to travel from there to see the exhibition at Exceat? Give the six-figure grid reference for Exceat.
   c The Park Trail route can be seen on the map in Figure 27.4 and the photographs in Figure 27.5 a and b were taken on the Trail. If you were to walk the entire length of the Trail at an average speed of 5 km per hour, how long would it take you? From the evidence on the photographs, what damage do you think people could cause to the ecosystem along the Trail?
   d Lightweight camping is allowed at Foxhole. Give the six-figure grid reference of the site and describe its particular physical characteristics from the OS map.

Figure 27.6 Chalk downland vegetation succession

Short springy turf with many varieties of flowering plants, e.g. cowslip

Taller, coarser grass with fewer herbs and smaller plants, e.g. upright broom

Scrub development, e.g. hawthorn, bramble, gorse

Woodland development, e.g. ash, sycamore, beech, yew

← Intense grazing → ← Low intensity grazing → ← Ungrazed, dependent upon time →

10 m

0 m

28 The tropical rain forest

The most ancient and complex ecosystem in the world is the **tropical rain forest**. It has been evolving over the past 50 million years to produce an enormous variety of plant and animal species. Scientists believe that many of the species that exist in the forest have yet to be discovered. However, despite its age, it is a very sensitive system. Once cleared it does not grow back, thousands of species are lost forever and the cleared area reverts to scrub or even desert. The heavy rainfall (about 2000 mm a year) comes from the moisture produced by the evapotranspiration of the vegetation. The soil is not very fertile and once the **nutrient cycle** is broken it soon deteriorates and gets eroded away. Figure 28.1 a, b and c show some of the features of the tropical rain forest.

Figure 28.1 Some features of the tropical rain forest

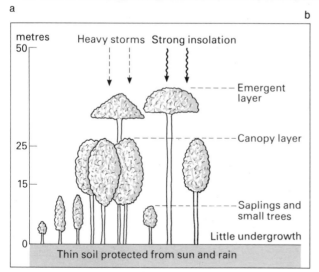

a

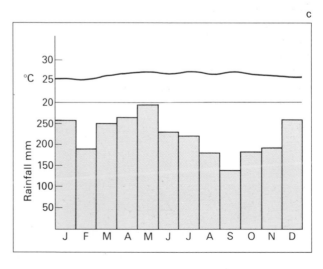

b

c

The newspaper report in Figure 28.2 describes some of the things that are happening to the forest in Brazil. Perhaps we should be concerned about such developments. Our industries in Europe need the valuable mineral resources that the Brazilian rain forest area can produce. However, medical and agricultural research rely on the forest species to supply new drugs and food crop varieties. Furthermore some climatologists believe that the rain forest gives to the Earth's atmosphere a vital proportion of the oxygen needed for life.

CASE STUDY

The Brazilian rain forest

Brazil warned of ecological 'catastrophe'

ECOLOGICAL catastrophe on a devastating scale is threatening the Amazon region of Brazil, according to a report released yesterday by Survival International and Friends of the Earth. If preventive action is not taken, the development of Greater Carajás, a hydro-electric, mineral, railway and agro-industrial complex, will turn an area the size of Britain and France combined, half of it tropical rain forest, into "a massive agro-industrial park, [and transform] its people into a destitute, landless labour pool".

The Greater Carajás programme began in 1980. It covers 10 per cent of Brazil, an area of 895,265 square kilometres (345,559 square miles) which contains bauxite, copper, cassiterite, nickel, gold and one of the world's largest reserves of iron ore. The exploitation of this mineral wealth was to be combined with agriculture, forestry, cattle-ranching and heavy industry and is planned to generate $17bn (£10.6bn) worth of exports annually by the 1990s.

The region is home to over 13,000 tribal people, theoretically protected under Brazilian law. Under the constitution, the Indians are guaranteed permanent possession of their lands and exclusive use of the natural wealth. But in Greater Carajás, over half the tribal lands have not been defined and have no legal protection from invasion by farmers, settlers, mining companies and gold-seekers.

Xikrin Indians, whose tribal lands have been taken over by a Brazilian development plan.

The report says much of the responsibility rests with the World Bank and the European Community, both major sources of finance, and accuses them of irresponsibly funding projects which ran counter to their policies on human rights and environmental protection. In spite of widespread protests, neither has been willing to insist on effective protective measures.

World Bank policy declares projects will only be funded if the resident communities have agreed to them. But because their tribal lands have not been defined, many communities were not consulted.

The EC is the biggest investor in the iron ore project, expected to supply 50 per cent of EC iron ore needs at favourable prices. The environmental cost, the authors argue, is devastating.

Eleven plants producing pig-iron, iron alloys and cement are to be installed along the Carajás railway, nine fuelled by charcoal from the destruction of 1.5 million hectares of forest.

In pursuit of development, the authors argue, even the organisations supposed to protect the Indians have subjugated Indian interests to what is seen as the national interest. The official attitude, says the report, is that the Indians will benefit from contact and integration into projects. But the authors say contact with Indian tribes brings them diseases they have no immunity to, the destruction of their environment brings hunger and social disintegration, leaving them dependent on government handouts or work on the projects which have destroyed their way of life.

The potential benefits to the nation, the report argues, are equally illusory. Because environmental advice has been ignored, prospects for sustainable development in Carajás are poor. The programmes are badly thought out, poorly administered and environmentally inappropriate. "The programme is a classic example of a mega-scale industrial fantasy which owes more to the dreams of politicians and businessmen than to rational planning for sustainable development," the report says.

The lushness of tropical rain forest gives an illusion of fertility but such forests are founded on poor soil. A hectare of Amazonian forest, which can contain 300 different species of tree, will not survive cattle farming or timber plantation for more than a few years before becoming desert.

The loss of the forest brings further problems: the microclimate changes, water retention is lost, rainfall drops, rivers become silted and polluted. Once the process is in train, even the vast hydro-electric schemes are at risk from silting and increased water acidity.

The report calls on those responsible—the EC, the World Bank and the Brazilian government, to urgently demarcate and protect tribal lands and to base development on ecological principles. Managed this way, the authors argue, the region offers sustainable, indefinite benefits.

The Independent 24 September 1987

Figure 28.2

ACTIVITIES

1 Study Figure 28.1 a, b and c which show some of the characteristics of the tropical rain forest environment. Answer the following questions.

   a Describe the climate of the rain forest. Why should such a climate encourage vigorous growth of natural vegetation?

   b How does the rain forest protect the soil beneath it?

   c What is likely to happen to the soil if the vegetation is cleared?

2 a With the aid of your atlas, draw an outline of South America.

   b On your outline mark on the following:
      i the tropical rain forest area;
      ii the border of Brazil;
      iii the River Amazon;
      iv the following cities: Brasilia, Rio de Janeiro, Recife, Belem;
      v the Carajas Region, centred at 51° W 5° S;
      vi make sure you give your map a title, key and scale.

   c From your map measure the area in km² covered by the tropical rain forest.

3 Read very carefully the newspaper article about the Brazilian rain forest in Figure 28.2. Then answer the following questions.

   a How large is the Greater Carajas Region?

   b What do the developers hope to gain from developing this region?

   c In what ways may the Xikrin Indians be affected if development takes place?

   d Why do Survival International and Friends of the Earth say that the World Bank and EC have to take most of the responsiblity for what could happen?

   e Why should the scheme be described as a 'megascale industrial fantasy'?

4 Design and draw a poster that might persuade the citizens of the EC to be concerned about the developments in the rain forest.

5 With the aid of the world vegetation map in your atlas, shade onto an outline map of the world the distribution of the tropical rain forest. Are most of the countries with rain forest developed countries or developing countries? How might this affect the future of the tropical rain forest?

Coursework ideas

1 Find out as much as you can about the tropical rain forest of a particular region. In particular:

   a find out its structure and draw an ecosystem diagram similar to the one in Figure 27.3 on p. 75;

   b discover how the native peoples traditionally live and why they should be regarded as being in tune with the natural ecosystem;

   c discuss whether the natural resources can be developed without destroying the natural environment.

2 Write a report on any ecosystem within reach of your home. There are many different types of ecosystem, and some of them may cover very small areas. You may have one of the following near you:

   a a woodland or forest (deciduous or evergreen);

   b a chalk grassland;

   c a marshland (freshwater inland or saltmarsh by the coast);

   d water meadows on a flood plain;

   e a piece of overgrown wasteland in a town or city;

   f a moorland.

To what extent is the ecosystem you have studied under threat from visitors or some kind of economic development?

3 Refer to Figure 27.1 on p. 74. Discuss ways in which, by taking actions locally (perhaps in your own neighbourhood), people can help the global environment.

Index

LONGMAN
CO-ORDINATED
GEOGRAPHY

Series editor - Simon Ross

Population

Sylvia M Wood

Acknowledgements

I should like to thank the following for the help they have given me in the preparation of this book:
Mrs Jill Norman — Librarian; Mrs D Hammond and the Office of Population Censuses and Surveys; Mr I Thomas, NRSC, Farnborough; Mr D Wilson, The Hunting Technical Services, Borehamwood; Mr L Bardou; Mr S Ross; Mr G Stone, Rhondda Borough Council; Mr G Harrison, Eastbourne Borough Council; Professor JP Cole, Nottingham University; Mrs KM Jarman; Miss AE Sutcliffe; Mrs B Saunders; Many Embassies, Councils and representatives of organisations like London Transport and Population Concern, Centre for World Development Education, Oxfam, World Family Foster Parents Plan, Save The Children.

We are grateful to the following for permission to reproduce photographs and other material:
J Allan Cash Limited, pages 35, 37 *below*, 41, 48, 58 (2); Chicago Tribune, 24; Colorific, page 47 (photo: Flip Schuike); Mary Evans Picture Library, page 6 *left*; Paul Forster, page 31 *above*; Geoscience Features, page 25; Sally & Richard Greenhill, pages 19 *above*, 27; Guildhall Library, London, page 2; Robert Harding Picture Library, page 6 *right*; Home Office, page 33; Hulton-Deutsch, pages 17, 57 *below right*; Hunting Technical Services Ltd, page 55; Hutchison Library, pages 12, 19 *below*, 36 *below*, 40; International Defence Aid Fund, page 57 *left*; IPPF, page 10; Library of Congress, page 11; Mansell Collection, page 49; Network, pages 7 *left* (photo: Denis Doran), 7 *right* (photo: Sunil Gupta), 32 (photo: Barry Lewis), 56 (photo: Barry Lewis); Northeast Studios Ltd, page 57 *above right*; The Observer, page 32 (photo: John Reardon); Office of Population Censuses and Surveys, page 3; Panos Pictures, pages 37 *above* (photo: Paul Harrison), 53 (photo: Sean Sprague) 54–55 (photo: Sean Sprague); Reflex Picture Agency Ltd, page 58–59 (photo: Caroline Penn); Stevenage Borough Council, page 22; Tony Stone Worldwide, page 38

Cover: Crowds, Oxford Street. London; Steve Sandon, Daily Telegraph colour library

Philip Allan Publishers Ltd. *Geography Review* Vol.1 No.1 — article by Ceri Peach. 'The Population of the United States'; British Rail; British Tourist Authority; English Tourist Board; Geographical Association, *Geography* 70 (2), 1985, p.160 Article by P Ogden 'France, recession politics and migration policy'; *Geographical Journal* Vol 150 (2) p.158; *Geography* 70 (2), 1985 p. 163 Article by Salt 'West German dilemma, Little Turks or Young Germans?; HMSO, CSO Annual Abstract of Statistics, 1984; HMSO *British Business* 7 December 1979; HMSO Ordnance Survey 1:25000 Pathfinder Map, Rhondda Valley (c) Crown Copyright; ISTAT-Roma; OPCS; Oxford Examination Board, Human and Regional Geography Paper for Oxford College Entrance Examination November 1987; (c) Pergamon Press plc,

Population Studies 11' 29 (1957–8) with the permission of OD Duncan and for *Population Geography* with the permission of J I Clarke; (c) George Philip Ltd; Rhondda Borough Council; Rhondda District Plan; Stevenage Urban Studies Centre; (c) Twentieth Century Fund, W S and E S Woytinsky, 'World population and production'; Tyne and Wear Passenger Transport executive; United Nations Population Fund; (c) World Bank, The World Bank Atlas, 1987

Ewan MacNaughton Associates for articles 'One Child only' slogan spreads across China' by Nigel Wade (abridged) from *The Daily Telegraph* 5.8.79. & 'France cheers le bébé boom' by John Izbicki from *The Daily Telegraph* 22.8.87. (c) The Daily Telegraph plc; The Observer Ltd for abridged articles 'My five-billionth baby' by Geoffrey Lean from *The Observer* 5.7.87, 'Waiting at death's door' by C. Watson from *The Observer* 3.1.88 & 'Rich and poor: Tale of two countries' by D. Willey from *The Sunday Observer* 15.11.87; Times Newspapers Ltd for an abridged extract from the article 'Romanians chafe at baby drive' by Peter Godwin from *The Sunday Times* 24.2.85. (c) Times Newspapers Ltd 1985.

We have unfortunately been unable to trace the copyright holders of the article 'Exodus in search of 'Good Life' reprinted in *The Independent* 8.1.88. and would appreciate any information which would enable us to do so.

Population

Sylvia M Wood

Contents

POPULATION STRUCTURE

1 The census

Look at Figure 1.1. It shows part of the *census* form used in England in early April 1981. A census has been carried out in England ever since 1801. They are usually carried out every ten years.

The aim of a census is to find out important information about the population. It is helpful to know how many children will require schooling in the years ahead, how many pensioners there are going to be, and so on. The census enables the government to see future trends and allows it to plan accordingly.

The census in England has to be completed by the **head of household** and the questions apply to those in the household on a given night. It is illegal not to complete the census.

As you can imagine, with every household completing a census, there is a huge amount of statistical information produced. This information is gathered together by the Office of Population Censuses and Surveys and processed. Nowadays all the information is stored on computers.

It is possible to obtain census information for different sized areas the smallest of which is the **enumeration district**. This is usually a few roads with a total population of about 500. Information is available for gradually larger and larger areas — **wards** (see Figure 15.7), **boroughs, counties** and **regions**.

Most geographers and **demographers** (people who study population) use census information to find out how aspects of population (like age, employment, car ownership) vary from one place to another. A useful document which gives summary information for each county is called the **County Monitor**. Your school or local library will have the Monitor for your county.

Further valuable sources of information on population are the **birth, death** and **marriage certificates**. Have a look at your birth certificate or that of another member of your family to see what information is recorded. Churches have **parish records** which have details of baptisms, marriages and burials (See Fig. 1.2).

Censuses are carried out in most countries of the world. However, in many developing countries they are not always very reliable for the following reasons:

1 a large number of educated people are needed to give out and collect forms;
2 people may be hostile to giving away personal information fearing taxes or military service, for example;
3 remote parts of countries can be difficult to reach;
4 the whole operation is expensive — the money might not be available.

ACTIVITIES

1 Study Figure 1.1, the 1981 census.
   a For three of the topics, suggest reasons why the information collected can be useful in planning for the future.
   b Can you suggest any other topics that should be covered in the 1991 census? Give reasons for your suggestions.
   c Although the statistics from each census are fed into a computer, names and addresses are **not**. Why do you think this is so?
   d Why is it important that all heads of household fill in the census?
   e Why do you think most censuses in Britain take place in the spring?

2 You have been asked to design the first census for a small country in the developing world. The people in the country are fairly well educated and have reasonable standards of living. However, the population is growing rapidly and the government thinks that it is about time some accurate population statistics were made available. Make up a simple census covering only those topics which you feel to be essential for future planning. Write a short introduction so that people know why the information is needed.

Figure 1.2 Parish records of deaths in London

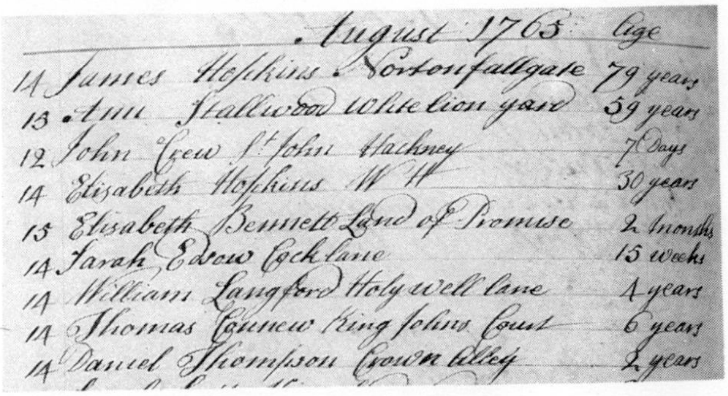

Figure 1.1

In strict confidence

1981 Census England

H Form for Private Households

A household comprises either one person living alone or a group of persons (who may or may not be related) living at the same address with common housekeeping. Persons staying temporarily with the household are included.

To the Head or Joint Heads or members of the Household

Please complete this census form and have it ready to be collected by the census enumerator for your area. He or she will call for the form on **Monday 6 April 1981** or soon after. If you are not sure how to complete any of the entries on the form, the enumerator will be glad to help you when he calls. He will also need to check that you have filled in all the entries.

This census is being held in accordance with a decision made by Parliament. The leaflet headed 'Census 1981' describes why it is necessary and how the information will be used. Completion of this form is compulsory under the Census Act 1920. If you refuse to complete it, or if you give false information, you may have to pay a fine of up to £50.

Your replies will be treated in STRICT CONFIDENCE. They will be used to produce statistics but your name and address will NOT be fed into the census computer. After the census, the forms will be locked away for 100 years before they are passed to the Public Record Office.

If any member of the household who is age 16 or over does not wish you or other members of the household to see his or her personal information, then please ask the enumerator for an extra form and an envelope. The enumerator will then explain how to proceed.

When you have completed the form, please sign the declaration in Panel C on the last page.

A R THATCHER
Registrar General

Office of Population Censuses and Surveys
PO Box 200 Portsmouth PO2 8HH
Telephone 0329-42511

Please answer questions H1 - H5 about your household's accommodation, check the answer in Panel A, answer questions 1-16 overleaf and Panel B on the back page. Where boxes are provided please answer by putting a tick against the answer which applies. For example, if the answer to the marital status question is 'Single', tick box 1 thus:

1 ☑ **Single**

Please use ink or ballpoint pen.

To be completed by the Enumerator

Census District	Enumeration District	Form Number

Name ....

Address ....

....

.... Postcode ⬚⬚⬚▨⬚⬚

Panel A
To be completed by the Enumerator and amended, if necessary, by the person(s) signing this form.

This household's accommodation is:

- In a caravan ☐ 20
- In any other mobile or temporary structure ☐ 30
- In a purpose-built block of flats or maisonettes ☐ 12
- In any other permanent building in which the entrance from **outside** the building is:
  - NOT SHARED with another household ☐ 10
  - SHARED with another household ☐ 11 ◄

H1 Rooms

Please count the rooms in your household's accommodation.
Do not count:

small kitchens, that is those under 2 metres (6ft 6ins) wide, bathrooms, WCs.

Number of rooms ....

Note
Rooms divided by curtains or portable screens count as one; those divided by a fixed or sliding partition count as two.

Rooms used solely for business, professional or trade purposes should be excluded.

H2 Tenure

How do you and your household occupy your accommodation? Please tick the appropriate box.

As an owner occupier (including purchase by mortgage):

1 ☐ of freehold property

2 ☐ of leasehold property

By renting, rent free or by lease:

3 ☐ from a local authority (council or New Town)

4 ☐ with a job, shop, farm or other business

5 ☐ from a housing association or charitable trust

6 ☐ furnished from a private landlord, company or other organisation

7 ☐ unfurnished from a private landlord, company or other organisation

In some other way:

☐ Please give details

....

Note
a If the accommodation is occupied by lease originally granted for, or since extended to, more than 21 years, tick box 2.

b If a share in the property is being bought under an arrangement with a local authority, New Town corporation or housing association, *for example, shared ownership (equity sharing), a co-ownership scheme*, tick box 1 or 2 as appropriate.

H3 Amenities

Has your household the use of the following amenities on these premises? Please tick the appropriate boxes.

- A fixed bath or shower permanently connected to a water supply and a waste pipe

1 ☐ YES – for use only by this household

2 ☐ YES – for use also by another household

3 ☐ NO fixed bath or shower

- A flush toilet (WC) with entrance inside the building

1 ☐ YES – for use only by this household

2 ☐ YES – for use also by another household

3 ☐ NO inside flush toilet (WC)

- A flush toilet (WC) with entrance outside the building

1 ☐ YES – for use only by this household

2 ☐ YES – for use also by another household

3 ☐ NO outside flush toilet (WC)

H4   Please answer this question if box 11 in Panel A is ticked.

Are your rooms (not counting a bathroom or WC) enclosed behind your own front door **inside** the building?

1 ☐ YES   2 ☐ NO

If your household has only one room (not including a bathroom or WC) please answer 'YES'.

H5   Cars and vans

Please tick the appropriate box to indicate the number of cars and vans normally available for use by you or members of your household (other than visitors).

0 ☐ None

1 ☐ One

2 ☐ Two

3 ☐ Three or more

Include any car or van provided by employers if normally available for use by you or members of your household but **exclude** vans used solely for the carriage of goods.

2 Population pyramids

A **population pyramid**, sometimes called an age-sex graph, uses horizontal bars to represent the proportion of people of either sex in different age groups. Put together, the bars often form a diagram that looks like a pyramid, hence the name. Figure 2.1 shows population pyramids for England and Wales in 1901 and 1981. Which one looks most like a pyramid?

Population pyramids tell us a lot about the population structure and aid planning. A number of typical shapes can be identified – see Figure 2.2.

True pyramidal shaped diagrams (Figure 2.2a) tell us that the population has a lot of young people but few elderly. Birth rates would be high as would death rates with few people living to old age. Such pyramid shapes are typical of developing countries where populations are **expanding**.

The graph in Figure 2.2b shows a fairly **stationary** population with fewer children and most of the population aged between fifteen and sixty years old. Figure 2.2c shows a **contracting** population where the birth rate is steadily dropping year by year and where there are a lot of elderly people. Both Figures 2.2b and 2.2c are common shapes for countries in the developed world.

The study of population pyramids is helpful for future planning. A 'baby boom' can be identified and then catered for accordingly as the 'boom' moves up the diagram over the years through childhood, to adulthood and into old age.

The following activities will further your understanding of population pyramids.

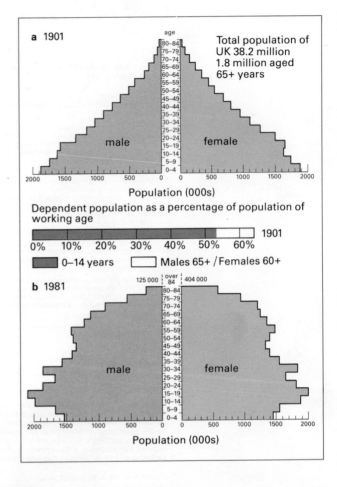

Figure 2.1 The age and sex structure of the population of England and Wales in 1901 and 1981. For 1981 the populations, male and female, over 84 years old are shown as numbers.

Figure 2.2 Population pyramids.

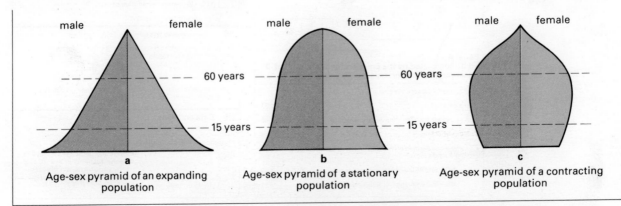

a	b	c
Age-sex pyramid of an expanding population	Age-sex pyramid of a stationary population	Age-sex pyramid of a contracting population

ACTIVITIES

1 a Make a copy of each of the pyramid shapes in Figure 2.2. Give each a title. Use three different colours to shade the sections 1–15 years, 15–60 years and over 60 years. Do this for each pyramid.

   b Which pyramid has the highest proportion of under 15s?

   c Which pyramid has the lowest proportion of under 15s?

   d In twenty years time, which pyramid will suggest a need to provide for the greatest proportion of elderly?

2 Study Figure 2.1.

   a For each pyramid, use Figure 2.2 to state whether the population is **expanding, stationary** or **contracting**.

   b Use both graphs to contrast the female population (thousands) in the following age groups:
      i 0–4 age group;
      ii 30–34 age group;
      iii 75–79 age group.

   c Suggest reasons for the changing trends you have identified.

   d Study Figure 2.1b. Explain why school closures are likely in the next few years and why the Government must plan to set aside more money for pensions.

   e Make a tracing of the outline of Figure 2.1b. Give it a title and add the following labels:
      i large proportion of elderly;
      ii decreasing proportion of infants;
      iii effect of the Second World War;
      iv effect of the 1960s 'baby boom'.
      Be careful with the last two labels!

3 a Use the information in Table 2.1 to draw a population pyramid for England and Wales in 1986. Give the diagram a title.

   b Comment on the changing trends between 1981 (Figure 2.1b) and 1986. Do you think these trends will continue into the first part of the 21st century?

4 a Trace the population pyramid for Nigeria from Figure 2.3a. Use the same three colours used in Activity **1a** to shade the three age groups 0–15, 15–60 and 60+. Add labels to highlight the contrasts with the 1986 pyramid for England and Wales drawn in Activity **3**.

   b Is this an **expanding, stationary** or **contracting** population?

   c What planning guidelines for the next ten years would you give the Government?

   d Repeat exercises **a, b** and **c** for West Germany using Figure 2.3b.

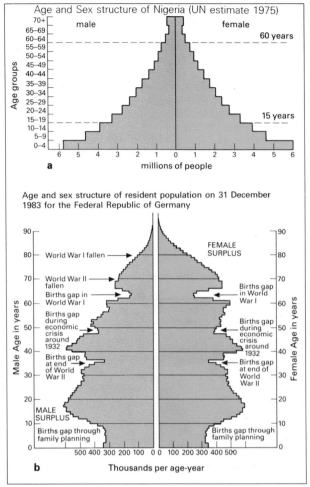

Figure 2.3

Table 2.1 Estimated resident population of England and Wales in mid 1986 in thousands

Age in years	Males	Females
0–4	1632.2	1551.0
5–9	1553.3	1471.3
10–14	1651.9	1561.6
15–19	2004.5	1906.7
20–24	2130.4	2072.2
25–29	1877.0	1847.0
30–34	1683.0	1658.3
35–39	1851.2	1848.7
40–44	1593.2	1570.1
45–49	1394.2	1382.0
50–54	1334.0	1332.7
55–59	1328.9	1374.0
60–64	1299.4	1415.1
65–69	1071.5	1279.3
70–74	897.2	1210.1
75–79	625.9	1014.0
80–84	325.9	688.6
85–89	113.6	343.1
Over 90	36.3	145.9

Source: OPCS Monitor

5 People below 15 and over 60 are sometimes referred to as **dependants**. They are generally looked after by those people aged 15–60. Work out the percentage of dependants for England and Wales from Table 2.1. To do this, add up the number of people below fifteen and over sixty. Divide this by the **total** population and multiply by 100.

Comment on any differences between the percentage of young and old dependants in 1901 (from Figure 2.1a) and your results for 1986. (See Figs. 2.4 and 2.5.)

6 Population pyramids can be drawn for towns. These often reveal interesting patterns.

a Use the information in Tables 2.2 and 2.3 to draw pyramids for Eastbourne and Stevenage 'New Town'.

b Use labels to compare the two graphs.

c Try to explain any differences.

Figure 2.4 An English family in 1901

Table 2.2 Eastbourne's population 1981 census

Age in years	Males	Females
0–4	1712	1600
5–9	1822	1783
10–14	2147	2091
15–19	2293	2460
20–24	2120	2274
25–29	1857	2017
30–34	2226	2312
35–39	1849	1895
40–44	1595	1731
45–49	1589	1768
50–54	1714	2126
55–59	1843	2472
60–64	1992	2728
65–69	2392	3477
70–74	2346	3688
75–79	1698	3184
80–84	923	2238
85+	495	1641
Total	32613	41485

Total population of Eastbourne 74098

Age in years	Total number of people	Percentage of the population
0–14	7731	10.4%
15–59	39565	53.4%
Over 60	26802	36.2%

Source: OPCS 1981 Census – small area statistics

Figure 2.5 An English family in the 1980s

Table 2.3 Population of Stevenage 1981

Age	Male	%	Female	%
0–4	2577	7.01	2409	6.38
5–15	6647	18.07	6380	16.91
16–24	6498	17.67	6345	16.82
25–34	6020	16.37	5681	15.01
35–44	4340	11.80	4627	12.26
45–54	4316	11.73	4522	11.99
55–59	2127	5.78	1960	5.19
60–64	1453	3.95	1557	4.13
65–69	1191	3.24	1356	3.59
70–74	837	2.28	1191	3.16
75+	777	2.11	1712	4.54
Total	36 783		37 730	

Source: Stevenage Urban Studies Centre

Figure 2.6 Retirement in Eastbourne

ACTIVITIES

7 Study Figure 2.7 which shows the distribution of people aged over sixty-five in the UK in 1981.

  a Which counties had over nineteen per cent of their population made up of elderly people?

  b Try to explain the pattern shown in Figure 2.7. Climate maps might help! Look in national newspapers for house prices in different areas of the country.

Figure 2.7 The elderly population of the UK in 1981

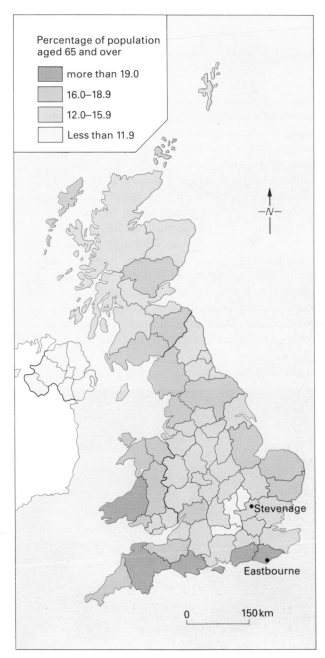

Percentage of population aged 65 and over

more than 19.0

16.0–18.9

12.0–15.9

Less than 11.9

–N–

•Stevenage

■Eastbourne

0   150 km

ACTIVITIES

8 Trace the triangular graph in Figure 2.8. Plot on it the position of Stevenage according to its population structure. In what way does its position differ from Eastbourne's?

9 Study the information in Table 2.4. It describes the age distribution of population in the UK by ethnic group. For each ethnic group draw a pie chart or divided bar to show the age distribution. On each diagram, use the same colour to represent each of the five age groups. (See Fig. 2.9.)

  Describe the main differences and try to suggest some reasons for them.

Table 2.4 Age distribution by ethnic groups in the United Kingdom for 1983−85 combined. Percentages of ethnic groups

Age	White	West Indian	Indian	Pakistani or Bangladeshi	Other
0−15 years	22	27	34	48	37
16−24 years	14	23	15	13	18
25−44 years	27	26	33	24	32
45−64 years	22	21	14	13	10
65+ years	15	3	4	2	3

Source: OPCS Monitor GH5 86/1, 1986

Figure 2.8 Triangular graph showing population structure in 1981

Coursework ideas

Use your local County Monitor to produce maps showing population characteristics for the different wards in your town. To do this you will need to obtain a base map showing the ward boundaries − this can usually be obtained from your Town Hall.

  The County Monitor will give you plenty of choice of population characteristics to map. Use a **choropleth** or shading technique to show the information − your teacher will help you here.

  If you mapped the proportion of elderly, it would be interesting to compare it with a map showing locations of old peoples homes and day centres. In the same way, does the location of, for example, mosques occur in wards with high proportions of ethnic minorities? There are all sorts of relationships you could look for.

Figure 2.9 British youth

3 *Introduction to population change*

The cartoon in Figure 3.1 shows how the number of people in an area or country changes.

All the time babies are born and people are dying. On average 3.9 children are born to each woman in the world. Sometimes during a war fewer children are born as families may be separated or people may not choose to have children at such dangerous times. During famines the number of births may fall as mothers may be ill-nourished. If women decide they want an education or a career they may decide to put-off having children. Sometimes people decide they do not want any children. In all these cases the **crude birth-rate** may be low.

Large numbers of births occur when people do not want or do not know how to prevent pregnancy. Read the story of Mamtaz in Figure 3.2

Figure 3.2

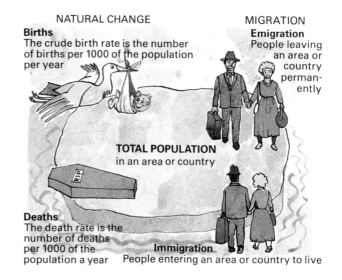

NATURAL CHANGE

Births
The crude birth rate is the number of births per 1000 of the population per year

MIGRATION

Emigration
People leaving an area or country permanently

TOTAL POPULATION
in an area or country

Deaths
The death rate is the number of deaths per 1000 of the population a year

Immigration
People entering an area or country to live

Figure 3.1 Components of population change

Mamtaz

Mamtaz is a small, very pretty woman in her early twenties. She lives in a typically poor, inaccessible, Bangladeshi village at Kaligonj, to the north of Dhaka. She was married at 13, and has had six children.

"My father's village is Dara Bakyti, about five miles from here. He was a farmer and I was the only daughter born to my mother: my stepmother had three.

I was married at 13 and had my first child when I was 14. I had no problems because my mother-in-law looked after him. It was an easy birth, after eight months, and he was very well.

Two years later I had a second boy, but he died after seven days from tetanus infection. My fifth child also died. My youngest daughter is three.

I did not use family planning because we did not have any facilities here. Recently I have tried Pills, but have had headaches. I also tried an IUD but I had problems with that. I don't want any more children but I don't know what to do.

Now I can look back and think what harm my parents did in getting me married so soon. If they had educated me maybe I would be in a better position now. My husband's income is not enough to maintain the family.

If I cannot afford to educate our two daughters, and have to get them married early, their position will be just like mine."

and see Fig. 3.3. In countries with no welfare provision for the elderly and few interests possible outside the home, people may need a large number of children for support and comfort. Look at Figure 3.4 which shows the reasons people give for having children. Some people like large families or belong to a religion like Roman Catholicism which forbids contraception. In these examples the crude birth-rate may be high.

Deaths may occur through old age, disease or accident. Through a natural or man-made hazard like famine or war, the number of deaths rises.

The relationship between the birth-rate and **death-rate** is constantly changing but the difference is known as the **natural increase or natural change**. Table 3.1 shows how birth and death rates have changed in the United Kingdom this century.

Of course variations in birth and death rates are not the only factors which explain population change. **Migration** or movements of people also help explain population losses or gains. **Immigrants** into an area to live increase numbers (see Fig. 3.5) and **emigrants** leaving an area or country permanently may reduce numbers.

Table 3.1 Birth and death rates and life expectancy in the UK

Year	Crude birth rate per 1000	Death rate per 1000	Average life expectancy
1901	30	17	Less than 55 years
1981	13	12	74 years

Source: OPCS

Figure 3.3 Mamtaz

Figure 3.4 Reasons for having children.

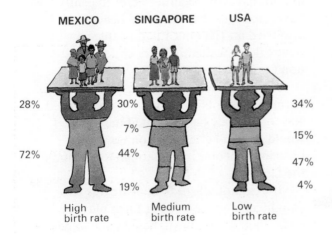

MEXICO   SINGAPORE   USA

28%   30%   34%

7%   15%

72%   44%   47%

19%   4%

High birth rate   Medium birth rate   Low birth rate

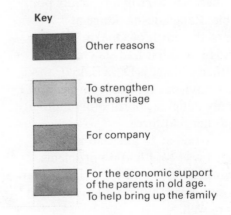

Key

Other reasons

To strengthen the marriage

For company

For the economic support of the parents in old age. To help bring up the family

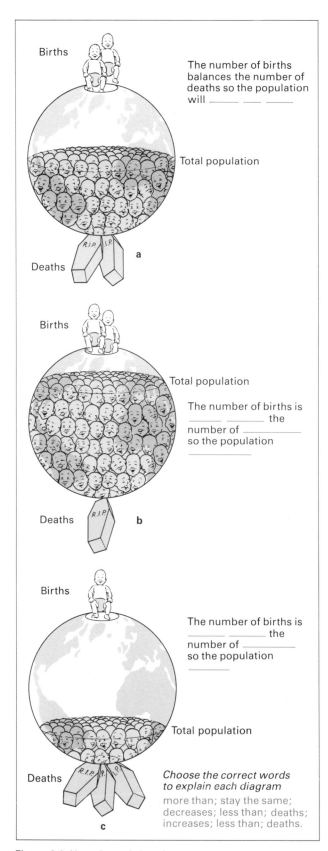

Births

The number of births balances the number of deaths so the population will ____ __ ____

Total population

Deaths

a

Births

Total population

The number of births is ____ ____ the number of ____ so the population ____

Deaths

b

Births

The number of births is ____ ____ the number of ____ so the population ____

Total population

Deaths

Choose the correct words to explain each diagram

more than; stay the same; decreases; less than; deaths; increases; less than; deaths.

c

Figure 3.6 Natural population changes

Figure 3.5 Immigrants to the USA last century

ACTIVITIES

1 Copy the diagram from Figure 3.1 showing how population in an area changes:

2 a Trace or copy Figure 3.6.

b Complete the captions by each diagram. (Migration may also be important.)

c Add labels to each diagram of reasons why the number of births may be high; the number of deaths may be high; the number of births may be low; the number of deaths may be low.

3 Work out the natural increase – the difference between the birth-rate and death-rate in the United Kingdom in 1901 and 1981 from Table 3.1.

4 Look at Figure 3.4. Describe in what ways the reasons for having children differ in Mexico, Singapore and the USA.

4 Global population change

Read the newspaper extract in Figure 4.2. It reports that in the summer of 1987 world population probably reached five billion people or 5 000 000 000. It has risen by 3 000 000 000 in only about sixty years!

Geographers are interested in the rate at which world population is growing and the location of the increases.

For centuries population did not increase very much. Death rates were high through food shortages and disease. People did not live very long. Their **life expectancy** – or the likely length of time on average a person could expect to live – was short. Notice in Table 3.1 that even in 1901 in the United Kingdom on average people could only expect to live fifty-five years.

However, through discoveries in medicine, curing and preventing diseases like smallpox and tuberculosis, people are living longer. Better hygiene and housing and diets for many have also caused the death-rate to fall in many parts of the World. Notice that Table 4.1 shows life expectancy is over seventy years of age for most people in developed countries.

Although many people have a comparatively short life in developing countries – it is on average only just over forty years in several African countries – many children are born and survive into adulthood (see Fig. 4.1).

Study Figure 4.3 which shows where the greatest rates of population growth occur. Compare the areas with the values in Tables 4.1 and 4.2. Every country with a population growth of three per cent or more a year can expect to double its population in twenty-four years!

Table 4.2 The rates at which population doubles

Percentage natural increase in population	Doubling time in years
1	70
2	35
3	24
4	17

Figure 4.1 Village life with many children, Nigeria

My five billionth baby

EARLIER this summer I was invited to a solemn celebration — a special United Nations conference to mark the imminent birth of the five billionth inhabitant of the planet.

The first 'billion baby' did not arrive until around the beginning of the last century — three million years or so after humans appeared on earth.

The second 'billion baby' arrived sometime in the 1920s, the third about 1958, the fourth in 1975. The acceleration is staggering. Four of the five 'billion babies' are probably still alive.

Demographers hope that the pace of increase will slow after that, but they do not expect world population to stabilise until it has reached 10 billion — double what it is today.

The greatest increases will be in the Third World, and the greatest strains, as ever, upon the world's poorest people.

Already three quarters of humanity live in the Third World, and 95 per cent of future growth will occur there. And the most rapid growth of all is in the poorest developing countries, particularly in Africa, which will become the second most populous continent.

In 1950 Africa's population was half the size of Europe's; before 2050 it will be three times as great. Kenya's population is doubling every 17 years; Nigeria is expected to become the world's most populous nation after China and India.

So the burden piles up intolerably on the weakest shoulders. If the five billionth baby has been born in Africa in 1987 it is nearly 20 times more likely to die before its first birthday in Sierra Leone, for example, than in Britain. It can expect to live to the grand old age of 34.

Even now only half the children in the world's poorest countries (mostly in sub-Saharan Africa) complete primary school; fewer than a quarter of the boys — fewer than a tenth of girls — even start secondary education. Only a quarter can get safe drinking water; less have any form of sanitation.

Trying to improve these woeful statistics would be hard enough, given the chronic poverty of the continent, even if the problem remained the same size. But the problem is doubling every 25 years along with the population.

As it is, enormous achievements have already been destroyed. Since independence, Africa has increased food production faster than Europe and the United States. But population has grown so fast that each African has a fifth less food to eat.

On the one side, there are those — who say that population growth is neutral, or even beneficial, in its effects and that nothing should be done. The United States and other countries are already reducing their support for population control.

On the other hand, there are those who attribute virtually all the world's problems to population growth.

To the rest of us, population growth is highly destructive, but it is neither the root of the world's problems, nor could it be simply controlled. It is a highly complicated social and economic issue.

In the seventh-century both birth rates and death rates were high — as for most of human history. Then health and sanitation improved in Europe in the nineteenth century, death rates fell, and population soared. Our great-grandparents did not start breeding like rabbits, but they did stop dying like flies. As they got richer people had fewer children, and population stabilised.

The same 'demographic transition' has begun, much more savagely, in the Third World. Death rates dropped from the 1950s, as life expectancy grew, and diseases like cholera, smallpox and typhus were beaten back. But birth rates are only beginning to turn down.

Medical technology provided death-control, so it was natural to look to it for birth control. But many early contraceptive programmes in the Third World proved disappointing — because people *wanted* lots of children.

They want them for good reasons. From the age of 10, children often produce more for the family by work in the fields than they consume. They provide security in old age. And infant mortality is still so high that an Indian couple needs to have more than six children to be statistically sure of having one surviving son.

As people become a little less poor and more secure, their desire for large families generally falls. And as women receive more education, better opportunities, and more rights the birth rate falls even faster.

At the same time, surveys have shown that many women would have fewer children if they could get effective contraception. So the best hope is by both providing the technology, and by fighting poverty. Where the two have gone hand in hand there have been astonishing successes. But attempts to short-circuit the process by compulsory sterilisation, as in Mrs Gandhi's India, have ended in disaster.

Providing the contraceptives is relatively easy. Fighting poverty is extraordinarily difficult, for it involves changing entrenched economic and political relationships both between and within countries.

Yet it must be done, not just to defuse the population bomb, but to address the appalling conditions of the billion poorest people already on earth, and to relieve the threats to farmland and forest, which are caused as much by poverty as by population growth.

Geoffrey Lean *Observer* 5.7.87

Figure 4.2

Table 4.1 Population totals: growth rates: life expectancy and Gross National Product for selected countries in 1985

Country	Population in millions	Percentage population growth rate 1973–1986	Life expectancy	GNP per capita in US$
*Algeria	21.9	3.1	60	2570
*Argentina	30.5	1.6	70	2120
*Australia	15.8	1.3	76	12370
*Austria	7.5	0.0	74	9100
*Bangladesh	100.6	2.5	51	150
Belgium	9.9	0.1	75	8320
*Bolivia	6.4	2.7	53	490
Botswana	1.1	4.2	57	810
Brazil	135.5	2.3	65	1660
Brunei	0.2	3.7	74	16890
*Burkina Faso	7.9	2.3	45	150
*Canada	25.4	1.2	76	14040
Chile	12.0	1.7	70	1440
*China	1041.0	1.4	69	320
*Colombia	28.4	1.9	65	1320
Denmark	5.1	0.1	75	11280
*Ecuador	9.4	2.9	66	1160
Egypt	48.5	2.7	61	660
*Ethiopia	42.2	2.6	45	110
France	55.1	0.5	78	9760
Gambia	0.7	3.5	41	230
German Dem. Rep. (East Germany)	16.7	−0.1	71	na
*German Fed. Rep. (West Germany)	61.0	−0.1	75	11040
*Ghana	12.7	2.7	53	370
Greece	9.9	0.9	72	3610
Guinea-Bissau	0.9	3.5	39	180
Hong Kong	5.4	2.3	76	6120
India	765.1	2.2	56	270
*Indonesia	162.2	2.2	55	530
Ireland	3.6	1.0	74	4770
Israel	4.3	2.1	75	6220
Italy	57.1	0.3	77	7700
Ivory Coast	10.0	4.3	53	650
Jamaica	2.2	1.2	73	910
*Japan	120.7	0.8	77	11270
*Kenya	20.4	4.0	54	290
Kuwait	1.7	5.5	72	14870
*Libya	3.7	4.3	60	7170
Luxembourg	0.4	0.2	74	14300
*Malawi	7.0	3.1	45	170
*Malaysia	15.6	2.4	68	1980
*Mexico	78.8	2.8	67	2100
*Netherlands	14.5	0.6	77	9380
*Nigeria	99.7	2.9	50	820
Norway	4.1	0.4	77	14510
*Pakistan	96.1	3.1	51	340
*Portugal	10.2	1.0	74	1960
Romania	22.7	0.7	72	na
*Saudi Arabia	11.5	4.8	62	8620
Singapore	2.6	1.3	73	7590
South Africa	32.4	2.4	55	2030
Spain	38.6	0.9	77	4280
*Switzerland	6.4	0.1	77	16330
Tanzania	22.2	3.4	52	280
*Trinidad and Tobago	1.2	1.6	69	6100
United Kingdom	56.5	0.0	75	8430
*United States	239.2	1.0	76	16730
USSR	277.4	0.9	70	na
*Venezuela	17.3	3.2	70	3080
Zambia	6.7	3.3	52	410

Source: World Bank Atlas 1987 and 1988 update.

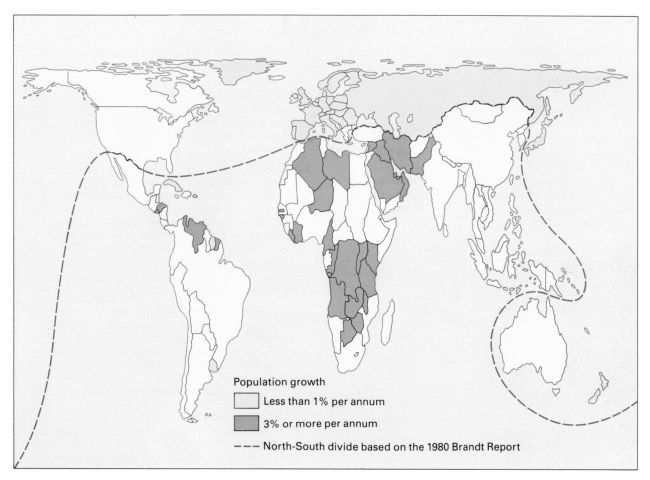

Figure 4.3 The world divided into the richer 'North' and poorer 'South', showing areas of population growth. (Arno Peters Projection)

As you can see from the map in Figure 4.3 and Table 4.1 the countries with the fastest rates of population growth are some of the poorest. In many cases their GNP is under US $1000 per person a year. (**Gross National Product** (GNP) means the total value of all goods and services produced in a country in a year, divided by the total population.)

Countries like The Gambia, Malawi, and elsewhere in Africa in particular are very poor and are finding it hard to develop and improve their economies.

Population growth rates are much lower in many developed, richer countries.

Table 4.3 Estimates of world population growth by region (in millions)

Year	Europe and the USSR	Asia	Africa	Latin America	North America	Oceania
1750	167	498	106	16	2	2
1800	208	630	107	24	7	2
1850	284	801	111	38	26	2
1900	430	925	133	74	82	6
1950	572	1381	222	162	166	13
Estimate 2000	880	3458	768	638	354	32

ACTIVITIES

1 a Draw a line-graph to show the increase in world population from 1650 when the population may have been 500 million. Use the information in the article in Figure 4.2 for your graph.

   Put time on the horizontal axis and numbers of people on the vertical axis.

  b Beneath the graph describe what is happening to the length of time it is taking for world population to double itself.

  c Discuss why you should be concerned that world population is increasing at a rapid rate.

2 Divide your paper into five columns. Head the columns as follows:

Areas of Population Change 1973—1985

Population loss	Population growth of 0 to 1% per annum	Population growth of +1% to 2% per annum	Population growth of +2% to 3% per annum	Population growth of +3% per annum

  a Using Table 4.1 insert each country into its correct column according to its percentage population change recently.

  b Look up the location of all the countries listed.
   i Which continent has countries where the population is not increasing?

   ii Which continent has most countries with a population increase of over three per cent per annum?

  c On a world outline of Figure 4.3 shade and name all countries with a population increase of three per cent or more per annum. Describe their position compared with the 'North-South' boundary.

3 a Draw a scattergraph, preferably on log/normal graph paper, to show the relationship between life expectancy and GNP per head for the thirty countries starred in Table 4.1. Put life expectancy on the horizontal, arithmetic scale and GNP on the vertical log scale. (Your teacher will give you the special graph paper from the *Teacher's Book* and discuss with you why it is used for this exercise.)

  b Write a few sentences describing the relationship the graph shows.

4 a Using the figures in Table 4.3 draw superimposed line graphs to show the rate of world population growth by region since 1750.
   i Put time on the horizontal axis and estimates of numbers of people on the vertical axis. (You will need to take great care choosing your vertical scale. Be sure you have a large sheet of graph paper!)
   ii Plot each region in turn and use a different colour for each region. Make a key or label each line.

  b Beneath the graph describe which continent has had the fastest rate of growth this century and which the slowest.

5 National population change

In Britain we do not know for sure what the population size was before 1801. In that year the first census gave the total as 10.5 million. Table 5.1 shows the growth of population in the United Kingdom this century.

Look at Figure 5.1. It shows how population grew mainly through natural increase rather than immigration. The death rate fell from about seventeen per 1000 in 1901 to twelve per 1000 in 1981. Better diet, clean drinking water and health care helped reduce the number of deaths. In 1868 an Act of Parliament allowed bad houses to be condemned as unfit for people to live in (see Fig. 5.2). Medical Officers and Health and Sanitary Inspectors were established after an Act in 1875. Both these Acts improved conditions especially for poor working people.

If you talk to your grandparents or elderly friends, they will probably tell you that they were from a big family with several brothers and sisters. As children survived, the birth rate slowed down. Look at Figure 3.1 which shows this. Family size in England and Wales has fallen so that on average there are now about two children per family (see Figs. 2.4 and 2.5). In countries like West Germany less than two children are born in each family on average.

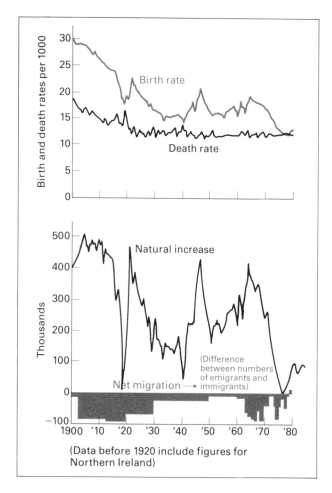

Figure 5.1 Factors affecting population growth in the UK.

Figure 5.2 Unhealthy conditions in nineteenth century Britain

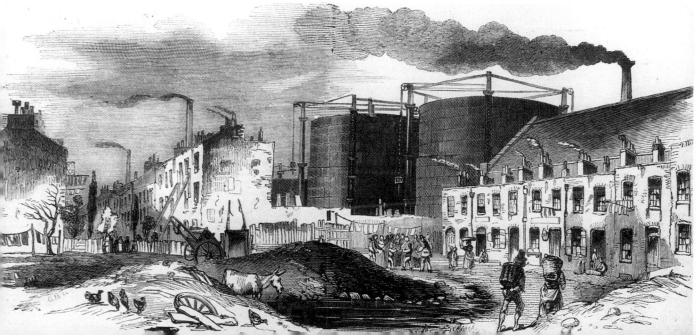

Table 5.1 Population of the United Kingdom

Year	Totals in thousands
1901	38237
1911	42082
1921	44027
1931	46038
1941	No census during World War II
1951	50290
1961	52807
1971	55928
1981	56352

Source: OPCS

Table 5.2 China's changing population

Year	Population total to nearest million	Birth rate per 1000	Death rate per 1000
1953	588	37	14.0
1964	695	39	11.5
1981	997	21	6.4

Of course population rates of change vary in different parts of a country. Figure 5.3 shows that, in Britain, counties like Berkshire and Cornwall have increasing populations. People are moving to areas where there is work and an attractive environment (see Fig. 5.4). Declining industrial areas like Cleveland and Merseyside are losing population.

Look at Table 5.2. China is the country with the largest population in the world. It has about 1000 million people. It had a population of about 542 million in 1949 when the Communists came to power. With high birth rates and falling death rates its population continues to grow (see Fig. 5.5) in spite of famine in some years.

In Nigeria, as in many developing countries, population is growing rapidly through very high birth rates − possibly seventy-four per 1000 in places. In eastern Nigeria the average age of marriage is 15. There is little use of contraception and more than eight children in a family is common in rural areas.

People in towns in African and Latin American countries have access to family planning clinics and education (see Fig. 5.6). Look at Table 5.3. These people tend to have fewer children than those in rural areas. In many Asian countries birth rates are high in towns and the countryside.

Figure 5.3 English counties showing the greatest change of population (projections 1983−1991).

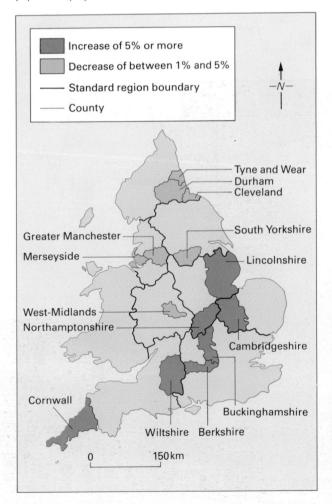

Figure 5.4 New homes in Milton Keynes, Buckinghamshire

Table 5.3 Urban and rural fertility for selected countries

Country	Average for continent	Average number of children per family for a five year period in the 1970s	
		Urban	Rural
Kenya		1.00	1.44
Ghana		1.02	1.11
	Africa	1.03	1.17
Mexico		1.17	1.29
Venezuela		1.00	1.50
	Latin America	0.96	1.19
Bangladesh		1.08	0.98
Indonesia		0.77	0.83
Malaysia		0.88	1.02
Fiji		0.80	0.93
	Asia and Oceania	1.04	1.05

Table 5.4 Average annual population growth rates in percentages in urban and rural areas of selected countries

Country	% Urban population	Average annual growth in % 1975–1980	
		Urban	Rural
Bangladesh	11.2	6.7	2.4
Brazil	67.0	4.0	−0.5
Egypt	45.4	3.4	1.9
Ethiopia	14.5	6.1	1.2
Indonesia	20.2	3.6	1.3
Kenya	14.2	7.3	3.5
Mexico	66.7	4.1	0.9
Venezuela	88.3	4.3	0.1

Figure 5.5 Some of China's people

Figure 5.6 A family planning clinic in a Nigerian town

$\boxed{\text{ACTIVITIES}}$

1 a Draw a line graph using information in Table 5.1 to show population change in the United Kingdom this century. Put time on the horizontal or x axis and population on the vertical or y axis.
   Give the graph a heading.

  b Comment on the trend and rate of change of the United Kingdom's population this century.

2 a Work out the natural increase – the difference between the birth rate and death rate – in the United Kingdom in 1901 and 1981 from Table 3.1.

  b Why did the United Kingdom's population increase rapidly in the early part of the twentieth century?

  c Interview and record a talk with an elderly person about what their home life was like when they were a child.
    i Try to find out whether they were part of a large or small family.
    ii Compare their memories with your family life today. You could produce a written or tape-recorded report of your investigations.

3 a Draw four proportional circles to show China's population in 1949, 1953, 1964 and 1981 from Table 5.2 and figures in the text. Use the square root of the population at each date to work out the radius of your circles. For example:

   1949 $\sqrt{(542 \text{ million})} = 23.3$ million

   Therefore 23.3 mm. will be the radius of your circle representing China's population in 1949.
   Keep the same scale of 1 mm to 1 unit for each circle so they are comparable. Label each diagram.

  b Describe what has happened to China's population since 1949.

  c What proportion of the world's population is Chinese? Assume China's population is 1000 million and that world population is 5000 million.

4 a Choose one Asian and one African and one Latin American country from Table 5.4.
    i For each, draw vertical bars to show average annual population percentage growth rates for urban and rural areas.
    ii Colour all bars showing urban growth rates in one colour and rural growth rates in another. (If you choose Brazil, think how you can show its rural population rate of change. Perhaps you could draw your bar beneath a base line.)
    iii Give your diagrams a heading.

  b Write a note beneath the diagrams, commenting on the different population growth rates between urban and rural areas in your selected countries.
   Suggest reasons for the differences – remember migration may be important too.

6 *Local population change*

People live where work is available. The population growth in towns often reflects available work and relative prosperity.

CASE STUDY *Merthyr Tydfil*

Look at Figure 6.1. See how the population has changed in Merthyr Tydfil, an industrial town in South Wales.

When iron ore and coal began to be mined in Merthyr Tydfil at the beginning of the nineteenth century, people moved there for paid work (see Fig. 6.2). The settlement grew from a village to a town by the twentieth century. Look at Figure 6.3 which shows such a change. Building a canal and a railway not only helped to move the iron and coal to markets but helped people move there too. This century the population has declined and labels on Figure 6.1 give some reasons for this.

CASE STUDY *Stevenage New Town*

Stevenage New Town in Hertfordshire, north of London, grew for a different reason. In 1947 the Government decided there should be several modern towns with good houses near places of work. Stevenage was one. This town's population grew from 6660 in 1951 to 74 513 in 1981.

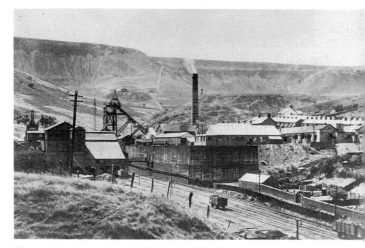

Figure 6.2 A coal mine in Merthyr Tydfil, 1913

Figure 6.1 The changing population and prosperity of Merthyr Tydfil 1750–1970.

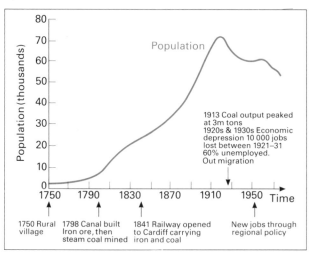

1913 Coal output peaked at 3m tons
1920s & 1930s Economic depression 10 000 jobs lost between 1921–31 60% unemployed. Out migration

1750 Rural village
1798 Canal built Iron ore, then steam coal mined
1841 Railway opened to Cardiff carrying iron and coal
New jobs through regional policy

THE RURAL PHASE – early 18th century

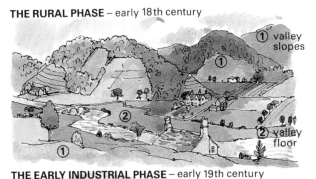

① valley slopes
② valley floor

THE EARLY INDUSTRIAL PHASE – early 19th century

① valley slopes
② valley floor

THE HEAVY INDUSTRIAL PHASE – early 20th century

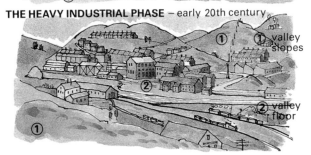

① valley slopes
② valley floor

Figure 6.3 Population growth in a mining town.

Figure 6.4 A new pensioners' club in Stevenage

At first, most of Stevenage's population were young married couples and their children who moved from slums or bombed areas of London. Recently natural increase explains most of the population growth in the town. The town now has more retired people than it had twenty years ago for its population is ageing. The public services in the town, like numbers of primary schools and pensioners clubs (see Fig. 6.4), have had to change too.

ACTIVITIES

1 a Look at Fig. 6.5, the Ordnance Survey extract on a scale of 1:25 000 of part of the Rhondda valleys in South Wales. Find the location of this area in your atlas.

 b On a piece of tracing paper placed over the Ordnance Survey extract shade in the areas of settlement in the Rhondda valleys. Name them.
  i In another colour, shade all quarries.
  ii Mark spoil heaps in another colour.
  iii Mark the location of mines (include disused).
  iv Make a key. Add the scale to your map. Show the direction of north. Give the map a title or heading explaining what it shows.

 c Write a few sentences describing the location of the population of the area shown in the map extract. Suggest why people settled where they did.

 d Draw a line graph to show population change in the Rhondda Borough from 1801 to 1986. Use the information in Table 6.1. You should put time on the horizontal or x axis but be careful after 1981! You must also think of a suitable vertical scale for the y axis on which to show population. Give the graph a title.
  Add labels to your graph commenting on trends in the population in the Rhondda Borough at different times.

 e Look at Table 6.2. Compare the causes of death in 1900 with those in the Rhondda Borough in 1980. What factors may explain the differences?

 f Why have many settlements in South Wales, like Merthyr Tydfil, and those in the Rhondda Valleys lost population since the 1920s?

Table 6.1 Population data for the Rhondda Borough

Year	Population
1801	542
1811	576
1821	647
1831	542
1841	748
1851	951
1861	3035
1871	23 950
1881	55 632
1891	88 351
1901	113 735
1911	152 781
1921	162 717
1931	141 346
1941	No data
1951	111 357
1961	100 314
1971	88 995
1981	81 725
1986	77 800

Source: Rhondda Borough Council

Table 6.2 Causes of death in the Rhondda Borough 1900 & 1980

1900		1980	
Cause	Numbers	Cause	Numbers
Measles	121	Infections & parasitic diseases	6
Scarlet fever	35		
Influenza	18		
Whooping cough	58	Cancer	230
Diphtheria	125	Endocrine, nutritional, metabolic and immunity disorders	4
Enteric fever	24		
Dysentery	10		
Enteritis	162		
Erysipelas	1		
Puerperal fever	9	Disease of the blood	1
Other septic diseases	4	Mental disorder	1
Tuberculosis	141	Disease of nervous system	5
Alcoholism	4		
Cancer	43	Disease of circulatory system	615
Premature birth	66		
Development diseases	230	Disease of respiratory system	254
Old age	64		
Softening of the brain	8	Disease of digestive system	24
Heart disease	65		
Bronchitis	152	Disease of genito-urinary system	10
Pneumonia	194		
Disease of the stomach	6	Disease of musculoskeletal system	4
Obstruction of intestines	16		
Cirrhosis of liver	10	Congenital abnormalities	2
Bright's disease	42	Perinatal conditions	14
Female tumour	1		
Accident or negligence	101	Injury and poisoning	28
Suicide	2	Other conditions	16
Other causes	11		

Source: Rhondda Borough Council

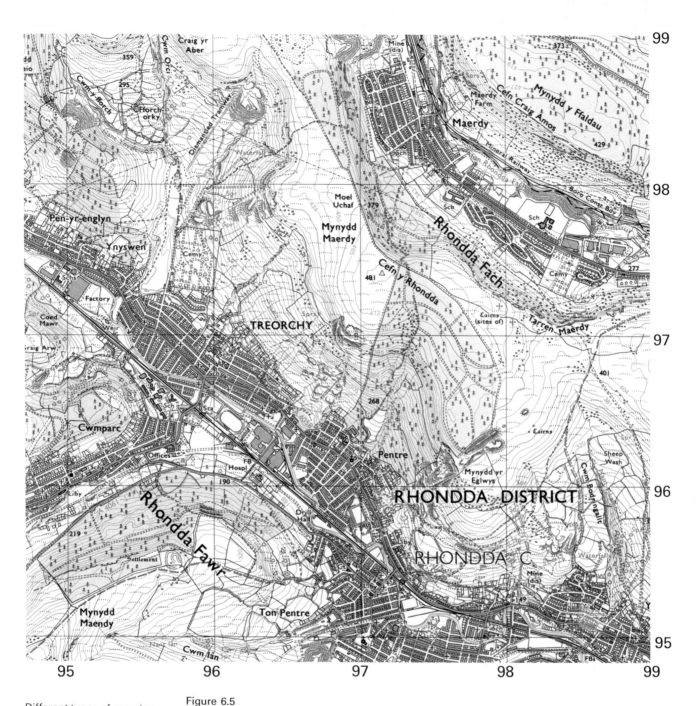

Figure 6.5

Different types of quarries

⌢ Disused pit or quarry

⌢ Refuse/slag heap

⌢ Chalk pit, clay pit
or quarry

Coursework ideas

Find out from your school or public library the
population changes in your home area such as
your ward or district this century. Discover some
key dates relating to changes in transport such as
the coming of the railway or the opening of
factories so that you can produce a graph rather
like Figure 6.1 of Merthyr Tydfil.

7 Population and resources

Growth of world population has led people to wonder about the future. Look at Figure 7.1.

The Rev. **Thomas Malthus** took a gloomy view in 1798. He argued that the number of people would increase faster than food supplies. Starvation, war and disease would follow when the land could no longer support all the people. In other words the land had reached its **carrying capacity**.

Malthus believed that population growth might be checked if people married later and so had fewer children. Eventually, however, the lack of **resources** would limit the size of the population.

Read the article in Figure 7.2. Some aspects of recent famines in Ethiopia reflect Malthus' ideas. However, it is important to realise that drought, civil war and povery have reduced agricultural output.

Figure 7.1 One view of the future

In 1972 people from ten countries met in Rome. The group was known as **'The Club of Rome'**. They fed information into a computer in order to predict what may happen if world population and the use of resources continued to grow. The results are shown in Figure 7.3. Remember these are only predictions. They do suggest that if we continue to use the world's resources at the

Figure 7.2

Waiting at death's door

CATHARINE WATSON
reports from Adigudum

Ethiopia's drought was noticed on the day the last raindrop fell last July.

But Ethiopia still hangs by a thread. Aid workers say it would be fatal to relax. 'If we don't call it a catastrophe, it will become one,' said one worker in Tigray.

There are good reasons to worry. This year the food shortfall is greater, and the civil wars more intense, than during Ethiopia's last famine. And only one third of the 1.3 million tonnes of food that Ethiopia needs to survive until the next harvest in November has been promised by donors.

Tigray and Eritrea are the regions worst hit by the famine, too. Farmed for 3,000 years, they are beautiful but bleak: windswept, tree-less and rocky. The dust which gums the children's eyes shut is Ethiopia's topsoil, blowing into the waters of the Nile at the rate of over one billion tonnes a year.

Semerer Abreham is a 40-year-old peasant in Eritrea. In 1984, 80 per cent of his crop failed. This year the figure is 100 per cent. Adis Selassie, 37, who lives in Tigray, lost 40 per cent of his crop in the last drought; this year he lost 75 per cent.

Both are fatalistic and deeply attached to their land. Abreham believes the rains failed because people in his village mistrusted each other. 'Rains come when God wants,' he said. Selassie does not want to move to a greener, wetter and more fertile part of Ethiopia. 'Our culture does not allow us to freely go away,' he said, leaning on his donkey and waving at a barren stretch of land. 'This is the place of my grandmother.'

Thomas Grannel, of the World Food Programme, says simply: 'History has caught up with Ethiopia.' Eighty-eight per cent of the people, and 67 per cent of the 40 million animals, live in the eroded and deforested highlands. The land is exhausted.

Grannel's agency has spent $250 million and 12 years on planting trees and terracing. This, he says, has 'stabilised' 2.5 per cent of the seriously eroded land.

But for Ethiopia's 47 million people, seven million of whom now risk starvation, that pace is too slow. It cannot go any faster because Ethiopia is starved of funds for development. Britain, the US and most other Western countries give no development aid because of the Government's Marxist policies. Ethiopia receives only $9 per person per year in development aid: other poor African countries receive $20.

The crunch will come in February and March, when one million Tigreans now subsisting on their meagre harvest will suddenly need help.

The *Observer*
3.1.88

HIGH

— Population
— Resources
– – Food per capita
·········· Pollution

– – – Industrial output per capita

LOW
1900 20 40 60 80 2000 20 40 60 80 2100AD
Year

Figure 7.3 The limits to growth model of population development.

present rate the limits to growth on this planet will be reached in the next one hundred years as resources run out and pollution increases.

Many people believe that adequate food supplies already exist in the world. It is the distribution which is to blame with some groups receiving more than their fair share, while some receive less.

Another geographer, **Esther Boserup**, believes that a growing population forces farmers to adapt and change their methods. By doing this, higher yields are produced. In the Middle Ages in England farmers left a third of their land fallow in turn each year to keep it fertile. Now, using intensive methods and growing crops under glass, several harvests can be gained each year from an area of land (see Fig. 7.4).

Figure 7.4 Fertile land by the River Nile allows intensive farming

ACTIVITIES

1 a Draw a graph to show how population might exceed food production over time according to Malthus. Put time on the horizontal x axis and growth rates for people and food supplies on the same vertical y axis. Assume the number of people increases at a geometric progression of 1, 2, 4, 8, 16, 32. The number doubles for each period of time such as every twenty years. Food supply increases over the same spans of time at an arithmetic progression of 1, 2, 3, 4, 5, 6. Plot a line to show how population grows over 100 years. On the same graph plot a line to show how food production grows over the same time period. Label each line. Label each axis. Give the diagram a title.

  b By the graph, write two or three sentences describing the relationship between population and food supply according to Malthus.

2 a In 1798 Malthus predicted only a slow increase in food production. In fact, production has increased greatly. In groups or on your own, suggest reasons for this rather more rapid rate of increase than Malthus had expected.

  b Collect pictures and articles from papers and magazines to make a display showing modern methods of food production.

3 a Read Figure 7.2. Which parts of Ethiopia are short of food and suffering from famine?

  b Make a list of the reasons why Ethiopia is short of food. Is rapid population growth the sole cause?

  c Discuss and then write down what you would do to relieve hunger in Ethiopia in the short term if you were in charge of a relief programme.

  d What long-term development schemes could you suggest which would benefit the people of Ethiopia in the future?

4 a Look at Figure 7.3 showing the results of 'The Club of Rome's' predictions. Discuss why the pollution curve follows the population curve.

  b Make a list of as many types of pollution that people cause as you can.

  c Write down examples of pollution of the air, water and land found between your home and school or in or near your school grounds.

8 The Demographic Transition Theory

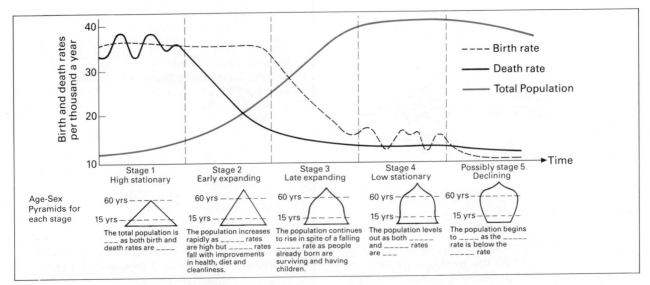

Figure 8.1 The Demographic Transition Theory.

A more hopeful prediction for the fate of the world's population is the **Demographic Transition Theory**. It is based on evidence of falling death rates and later falling birth rates in Western Europe since the early nineteenth century. Look at Figure 8.1. It suggests that the population of a country or area goes through four or even five stages. Eventually population growth stabilises or decreases.

At first, although many are born, the population remains low as there is a high death-rate. Secondly, when the death rate falls through better food, health and hygiene, the population increases rapidly. In time, realising children are surviving, people have fewer babies. Contraception helps this. Although the birth rate falls, the population total continues to increase as many people are in the reproductive age-group. The fourth stage of the theory is reached when the total population stabilises. A possible fifth stage is when the total population declines with birth rates below death rates.

Although this is an optimistic theory it may not be true and may not apply to all countries such as in the developing world. Time will tell!

ACTIVITIES

1 a Make a copy of Figure 8.1 showing the Demographic Transition Theory. Complete the description under each stage by filling in the blanks.

  b Shade in the natural increase of population on the diagram — (the gap between the birth rate and death rate.)

2 Look at the population pyramids for England and Wales in 1901 and 1981 in Figure 2.1.
  a In which stage would you put the pyramid for 1901?
  b In which stage would you put the pyramid for 1981?
  c Is it easy to fit a pyramid shape for an actual country to a stage in the theory? What other information might you need?

3 Look at Figure 2.3.
  a In which stage of the theory would you place the population pyramid for Nigeria?
  b Explain your reasons.
  c In which stage of the theory would you place the population pyramid for West Germany?
  d What do you predict may happen to the total population of West Germany in the next twenty years?

9 Policies of population control

'One child only' slogan spreads across China

By NIGEL WADE in Peking

THE slogan, "It is best to have only one child and two at most," is spreading across China as more and more provinces adopt birth control incentives and economic sanctions against couples having more than the approved number of children.

A series of strict birth control measures has been announced in several parts of China.

Coupons awarded

It is now official policy to reward couples who have only one child and undertake not to have more. Some provinces give these couples a "planned parenthood honour" coupon, which ensures priority for their child in education and medical care.

Couples having only one child will receive the same housing floorspace as couples having two, and an only child will receive an adult's grain ration.

To placate couples in conservative peasant areas where male children are traditionally preferred the new birth control measures provide preferential pension terms for peasants having only daughters. This is to stop them going on having children until they get a son to support them in their old age.

Couples in some provinces who produce a second child after being rewarded for undertaking to have only one will have their rewards revoked and will also be required to pay a surcharge for their children's health expenses.

Too many children

The surcharge will rise to seven per cent of the parents' combined monthly income if they have as many as five children.

Couples having three or more will be denied extra living space and extra food ration tickets.

A Peking birth control official said recently that the old belief that every family should have a son was still common, and that about 15 per cent. of all rural households had too many children.

China wants to cut its population growth rate to less than one per cent. a year by 1980.

The average annual population growth rate in recent years has been about two per cent., which is higher than the rate of growth in grain production.

Figure 9.2
The *Daily Telegraph*, **June 5 1979**

Lu Bo is a twenty-nine year old cashier in a hotel in Jinxian county in Liaoning Province in China. Her child has just had her first birthday. Lu said that last year she only worked for seven months. The rest she took off to have the baby. As a result her income was reduced. As her husband's wages are low she does not want another child.

In the past, many Chinese like Lu Bo had more children in the hope of having a boy. On marrying she would have gone to live with her husband's family.

Lu Bo and others like her now want to work and have their own homes. They want to buy furniture and improve their homes. The Chinese Government is encouraging directly people like Lu Bo to have small families.

Governments may introduce policies to control population growth in their countries. China and India are both trying to limit their populations: they have **anti-natalist policies**.

Lu Bo and other young Chinese like her have been encouraged since 1979 by the Government to have only one child.

The Chinese Government was alarmed that it might not be able to feed and support its population. Look back to Table 5.2. You learnt in the section on 'Population Growth and Change' that China's population went from 542 million in 1949 to nearly 1000 million in 1981. This growth was due to a high birth-rate and a falling death-rate.

Figure 9.1 Advertising a one child per family policy in China

In 1979 China began an anti-natalist policy. A man cannot now marry until he is twenty-two and a woman cannot marry until she is twenty. The couple are urged to have only one child (see Fig. 9.1). Remember Mamtaz whom you read about in Chapter 3. She was married at thirteen and had six children by her early twenties! Read Figure 9.2 to find out the rewards and penalties linked to China's policy.

By 1984 the unpopular policy of one child per couple was relaxed for some people. Those allowed to apply for permission to have a second child include couples who were both only children, rural couples where one or the other has a parent who was an only child and only sons who have fathered a daughter.

The official reason for the change in policy is that many females died in a famine from 1959–61. Therefore there are fewer women to have children than had been expected.

Look at Figure 9.3. It shows computer predictions of China's future population structure if the birth rate falls. Notice the population pyramids A and C for the year 2000. If each female has one child many old people will need to be looked after by fewer young people.

India has had family planning services since 1930. Between 1972 and 1977 sterilizations (especially of males since the operation is simpler than for females) were carried out. Sterilizations, making people infertile, were at first on volunteers. By 1975 many were forced to be sterilized, it is claimed, against their will. The outcry against these acts helped lead to a change of Government in 1977. Perhaps more importantly, however, it has been a severe set back to the cause of population control.

Figure 9.3 Projected population pyramids for China if population policies are continued for 100 years.

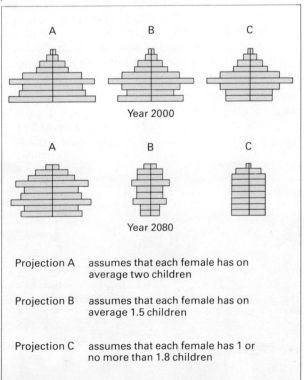

A   B   C

Year 2000

A   B   C

Year 2080

Projection A	assumes that each female has on average two children
Projection B	assumes that each female has on average 1.5 children
Projection C	assumes that each female has 1 or no more than 1.8 children

ACTIVITIES

1 Form discussion groups of three or four pupils each representing the Government of a developing country. Each group should appoint one person to present their views to the class at the end. You are to discuss the following proposals and decide which *five* in rank order should be introduced to improve the standard of living of the population. Most are very poor peasant farmers with large families. You cannot afford to introduce more than five schemes at once. You must also consider the likely long-term effects of your actions.

a Send abroad fifty of your best students to be trained as doctors, engineers, teachers and nurses.

b Provide free education for all children up to the age of fourteen.

c Purchase and run ten mobile health caravans to reach the more remote parts of the country.

d Immunise all children against polio, malaria, whooping cough, measles and tetanus.

e Introduce and run mobile family planning services to reach women in rural areas.

f Provide a small pension for all elderly people over the age of sixty-five years.

g Lay a clean, piped water supply throughout the capital city.

2 Watch television advertisements involving families. Record the number of children shown in each family. What do you find advertisers regard as the typical family size in the United Kingdom?

3 a Read Figure 9.2. Make a list of the benefits the Chinese were offered after 1979 if they agreed to have only one child.

b Make a list of what penalties would be imposed if they had more than one child.

4 Imagine you are the third child in a large family of seven children in a developing country. Make two lists:
  i giving the advantages of being part of a large family;
  ii the disadvantages of being part of a large family.

10 Policies of population growth

Chirac : « La famille est prioritaire »

Le premier ministre veut poursuivre en ce sens une politique «résolue et ambitieuse» pour endiguer le déclin démographique.

Dans son projet de gouvernement, Jacques Chirac a aussi des priorités et, pour lui, il l'a dit hier, la priorité des priorités c'est une politique « résolue et ambitieuse en faveur de la famille ». « Toute l'ex-périence de l'histoire de nos civilisations, a-t-il déclaré, montre que cha-que fois que la cellule familiale s'est dégénérée, cela s'est traduit par un déclin de la civilisation. Nous sommes en période de déclin démographique. C'est dangereux et c'est mortel. Nous avons une responsabilité à l'égard de nos petits-enfants. On ne peut pas, après avoir hérité une France belle et forte, qu'elle l'a été dans les années soixante, nous désintéresser de la France que nous allons léguer à nos enfants et à nos petits-enfants. Une France ou une Europe en voie de décomposition serait alors colonisée par d'autres... »

Le Figaro, 6.3.88

Figure 10.1 ▼   Figure 10.2 ▲

France cheers le bébé boom

For centuries the French have worried about their birthrate. Now things are looking up at last, reports JOHN IZBICKI

For the third year in succession, the French population is on the increase.

By the year 2020, France will become the most populated country in the EEC, with 58.6 million, compared with 55 million in 1985. The UK, believe it or not, follows closely, rising from 56.5 million in 1985 to 58.5 million in 2020. Germany, however, over the same period, is set to plummet from 61 million to 51 million.

Now why, you might well ask, should this mini-explosion in France's birthrate alongside West Germany's clear dip, bring joy to the French? According to M. Michel Lévy, the French "believe subconsciously that a small population makes a country weak in every sense".

He goes further: "Many Frenchmen and women are convinced that France suffered disastrously in battles with and under the occupation of the Germans in both the first and second world wars because the German population was bigger than the French and that the failure to produce more children led to the nation's punishment."

This obsession with the number of children each woman should bear has been a French trait for many centuries. When the Black Death stalked Europe, Britain's population was about three million; France had some 20 million and was able to face the killer disease more easily than its neighbours.

The French are by no means alone in holding the view that there is strength in numbers and that a country's defence as well as its economy will prosper once its population has grown. This week, Lee Kwan Yew, Singapore's prime minister, appealed to the city-state's womenfolk to produce an additional 40,000 babies within 10 years to stop the population from growing "dangerously low".

Can a population not also grow "dangerously high"? China might have views to offer on that question. So might Miss Yu Xiao Zhan, a young Chinese woman who turned up in a Lille court on Wednesday charged with working without a permit. She claimed she had come to France to start a *famille nombreuse* (large family) — an ambition not permissible in China.

Mme Michèle Barzach, the Minister of Health and Population (the very fact that there is a Minister of Population says something ...) feels that France could and should do better. "The increase is still too small to allow for the renewal of the population," she says.

The birthrate among Frenchwomen of child-bearing age has increased from 1.82 children per woman to 1.84 over the past three years. For the population to remain stable, that birthrate should rise to at least 2.1 children apiece.

How do these figures compare with the rest of Europe? In Holland it is 1.51, in Norway 1.66, in Sweden 1.74, in Italy 1.4, in Portugal 1.8, and in Spain it has shown another steep slide from 2.35 in 1979 to 1.65 in 1984. West Germany has already reached what is considered to be its lowest individual birthrate figure with 1.28 in 1985, while the figure in East Germany, whose ideas on population resemble those of France and Singapore and whose policies encourage child-bearing, has risen from 1.54 in 1975 to 1.94 in 1980.

Well ahead of all its EEC partners, with 2.54 (in 1984), is Ireland.

Telegraph 22.8.87

Figure **10.3** *Sunday Times* 24.2.85

Romanians chafe at baby drive

by Peter Godwin
Bucharest

THE PEOPLE of Romania were last week given a pat on the head by President Nicolae Ceausescu for their progress towards yet another target: the production of more babies. Worried by Romania's flagging population, he chose the occasion of international women's day last year to order his citizens to go forth and multiply. Every couple should have at least four.

Last week he announced that the population growth rate was up to 5.2 per 1,000 last year, compared with 3.9 in 1983, the lowest since the war. Population has increased to 22,687,374.

However, the achievement was not altogether voluntary.

Artificial contraceptives have been banned and abortion, once the most common form of birth-control in Romania, is denied to most women.

Initially the president was rewarded with a rise in the birth rate from 6.1 to 18.1 per 1000, but it soon slid down again as doctors ignored the regulations.

By 1983 the authorities estimated that, of 742,882 registered pregnancies, 321,598 resulted in live births and the other 421,384 were aborted, only 9% of them for legal reasons. Including unrecorded pregnancies and backstreet abortions, the ratio of live births to abortions was estimated at about 1:4.

The president warned doctors that abortion legislation must be followed to the letter. Doctors who fail to justify abortions they perform are imprisoned for up to three years and struck off the register. Patients face heavy fines and up to two years in jail.

Under the population drive, young women are obliged to undergo regular pregnancy tests and, once registered as pregnant, are watched carefully to ensure they do not get rid of the baby. Any termination of pregnancy is closely investigated.

Contraceptives have been banned and can be obtained only on the black market.

Those not married by 25 pay an extra 8% to 10% in penalty tax. They cannot apply for housing and in practice their careers suffer through lack in promotion. They are not considered for trips abroad.

Television campaigns have been screened and government doctors preach the virtues of big families in factories, offices and schools. The legal marrying age for girls was dropped from 18 to 15 and attempts made to remove the stigma of unmarried motherhood. Special classes for pregnant schoolgirls have been organised to permit them to continue their schooling after the birth. Fertility clinics have been set up around the country to investigate failure to conceive.

As incentives, women are given generous maternity leave on full pay, a reduction in working hours with no salary loss after the birth, and special paid leave to look after sick children. They also get a lump sum bonus as a reward. Non-working mothers also get a lump-sum for each baby, plus a monthly payment for each child.

This extensive package of measures, however, has done little to tackle the underlying causes of Romania's population decline. For many Romanians, huddled in freezing, overcrowded apartments, queuing for food and struggling to make ends meet, more children is the last thing they need. Overhasty industrialisation and the rush to towns has resulted in one of the lowest standards of living, in real terms, in eastern Europe.

As with most things, the Romanian government has grand five year plans for population growth: by the end of this year the population target is 23.4 m, rising through 25 m in 1990 to 30 m by the year 2000.

Six-year old Peter in Bavaria, West Germany, was looking forward to his first day at school. Football was his favourite game. However, he came home crying from school at the end of the day. There were not enough in his class to form a team of eleven.

Some countries like West Germany, France and Romania in the developed world and Singapore and Malaysia in the less developed world are trying to encourage the growth of their populations or sections of their populations. They have **pro-natalist policies**.

One reason Peter had so few class mates is because few children are being born in West Germany. Since 1974 there have been years when the birth rate is lower than the death rate.

Chancellor Kohl stated in 1983 that the Republic of West Germany 'must again foster a positive attitude towards children. Families will continue to receive assistance in the form of child benefits and tax relief.' To encourage larger families in 1987 the state paid DM 50 per month for a first child, DM 100 for the second, DM 220 for the third and DM 240 for the fourth and every child after that until the age of eighteen years. Benefits and tax relief continue to the age of twenty-seven years if that person is in training. Rent allowances, reduced fares on public transport and student loans are financial supports for large families.

Look at Figures 10.1 and 10.2. Read them carefully to find out why France believes it is necessary to have a growing population.

Romania, as a European communist state, had a pro-natalist policy from 1965. Read Figure 10.3. It describes why and how Romania encouraged more births under Ceausescu.

Singapore in the developing world tried in 1966 to limit the size of families to two children. Since 1983 the state has wanted graduates to marry and have children. The Government thinks this might improve the intelligence of the population! A slogan encourages the intelligent with the words: 'At least two. Better three. Four if you can afford it.' It is believed a larger population would stimulate economic growth (see Fig. 10.4).

The Prime Minister of Malaysia has gone further by encouraging Malaysian families since 1982 to 'Go for five!' children. Malaysia's population is less than 16 million people. The authorities think the country could support seventy million people! They say 'seventy million people working very hard and producing a lot of goods could easily live in Malaysia — and even produce sufficient food.'

More people would make more goods, perhaps for export. More people would need more goods and services and so stimulate industries. More jobs would be created. In this way the economy might spiral upwards.

To encourage larger Malaysian families, maternity benefits and tax concessions are granted. It has been suggested that the bumiputera (the local Malays) are encouraged to have larger families than the immigrant Chinese and Indians. In this way most of the population would be Malay.

Figure 10.4 A young worker in Singapore

ACTIVITIES

1 Imagine you are an only child aged fourteen of parents in their late forties in West Germany. Make two lists:

   a giving the advantages of being in a small family;

   b giving the disadvantages of being in a small family.

2 Using Figures 10.2 and 10.5, explain why the French feel they need a big population.

3 a Read Figure 10.3. Write down why Romanians were urged to have at least four children.

   b How did Romania try to increase the number of births in recent years?

4 Make a poster urging people to have more babies in West Germany or France or Singapore or Malaysia. It should suggest why population growth is needed in your chosen country.

Figure 10.5 France wants more babies

11 Controlling immigration

Figure 11.1 Party-going in Brixton

Figure 11.2 USA citizens are of many ethnic groups

Look at Figure 11.1. A street party was held in Brixton in South London in June 1988. It was held to celebrate the arrival, forty years ago, of West Indian workers to Britain. At that time in the late 1940s there was a shortage of unskilled workers. Organisations like London Transport needed workers. Men were recruited in the Caribbean where there was high unemployment to work in Britain. Many brought their families.

In the last forty years many other immigrant groups such as European and Kampuchean refugees settled in Britain. Asians from East Africa and India, Bangladesh and Pakistan have come to Britain. Many could settle here because they were Commonwealth citizens. They held a British passport. Some had been forced to leave some African states. In 1972 about 25 000 British Asians came to Britain, expelled from Uganda. Look at Table 2.4. You will see that now in Britain large percentages of children aged under fifteen years are of immigrant origin.

Governments place **controls on migration** sometimes within states or across their borders. If you have been abroad, you will know that on leaving or entering a country passport checks are made. Sometimes a visa or work permit is needed before entry is possible. Countries like the USA and the USSR have strict controls on who enters or leaves.

In 1962 the Commonwealth Immigrants Act was passed in the United Kingdom. That Act and those passed afterwards in 1968, 1971 and 1982 limited the numbers and types of immigrants into Britain. People allowed in had jobs to come to or particular skills or rights of citizenship. A **citizen** is a native or member of a state with certain rights such as to vote in elections. Now, if children are born abroad even though their parents are British they may not be able to be British citizens.

113 000 British citizens and 130 000 **aliens** or non-British people entered the United Kingdom in the year ending June 1986. In the same year 106 000 British citizens and 70 000 non-British people left the country. There was therefore a **net gain** of 67 000 people through migration. A net gain means there were more entering than leaving the country.

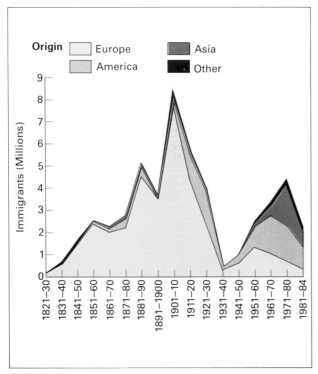

Figure 11.3 The origin of immigrants to the USA 1821–1984.

CASE STUDY *Immigration in USA*

President Kennedy in the 1960s, who was himself the grandson of an Irish immigrant, said that the United States of America was 'a society of immigrants' (see Fig. 11.2.).

In the nineteenth century millions of immigrants entered the USA. The population was seventy-six million in 1900. It rose to 151 million in 1950 and was about 240 million in 1986. Look at Figure 11.3 and see the numbers and origin of immigrants to the USA. Probably over fifty-one million people entered the USA between 1820 and 1984. Between 1925 and 1965 a quota system existed. It meant that only limited numbers of people from certain countries could enter. The numbers were in proportion to the nationalities in the USA in 1920. You will see in Figure 11.3, eighty-seven per cent of all immigrants to the USA at that time were from the United Kingdom, Ireland, Germany and Scandinavia. Refugees were allowed in as a special group after World War II.

In 1965 the quota system was abolished. The Act replacing it permits 270 000 people into the USA a year. Priority is given to single adults who are children of American citizens. As in Britain, people with special skills like scientists, doctors and nurses are also given priority.

Figure 11.4

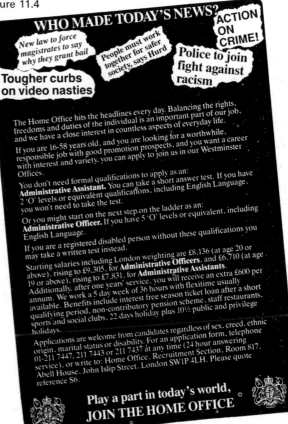

12 Factors influencing population distribution

Look at Figure 12.1, the Map extract. It shows part of the Hardanger Fjord in Western Norway. Locate the fjord in your atlas. Notice that settlement is unevenly spaced across the map.

People have always tried to live where **resources** provide them with fresh water, food, shelter and work. (See 'Settlement site' in the *Settlement* book).

Resources and other geographical advantages are unevenly spread across the earth. As a result some areas are more thickly populated than others. Eighty per cent of the world's population live on less than twenty per cent of the land surface.

The Norwegian map extract shows that settlements tend to be at a low altitude, by the edge of the fjord and on flat, fertile, alluvial fans such as at Ullensvang. Transport by boat along the

Figure 12.1 Part of the Hardanger Fjord, 1:50,000

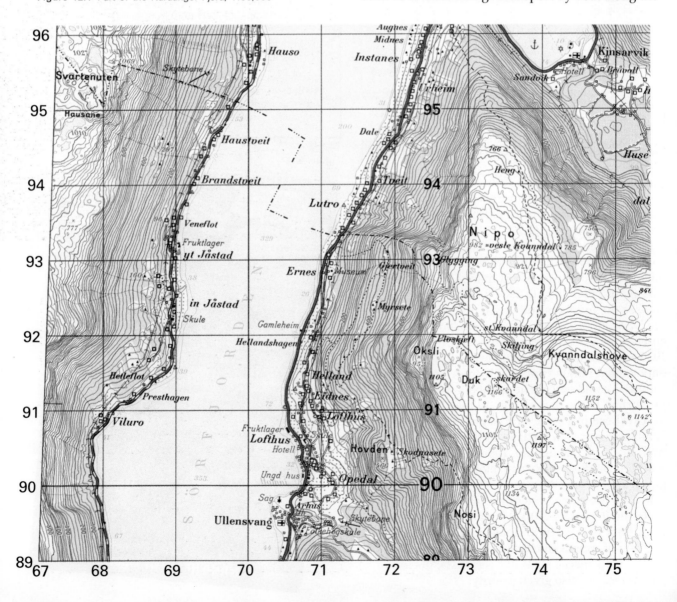

fjord and by the coast are important links between settlements. A few cabins and huts are on slopes up to about 400 or 500 m above sea level, but above that, there is little settlement (see Fig. 12.2). Favourable settlement sites for the 4.1 million Norwegians are in short supply. Over seventy per cent of the country is glaciers, lakes or barren areas, fifty per cent is north of the Arctic Circle. Twenty five per cent is over 1000 m above sea level.

Population density is a useful expression for describing and comparing population distribution. Imagine that the squares in Figure 12.3 represent a deserted island to which the settlers, seen arriving by boat, spread evenly over the land. It is easy to see that on average there would be two people on every square kilometre of territory. Population density can be calculated as:

$$\textbf{Population density} = \frac{\textbf{Population}}{\textbf{Area}}$$

Figure 12.3 Population density.

0   1 kilometre

Figure 12.2 Part of the Hardanger Fjord

ACTIVITIES

1 a Draw a cross-section from the Norwegian map extract from 671897 to 738897. You need only mark the contours at every 100 m. Use a vertical scale on your cross-section frame of 1 cm to 200 m and a horizontal scale of 2 cm to 1 km or 1000 m.

   b Mark and name the location of the fjord, the area of settlement, the road, the forested slopes, the plateau. Remember to make the scales, label each axis and put a heading on your section.

2 a Describe and suggest reasons for the location of Kinsarvik in the Kinso Valley on the Norwegian map. Try to explain why more houses are on the north-east side of the valley. How may aspect affect settlement?

   b Describe and explain the pattern of communications. Why are ferry routes important (742958, dotted line)?

13 Global distribution of population

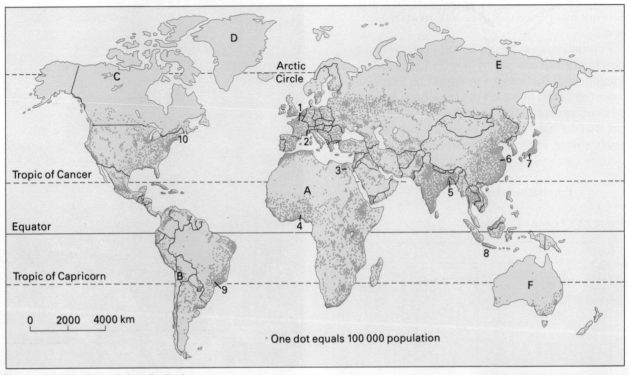

Figure 13.1 World population distribution.

Look at the **dot map** on Figure 13.1. It shows, using dots, where most people live in the world. One dot represents 100 000 people. The map can only give a very general impression of where people live.

It is clear that much of the land is virtually uninhabited. In fact about sixty per cent of the land has no people on it. Most of these areas are hostile to settlement. The hot desert areas (see Fig. 13.2) like the Sahara Desert have few people. Cold desert areas too like north of the Arctic Circle in Canada (see Fig. 13.3) have few people living there. High, steep slopes, such as on the Andes and Himalayas have few people. Small clusters of people may be found where resources like oil or minerals are extracted, but they do not show up on a map like this.

Figure 13.2
Low density settlement in the Sahara

About seventy per cent of the world's population live in fertile lowlands in:

1 Western Europe, like the River Rhine delta area in the Netherlands (see Fig. 14.3);
2 Asia, like the River Ganges and Brahmaputra deltas in India and Bangladesh and in valleys in eastern China;
3 North eastern North America near New York and Boston.
   Smaller concentrations of people live:
a near the coast in parts of South America (see Fig. 13.4) and
b Australia;
c in the Nile valley in Africa (see Fig. 7.4); by water and other resources in Central Africa.

You will notice that most people – in fact over ninety per cent – live in the northern hemisphere where most of the land is found.

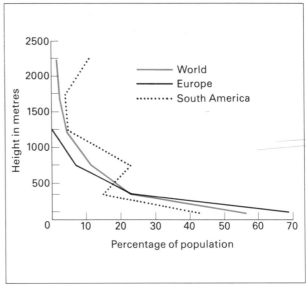

Figure 13.5 The vertical distribution of population.

Look at Figure 13.5. Most of the world's population live at altitudes below 500 m. A height of about 5000 m is generally the limit of comfortable human occupation. Above that height, low oxygen levels and low temperatures make life difficult.

Figure 13.4
High density Settlement in Rio de Janeiro

Figure 13.3
Low density Settlement in The Canadian Arctic

ACTIVITIES

1 a On a world outline map mark the main areas of population from Figure 13.1. Insert the equator, the tropics, the Arctic and Antarctic circles and the 0° line of longitude.

  b Name the areas of dense population labelled 1–10 using your atlas. Mark these on your map.

  c Name the areas of sparse population labelled A–F using your atlas. Mark these on your map.

  d With the help of this section and an atlas try to explain why the following areas have many people:
   i the coastal areas of the Netherlands and South East England;
   ii the Nile delta and valley in Egypt;
   iii the Ganges and Brahmaputra deltas in India and Bangladesh;
   iv the island of Java in Indonesia.

  e With the help of this section and an atlas try to explain why the following areas have few people:
   i a large area of Africa on the Tropic of Cancer;
   ii the northern coastal areas of the USSR;
   iii the Andes in South America.

2 What are the disadvantages of dot maps in showing population distribution or location?

14 Population distribution in Europe

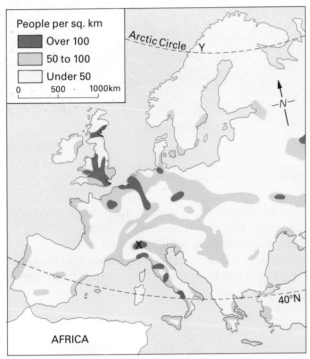

Figure 14.1 Where people live in Europe.

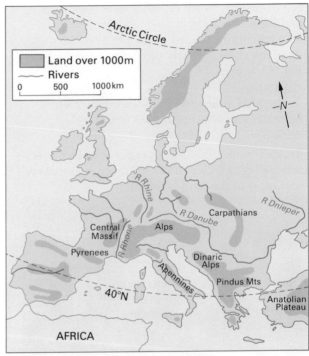

Figure 14.2 Physical geography of Europe.

Figure 14.1 shows where people live in Europe. As you can see, there are areas where there are over 100 people on every square kilometre of land. There are larger areas with fewer people to each square kilometre.

Compare the map of where people live in Figure 14.1 with the physical map of Europe Figure 14.2. The areas of high population density of over fifty people per kilometre square occur in strips mostly running west to east through comparatively low land in central Europe (see Fig. 14.3). Some river valleys, like the Rhine, where resources of water, fertile silt, minerals and good transport occur, have also attracted settlement. Coastal areas of the English Channel and the Mediterranean where trade has occurred have pockets of high population density.

The areas of lowest densities are on highland above 1000 m and in high altitudes where winters are long and cold (see Fig 14.4).

Figure 14.3 Concentration of settlement in the Randstadt near the mouth of the Rhine

Look at Table 14.1. You will see that the area and population size of each country in Europe vary very much.

ACTIVITIES

1 a Trace Figure 14.2 of the physical map of Europe. Use an atlas to name all rivers and mountain areas shown. Add a title. Include a key if necessary.

  b On a piece of tracing paper copy the outline of Europe from Figure 14.1 and shade the areas where over fifty people per kilometre square live. Add a title. Make a key.

  c Fasten the traced map of where people live in Europe exactly *over* the map of the physical geography of Europe. (Only fasten one side of the traced map so that you can use the two together, or the physical one alone.)

  d With the help of your maps drawn in exercises a, b and c, an atlas and other reference books, try to explain:
   i why many people live in area X on Figure 14.1;
   ii why few people live in area Y on Figure 14.1.

2 Look at Figure 13.5 in the previous chapter.

  a What percentage of Europe's population live above 1000 m?

  b What percentage of Europeans prefer to live in areas under 1000 m? Suggest reasons for this.

3 Look at Table 14.1 showing the area and population of European countries.

  a Copy and complete Table 14.1 by working out population densities. Some have been done for you. Look back to the text on page 35 to find out how to calculate population density.

  b On an outline map of European countries, (one is in the *Teachers' Book*), show by four types of shading the average population density for each country.
   i Use a light shade for countries with average densities of under 100 people per kilometre square.
   ii Use a slightly heavier shade for countries with average densities of 100 to 199 people per kilometre square.
   iii Use heavier shading for countries with average densities of 200 to 350 people per kilometre square.
   iv Use the heaviest or darkest colour for countries with average densities of over 350 people per kilometre square.

     Remember to give your map a title and a key.

  c Choose one country with a low density and one with a high density. Try to suggest reasons for the difference in density. Use your atlas to help you.

  d Why does your map of population densities in each European country give a misleading picture of where people really live?

Figure 14.4 The empty, high plateau of Norway

Table 14.1 Area and population of European countries (1984)

Country	Area in 000s of km^2	Population in millions	Population density per km^2
Albania	27	2.9	107
Austria	83	7.6	92
Belgium	30	9.9	330
Bulgaria	111	9.0	81
Czechoslovakia	125	15.4	123
Denmark	42	5.1	
Finland	305	4.9	
France	546	54.9	
German Democratic Republic	106	16.6	
German Federal Republic	244	61.1	
Greece	131	9.9	
Hungary	92	10.0	
Iceland	100	2.4	
Ireland (Irish Republic)	69	3.5	
Italy	294	57.0	
Luxembourg	2.6	0.3	
Malta	0.3	0.4	
Netherlands	34	14.4	424
Norway	308	4.1	13
Poland	305	36.9	121
Portugal	92	10.1	
Romania	230	22.9	
Spain	499	36.7	
Sweden	412	8.3	
Switzerland	40	6.4	
USSR	22272	275.0	12
UK	242	55.6	230
Yugoslavia	255	23.0	
Data for the smallest states (early 1980s)			
Andorra	1	0.034	34
Monaco	0.002	0.026	13000
San Marino	0.061	0.021	344
Vatican City State	0.004	0.001	250

Source: UN from Philips Modern School Atlas

(N.B. Due to the different sources involved, some of the figures in this table are different from those in Table 4.1.)

15 National population distributions

In all countries, people are spread unevenly. Study Figure 15.1. It shows population density in Nigeria. Concentrations of population are influenced by physical and economic factors. Medical, social and historical reasons are important too. In parts of northern and southern Nigeria mineral and agricultural resources occur. Trade links are important. Economic advantages have attracted people to these areas. The labels on Figure 15.1 show the attractions of North and South Nigeria.

Several factors have made the central belt of Nigeria unattractive to man:

1 The area has been infested with parasitic tsetse flies which cause **Trypanosomiasis** or sleeping sickness in man. The flies, or vectors since they transmit the disease, also affect domestic animals like cattle. There are many species of tsetse fly which affect animals. About five species damage man. The flies like the bushy vegetation of central Nigeria where they shade from the hot sun. The flies, about the size of our blue bottles, suck blood from man and animals. In this blood are small eel-like organisms which lie dormant in the fly. Later when the tsetse fly feeds on another creature the trypanosomes are transferred to that creature or host. The poisonous parasites grow to maturity, killing or weakening man and animals.

2 **Onchocerciasis** or river blindness is another condition caused by parasitic worms. Within twenty km of some central Nigerian rivers many people are blind through this worm (see Fig. 15.2). In this part of Africa people say 'nearness to rivers eats the eyes'.

3 Wars between tribes in central Nigeria reduced the population.

4 Slave traders from the savanna in the north and from the coast in the south captured people from this area. People were taken from these areas until the end of the nineteenth century.

5 Forest and game reserves such as Yankari east of the Jos Plateau have been made this century. They too discourage permanent settlement.

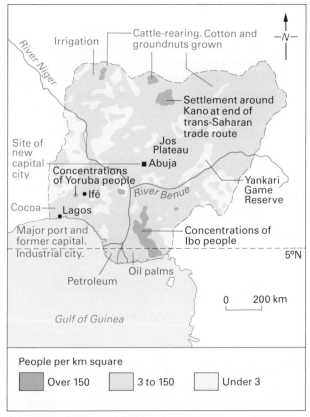

Figure 15.1 Population distribution in Nigeria.

Figure 15.2 River blindness, Nigeria

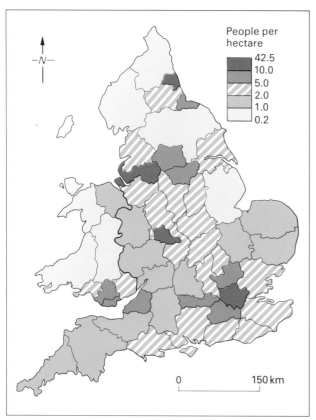

Figure 15.3 Population density in English and Welsh counties 1981.

The map in Figure 15.3 shows average population densities in English and Welsh counties in 1981. It may appear that population densities change suddenly at a county boundary. In reality changes usually occur gradually.

Figure 15.4 An area of high density population in London

You have already learnt that population is **dynamic** that means it moves and changes. Figure 15.5 maps the results of finding the **population centre** for the USA between 1790 and 1980. Gradually there has been a westward shift of the population.

Figure 15.5 The changing centre of the United States' population 1790–1980.

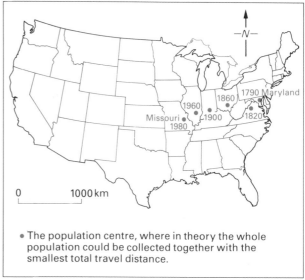

• The population centre, where in theory the whole population could be collected together with the smallest total travel distance.

Figure 15.6 shows a **Lorenz curve**. It is a special diagram showing what percentage of the population is found on a given percentage of the land. In 1950 in the USA over sixty per cent of the population lived on about ten per cent of the area of the country. People were concentrated into clusters especially on parts of the east and west coast and near the Great Lakes.

Figure 15.6 The population concentration in the USA in 1950 shown by a Lorenz curve.

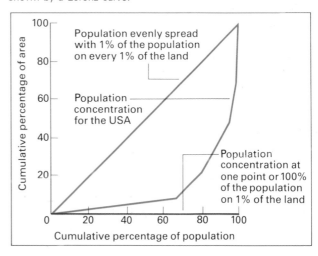

Figure 15.7 District Plan Proposals for some wards in the Rhondda Borough (pre 1984).

Table 15.1 Population changes in the wards covered by the 1:25 000 OS map extract in the Borough of the Rhondda for 1922 and 1981. (The ward boundaries were redrawn in 1984)

Ward	1922	1981
Treherbert	15 082	7450
Treorchy	17 570	7668
Pentre	12 238	5459
Ystrad	13 009	6398
Ferndale	19 360	9368

Source: Rhondda Borough Council

To find out by what percentage the population has declined in each ward from 1922 to 1981:

$$\frac{1981 \text{ population for the ward}}{1922 \text{ population for the ward}} \times 100$$

For example, in Treherbert Ward the population has declined to:

$$\frac{7450}{15\,082} \times 100 = 49\% \text{ of what it was in 1922}$$

The population has therefore declined by 51% since 1922.

ACTIVITIES

1 Look at Figure 15.1 (and the picture of part of Nigeria). With the help of Figure 15.1 and an atlas:

a Give five reasons why many people live in Southern Nigeria. The following list will help you:
  i Look for advantages for trade and transport.
  ii What economic resources occur there? What happens to them?
  iii What is the shape and fertility of the land like?
  iv Does the climate help farmers?

b Give as many reasons as you can why there are areas with over 150 people per square kilometre in northern Nigeria.

c What is the average density of population in central Nigeria? Give as many reasons as you can to explain why this region has low densities of people.

d What are the advantages of the location of Abuja the new Nigerian capital city?

e What are the disadvantages of Abuja's location?

2 Using Figure 15.3 and an atlas:

a Name four conurbations in England where on average population densities are over ten people per a hectare.

b Make a study of one of these urban areas to find out why people clustered in that area.

c What types of dwellings in your chosen urban area allow high densities of population to be housed?

3 Using Figure 15.3 and an atlas:

a Name all the counties in England and Wales which have less than one person per hectare on average.

b Describe the location of these counties. Do their locations have anything in common?

c Choose one county which has less than one person per hectare on average.
  i Make a study of this county with the help of your atlas and books from your library. Try to find out reasons why so few people are supported by the area.
    • Are there few economic resources like workable minerals?
    • Is the quality of the farmland poor?
    • Is there much high, steep, exposed land?
    • Are communications poor?
    • What is the weather like? Is it very wet? Is it very cold? Give climatic figures to support your opinions.
    • Are there many jobs?
    • Do young people want to live there? Are homes available?
    • Is the area near a major city?

  ii Make a report on your chosen county. Give the reasons why you think its average population density is low. Your report could be written; could be as a tape recording or on a wall poster.

Different groups in the class could work on different counties making different types of reports.

4 Look at Figure 15.7 of some wards in the Rhondda valley.

a On tracing paper mark the ward boundaries and the edge of the District Plan. These wards are on the Ordnance Survey extract on page 23. Place the tracing over the Ordnance Survey extract using the grid references to help you locate the wards.

Table 15.1 shows that population totals and therefore densities have declined between 1922 and 1981.

b With the help of the information on Table 15.1 work out and list for each ward by what percentage the population has declined between 1922 and 1981.

c In 1922 many people in the Rhondda valley were living in very over-crowded homes. The Medical Officer of Health reported in that year that nearly 5000 families lived in only two rooms per family. One family of twelve lived in only two rooms!

The present Rhondda District Planners want to provide housing 'better suited to the needs of the present and future population'. (see Fig. 15.8)
  i On your tracing and with the help of the Ordnance Survey extract of the Rhondda valleys, mark where you would put new houses.
  ii Mark where and what type of recreational facilities you would provide. Keep in mind the population trends, the shape of the land and the existing land use.
  iii Add to the map any other development – perhaps shops or parks which you think would improve the area.

d Explain why it is not easy to provide at once better homes and leisure facilities in an area like the Rhondda. (See Fig. 15.8.)

e Describe how population changes in some of the Rhondda valley wards have reduced some problems, but made others worse.

5 a From an atlas draw or trace a map showing the distribution of the population in the USA.

b Look at Figure 15.5. Try to find out why many people in the USA have moved westwards.

Figure 15.8 New homes in the Rhondda Valley

Coursework ideas

1 a From your school or public library or Council or Planning Office obtain a map of the wards − or some wards − in your town or nearest town. You may have to make a tracing from a borrowed map.
   b Find out the population in a recent year for each ward. Find out the population for the same wards ten, twenty or more years ago. Calculate, as in Table 15.1, the percentage population change.
   c Visit the areas concerned. Observe, map and record, using sketches or photographs the effects of population change on the area.
2 Conduct local field work to see if different ages of housing have different densities.
   a Walk through the settlement recording on a map where different styles and ages of homes occur.
   b On a larger scale map of your town rule lightly in pencil small squares of equal size. Draw the squares − perhaps of one centimetre square for a 1:25 000 scale Ordnance Survey map − where you know different ages of homes are found. Go to those areas and count the number of homes or separate dwellings found there. Record the information.
   c Transfer the results of your investigation into a report. Explain fully how you collected your information. Use maps, sketches and other illustrations to help you reach your conclusion.
3 Test the hypothesis that fewer dwellings are found on higher land, steeper slopes or shaded slopes.
   a Select a suitable area for your study such as across a valley. Draw a transect or cross-section of the area from a large-scale Ordnance Survey map or make your own transect. (Your teacher will be able to help explain how you can do this.) Make sure you take a line across land you may walk over or can see.
   b Following the transect record on it the location of numbers and types of dwellings.
   c Consider what factors, other than physical geography, may account for your observations.

16 The nature and causes of migration

Perhaps you have lived in more than one place since you were born. Possibly you have had to travel to school. You may have travelled further on holiday. All these movements are called **migrations** and are of interest to a geographer. Look at Figure 16.1. It shows why some people moved recently.

Permanent migration involves people moving for good to a new area (see Fig. 16.2). Sometimes the new area is in another country or continent. **Temporary migration** is when people move from one place to another for a short time. They usually return home.

Figure 16.1 Causes of migration

TAMIL refugees seeking asylum in Britain claim they are persecuted in Sri Lanka.

Ethiopia's Marxist Government estimates the drought has affected life for 7.7 million people in this country of 33 million, forcing 2.2 million from their homes and threatening 5.5 million with starvation.

LEIGH offers to success oriented individuals, a career growth opportunity in Canada's high technology industry coupled with competitive salaries, excellent benefits and a comprehensive "Relocation to Canada" package.

They have fled in fear from widespread fighting between Government and rebel forces. The rebels of the Mozambique National Resistance (MNR) are burning villages, destroying schools and health posts, and making it unsafe to use roads or grow food in many areas.

Houses crushed
A landslip in the north-west Yugoslav mining town of Zagorje ob Savi destroyed eight houses and a factory. Residents and machinery had been evacuated. —Reuter.

RETIRE IN STYLE
From £46,850. Fabulous Regency-style cottages and apartments purpose built for perfect retirement in elegant Cheltenham.

Pozzuoli has the rare distinction of being founded on the roof of a volcano. The worst, and most recent, disturbance ended in January 1985, after 18 months in which a bulge, 1·8 metres high at its peak, appeared over some 80 square kilometres. In Pozzuoli, sitting astride the developing summit, the immediate repercussions were ruinous: of the town's 72,000 inhabitants, over half were obliged to abandon homes made unsafe by the movement of the ground. As the sheltered harbour became shallower, so more and more of the local fishing fleet had to find anchorage elsewhere. Tourism slumped.

Figure 16.2 War time refugees

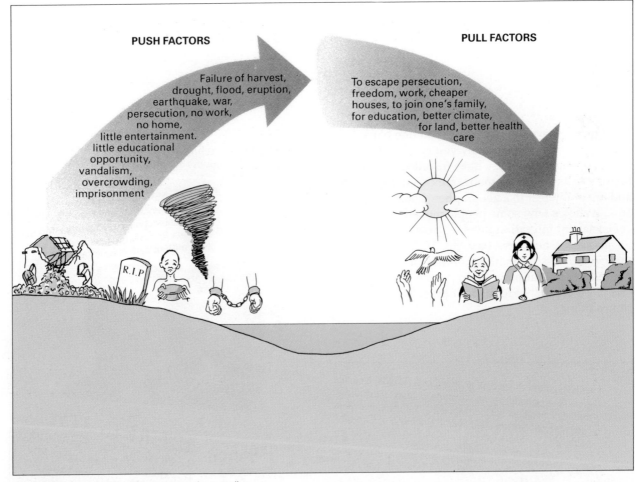

PUSH FACTORS

Failure of harvest, drought, flood, eruption, earthquake, war, persecution, no work, no home, little entertainment, little educational opportunity, vandalism, overcrowding, imprisonment

PULL FACTORS

To escape persecution, freedom, work, cheaper houses, to join one's family, for education, better climate, for land, better health care

Figure 16.3 Push and pull factors on where to live.

There are two sets of reasons which often combine to explain migrations. They are called **push** and **pull factors**. Push factors force people from an area. Pull factors attract people to an area. Figure 16.3 shows some push and pull factors.

ACTIVITIES

1 Look at Figure 16.1. Make two lists, one giving examples of 'pull' factors and the other examples of 'push' factors. Try to add more of your own.

2 a On two maps of the world use Table 16.1 to draw arrows to show the direction of migration to and from the United Kingdom recently.
   You could make the width of the arrows proportional to the number of migrants.
   Use one map to show the number and origin of people entering Britain. On the other map show the number and destination of migrants leaving Britain. Give your maps titles and keys.
  b Where do most immigrants come from? Why?

Table 16.1 The origin and numbers of migrants to and from the United Kingdom – mid 1985–mid 1986

Country of last or next place of residence	Numbers entering the UK in thousands	Numbers leaving the UK in thousands
Commonwealth countries		
Australia	16	23
Canada	4	6
New Zealand	7	8
African countries	13	6
Bangladesh, India, Sri Lanka	15	3
Caribbean	5	2
Other Commonwealth countries	20	15
Non-Commonwealth countries		
European Community	55	33
Rest of Europe	18	17
United States of America	25	27
Other American countries	2	4
Republic of South Africa	19	3
Pakistan	10	2
Middle East	16	17
Other Foreign countries	16	10

Source: OPCS Monitor MN 87/3 1987

17 Permanent migration

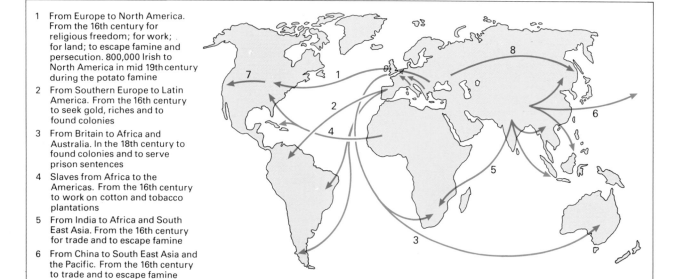

1. From Europe to North America. From the 16th century for religious freedom; for work; for land; to escape famine and persecution. 800,000 Irish to North America in mid 19th century during the potato famine

2. From Southern Europe to Latin America. From the 16th century to seek gold, riches and to found colonies

3. From Britain to Africa and Australia. In the 18th century to found colonies and to serve prison sentences

4. Slaves from Africa to the Americas. From the 16th century to work on cotton and tobacco plantations

5. From India to Africa and South East Asia. From the 16th century for trade and to escape famine

6. From China to South East Asia and the Pacific. From the 16th century to trade and to escape famine

7. From East to West Coast in North America. From the 19th century for land; to escape drought; to work in films; to live in a sunny, dry climate

8. From west to east in Russia. From the 18th century to colonise land and develop resources for trade

Figure 17.1 Major intercontinental migration from the sixteenth century on. See Figure 17.5 also.

Moving to a new place permanently may be very disturbing even if people move willingly. Much has to be arranged even to move a short distance from one home to another in the same town. At times, large numbers of people have been involved in major migrations. They may have gone overseas or to completely strange areas. Sometimes they have been moved against their will. Look at Figure 17.1 which shows some intercontinental movement from the sixteenth century.

During the Second World War, 1939–1945, about twenty-five million people migrated. Some moved as refugees away from invaded or battle areas (Fig. 16.2). Some were put in concentration camps. Some were soldiers or prisoners. After the war over fourteen million people moved into West Germany from lands to the east. In 1961 the Berlin Wall was built to stop people from East Germany escaping to the West (see Fig. 17.2).

Israel since its foundation in 1948 has received immigrants from seventy countries. The Law of Return grants every Jew the right to settle there and become an Israeli citizen. The population of 806 thousand in 1948 has increased to 4.3 million. 1.4 million of these have been migrants with about half from Europe and the Americas.

Figure 17.2 The Berlin Wall

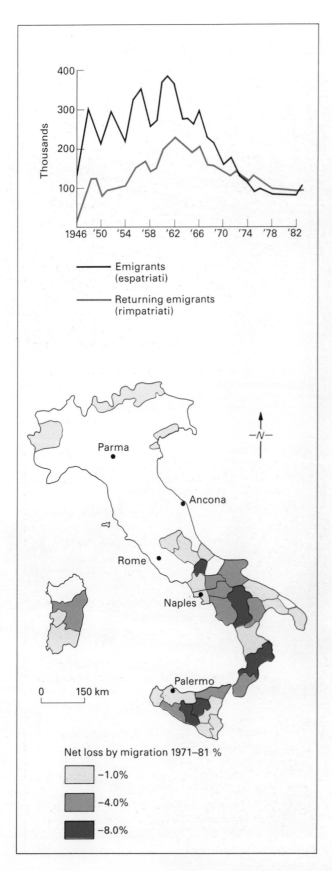

Net loss by migration 1971–81 %

- ☐ −1.0%
- ☐ −4.0%
- ■ −8.0%

Emigrants
(espatriati)

Returning emigrants
(rimpatriati)

Figure 17.4 Impoverished Southern Italy

In 1961, 113 000 Commonwealth immigrants mainly from the Caribbean came to the United Kingdom to work or to join their families.

Look at Figure 17.3. It shows that people have left some northern and many southern provinces in Italy (see Fig. 17.4). It has been estimated that twenty-six million Italians have emigrated since 1860.

Permanent migration also occurs within countries like Italy and the United Kingdom. A recent trend in many large European cities such as London has been a loss of population, particularly from central areas. People have been moving from **urban** to **rural** areas. Interestingly, however, in the three years between 1984 and 1987 Greater London showed a slight increase in total population.

In Norway in recent years there has been a shift in population especially from the northern counties in the Arctic. Look at Figure 17.10. Norwegians have been moving to the more southerly counties near Oslo, the capital.

Figure 17.3 Italian migration 1946–1982.

Figure 17.5 Migrating to North America from Ireland in the mid-nineteenth century

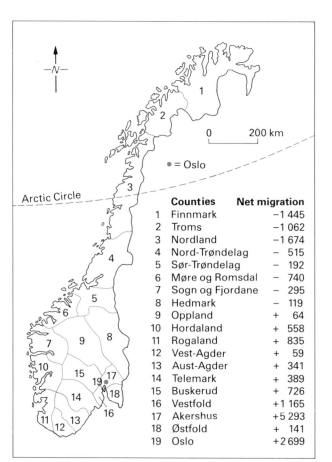

Counties	Net migration
1 Finnmark	−1 445
2 Troms	−1 062
3 Nordland	−1 674
4 Nord-Trøndelag	− 515
5 Sør-Trøndelag	− 192
6 Møre og Romsdal	− 740
7 Sogn og Fjordane	− 295
8 Hedmark	− 119
9 Oppland	+ 64
10 Hordaland	+ 558
11 Rogaland	+ 835
12 Vest-Agder	+ 59
13 Aust-Agder	+ 341
14 Telemark	+ 389
15 Buskerud	+ 726
16 Vestfold	+1 165
17 Akershus	+5 293
18 Østfold	+ 141
19 Oslo	+2 699

Figure 17.10 Population movement by migration in Norwegian counties in 1985.

ACTIVITIES

1 On an outline map of the world, mark the major migrations since the sixteenth century and add examples of recent migrations described in this section or known by your class.

2 Trace a large-scale map of North America. Using an Atlas, on your map:

  a mark all settlements named after places in Britain in one colour;

  b mark all settlements named after places in France in another colour;

  c mark all settlements named after places in Spain in another colour;

  d which parts of North America were settled by British, French and Spanish migrants?

3 a Trace the map of Italy in Figure 17.3 showing areas of **out-migration**. With the help of information in an atlas, label your tracing to describe the nature of the physical geography of the areas losing population.

  b Read Figure 17.6. List the main reasons why many people have left southern Italy. Where have they gone?

  c What do you think could be done, if anything, to help people stay in southern Italy?

  d Try to find out from your library what improvement schemes have been introduced in southern Italy since 1950.

Rich and poor: Tale of two countries

David Willey

Despite the pouring of thousands of millions of pounds of taxpayers' money into the economically less developed South during the past 40 years to implant new industries and to improve communications, the image of two Italys — the go-ahead North and the backward South — seems more, not less real. Although the citizens of Palermo and the citizens of Parma may watch the same television programmes, and go to the polls on the same day, they are living in different worlds.

While the big cities of northern Italy, and smaller wealth-generating industrial centres in the North like Parma forge ahead, achieving living standards for many Italians that exceed those of their social counterparts in Britain, the South remains stuck with ancient and apparently insoluble problems.

The imbalance shows up in practically all the statistics, beginning with the key index, unemployment. Whereas the overall percentage of unemployed for the whole of Italy in 1987 is 12 per cent, the figure for the South rises to 20 per cent, while in the North it falls to 7.5 per cent.

The main reason for the chronic lack of jobs in the South is the lack of industry; 80 per cent of Italian heavy and light industry is located in the North. Six per cent of southerners are still classified as illiterate, in comparison with only just over 1 per cent in the centre and North.

Consumer statistics point up gaps in living standards. In Calabria, family spending is 30 per cent below the national average. Calabrians own fewer cars, fewer television sets, make fewer calls from fewer telephones.

In prosperous Lombardy the local per capita gross product is 130 per cent above that of the Calabrian worker. A Milanese worker can afford to dress better, eat more expensive foods, can go out more often to sporting events, theatre, cinema or discos.

In the South, jobs in the public sector are regarded as pure gold for, once obtained, they offer security for life. So public competitions for even humble posts as janitors or secretaries attract many thousands of applicants and are fought with the powerful weapon of recommendations — letters of support from politicians or bishops or civil servants, the higher ranking the better.

Although Italians are marrying less and having fewer children (the population has actually begun to decline), the demographic picture varies radically between North and South. Southerners have an average two or three children, while the typical family in the North has one child. Divorce and abortion, which are both now legal in Italy, are practised much more in the North than in the more Catholic and traditionally minded South.

In Naples the family remains virtually the only stabilising force in a society which is falling apart at the seams through increasing drug addiction, organised crime and administrative neglect. The taps in the biggest city in southern Italy frequently run dry. Two thousand crimes of violence are reported each day to the police.

The family simply has to fill in where the State or local government fails. There is growing political disaffection in the South.

The North-South gap remains stubbornly there, and seems destined to grow.

The *Observer* 15.11.87

Figure 17.6

4 Read Figure 17.7 and with the help of Figure 17.8 describe and attempt to explain the direction of migration of population in the United Kingdom since 1971.

Exodus in search of 'good life'

AN INCREASING number of people are deserting Britain's cities in search of a more relaxed lifestyle in the country.

The exodus, particularly from London, now far exceeds the trend of rural depopulation in many areas, research findings show.

David Cross, of King's College London, used census information and the Registrar General's annual estimates of population to determine how much of the population growth in rural areas is due to long distance migration.

Of the 529 000 rise in population between 1971 and 1986 in the "top 33" rural growth areas, 97 per cent was due to migration, Mr Cross said.

And between 60 and 70 per cent of the gains were due to long distance migration between regions.

Leading the list of favourite destinations were: Wimborne, Dorset; Huntington; Caradon, near Plymouth; north Cornwall; Radnor in mid Wales; Forest Heath and Babergh in Suffolk; Breckland in Norfolk; Ryedale and Selby in north Yorkshire; east Yorkshire; and Holderness in Humberside.

The main trend, Mr Cross told the conference, was that while the more remote rural areas were often losing population through natural change as old people died and young people left to find work in the cities, this was being far outstripped by the new gains.

And more than a quarter of the new arrivals were in the 15 to 29 key worker age group.

Mr Cross said the reasons for the exodus included an increasing deconcentration of employment opportunities, improvement of travel facilities — particularly the high-speed rail links between East Anglia and London — and high property prices in London which enabled people to sell their homes at a large profit and move out.

The *Independent* 8.1.88

Figure 17.7

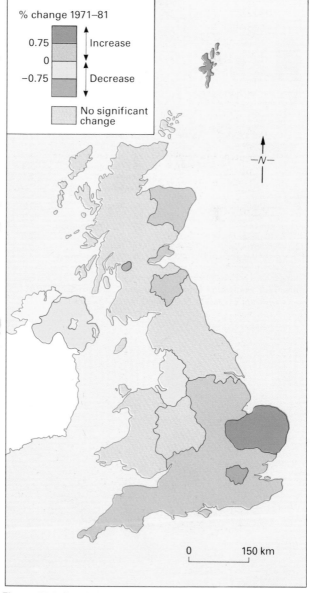

% change 1971–81

0.75 — Increase
0
−0.75 — Decrease

No significant change

0   150 km

Figure 17.8 Population change in the UK 1971–1981 through migration.

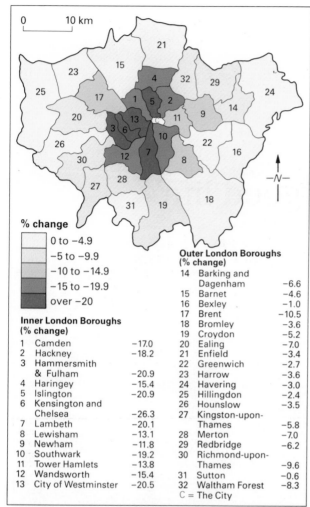

% change

☐	0 to −4.9
	−5 to −9.9
	−10 to −14.9
	−15 to −19.9
■	over −20

Inner London Boroughs (% change)

1	Camden	−17.0
2	Hackney	−18.2
3	Hammersmith & Fulham	−20.9
4	Haringey	−15.4
5	Islington	−20.9
6	Kensington and Chelsea	−26.3
7	Lambeth	−20.1
8	Lewisham	−13.1
9	Newham	−11.8
10	Southwark	−19.2
11	Tower Hamlets	−13.8
12	Wandsworth	−15.4
13	City of Westminster	−20.5

Outer London Boroughs (% change)

14	Barking and Dagenham	−6.6
15	Barnet	−4.6
16	Bexley	−1.0
17	Brent	−10.5
18	Bromley	−3.6
19	Croydon	−5.2
20	Ealing	−7.0
21	Enfield	−3.4
22	Greenwich	−2.7
23	Harrow	−3.6
24	Havering	−3.0
25	Hillingdon	−2.4
26	Hounslow	−3.5
27	Kingston-upon-Thames	−5.8
28	Merton	−7.0
29	Redbridge	−6.2
30	Richmond-upon-Thames	−9.6
31	Sutton	−0.6
32	Waltham Forest	−8.3
C	= The City	

Figure 17.9 Population change in Greater London 1971−1981. Population total 6 696 008 − declined by 10.1% from 1971.

5 a Make a tracing of Figure 17.9 which shows the Boroughs of Greater London. Use your own colour scheme to shade each borough according to its population change. The groups are:

Boroughs showing 0 to −4.9% population change or loss; −5 to −9.9%; −10 to −14.9%; −15 to −19.9%; −20% or more.

Remember to use the darkest shade for the over −20% category.

b Describe the map. Which areas lost most people? Discuss as a class what the reasons for this might be.

c Make a collection of estate agents' descriptions of similar houses for sale in different parts of London from a national newspaper − weekend papers often have a selection. Where are houses most expensive? Try to explain this.

d As a class, make a wall display of current newspaper cuttings about London. Do this for a month. At the end of that time what items have been featured? Which of the items help explain why people are moving from Greater London?

e How has the redevelopment of London Docklands encouraged people to move recently − especially since 1984 − into some Inner London Boroughs like Tower Hamlets? (Inner city redevelopment is dealt with in the *Settlement* book.)

6 a Trace the counties of Norway from Figure 17.10.
 i Shade in one colour areas which have lost population and in another colour counties which have gained population.
 ii Make a key. Give the map a title.

b Using your atlas and your library, find out what the weather and length of daylight are like north of the Arctic Circle in Norway in January. In what way are conditions in Oslo different?

c From your atlas count the number of towns in Arctic Norway. Are there many **central places** providing high order goods and services?

d Describe what your atlas shows about the type and direction of communications in Finnmark, Troms and Nordland in Norway.

e Write a short report suggesting reasons why many people choose not to live in Arctic Norway.

f From your atlas look at the countries which border Norway in the Arctic Circle. Why are the Norwegian Government and NATO members keen that the region remains populated?

g Imagine you are a Norwegian Government minister responsible for stopping out-migration in Arctic Norway. Outline, in a brief report, what the Government could do to attract more people to the area.

18 Case study: Migration in Indonesia

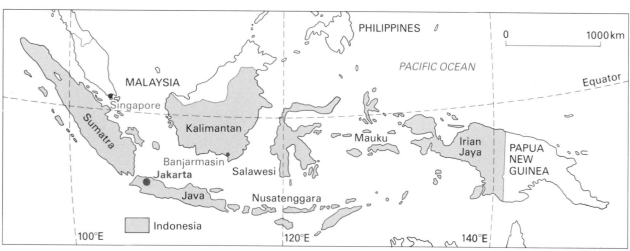

Figure 18.1 Indonesia

Table 18.1 Population distribution in Indonesia 1985 (estimated)

Province or island	Area in km²	Population in 000s	% of Indonesia's population	Population density per km²
Sumatra	473 606	32 922	19.93	70
Java	132 187	100 279	60.72	759
Nusatenggara (Bali, West Nusatenggara, East Nusatenggara, East Timor)	88 488	9411	5.70	106
Kalimantan	539 460	7842	4.75	15
Sulawesi	189 216	11 688	7.08	62
Maluku	74 505	1646	0.99	22
Irian Jaya	421 981	1368	0.83	3

Source: *Facts on Transmigration* Dept. of Information, Rep. of Indonesia

Indonesia consists of nearly 14 000 tropical islands in the west of the Pacific Ocean between Malaysia and Australia. Figure 18.1 shows the location of the main islands and of Jakarta, the capital.

The estimated population of Indonesia in 1986 was 169 million. Look at Table 18.1. Compare the population distribution on the principal islands. Population distribution is very uneven in Indonesia. Over sixty per cent of the people live on Java (see Fig. 18.2).

Figure 18.2 Squatter homes of rural to urban migrants in Jakarta

Two types of permanent migrations are occurring in Indonesia. Firstly, many people are moving from **rural to urban** areas like Jakarta. In 1971 eighty-three per cent of Idonesians lived in the countryside with only seventeen per cent in urban areas. By 1980, the percentage of urban dwellers had risen to twenty-two per cent. Migration as well as a high birth-rate account for this increase.

Secondly, since 1905 there has been Government sponsored **transmigration**. This involves the voluntary resettlement of families from the over-populated islands with high unemployment like Java to islands like Sumatra, Kalimantan, Sulawesi and Irian Jaya. These islands have plenty of land and need labour for development. Figure 18.3 shows the numbers of families who have moved in successive five-year plans (see Fig. 18.4).

Iskani is a fifty year old farmer. He moved with his parents to one of sixty-three transmigration sites at Takisung near Banjarmasin on Kalimantan. They moved from east Java in 1953. At first the Government only cleared the land. They did not offer expert guidance or fertilizers for the poor soil. It took all day on bad roads to get to market. Now, schools, a health centre, mosques and better roads serve the residents.

More recently the Government has provided each migrant family with at least two hectares of land. They are given a two room house and help in obtaining tools, seeds and seedlings of cash crops like coconuts, cloves and coffee.

Sukarman went to Kalimantan from east Java in 1986. On Java he owned 25 m^2 on which his house stood and his family income was about £15 a month. He has prospered from transmigration. He now owns 25 ha of land and his monthly income is over £30 a month.

With the help of funds from the World Bank, Indonesia has built health centres, schools, roads and sugar mills in the reclaimed areas of settlement. It is feared, however, that new transmigration schemes will slacken. This is because the price for Indonesia's oil fell on world markets in 1986. The Government has had to cut the budget for transmigration by fifty-four per cent. Indonesia will now improve areas already settled.

Some people would welcome less transmigration. They believe it destroys the lives of tribal people, damages fragile forest environments and does nothing to reduce the high birth rate on Java.

Figure 18.4 Transmigrants on Kalimantan

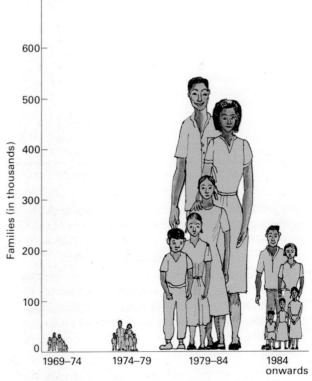

Figure 18.3 Numbers of transmigrant families in Indonesia in successive five-year plans.

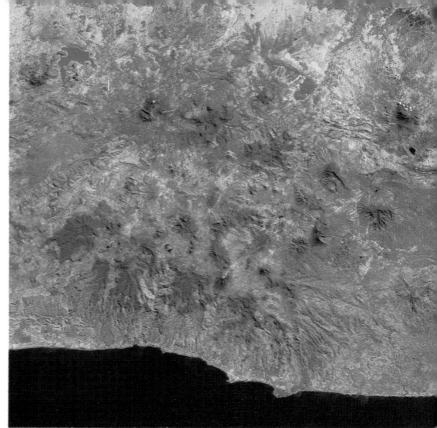

Figure 18.5 A satellite image of part of Java. Blue and black indicates water; pale green indicates limestone; green indicates rice growing; pink shows tea growing; the central green area is the Bandung Plain surrounded by volcanoes; dark red indicates very thick vegetation.

ACTIVITIES

1 Draw a map of Indonesia from Figure 18.1. Use information in the text to show by arrows the direction of the two types of migration in the country. Use different colours for each type of migration.

2 a Draw a bar chart to show the population of each province or main island in Indonesia from Table 18.1.

b From Table 18.1 write down the Indonesian island with the largest population and give its population density.

c Which two islands have the lowest population densities?

d What does transmigration try to do?

3 Look at the photograph of the home and land of a transmigrant in Kalimantan (Figure 18.4).
i Draw a labelled sketch of the picture. Your labels need to show the style of the house; what it is made of; why the roof overhangs the walls and why it has shutters.
ii Add labels describing the use of the surrounding land.

4 a Draw a climatic graph for Indonesia using the following figures:

b In what way does the climate influence:
i the style and building materials of the houses of transmigrants?
ii the crops grown by transmigrants?

5 Draw a pie-chart or pie-diagram from Table 18.1 to show what percentage of Indonesia's population live in each area shown in Figure 18.1. (To find the number of degrees, multiply the percentage figures by 3.6.)

6 Look at the satellite image of part of Java. (Fig. 18.5)

a What natural hazard could affect the area?

b How can the products of this hazard prove valuable to farmers?

c What protects the soil from erosion by the tropical rains?

d Does the image confirm your knowledge that Java is one of the most densely populated islands in the world?

	J	F	M	A	M	J	J	A	S	O	N	D	
Temperature °C	26	26	26	27	27	26	26	26	27	27	26	26	Average 26
Rainfall in cm	30	30	21	15	11	10	6	4	7	11	14	20	Total for year 179 cm

19 Temporary migration

Journeys to and from work and on holiday are examples of **temporary migration**.

The journey to work

In Britain many people have chosen to live away from their jobs. More and more people now live on the fringe of urban areas or in villages and **commute** daily to the cities. Increasing car ownership since the 1960s has been largely to blame for this trend. Read Figure 17.7 to discover which rural areas in England and Wales are gaining most people.

Improved rail services through electrification to London has encouraged the growth of commuting (see Fig. 19.1). Electrification can be said to reduce the **friction of distance** (see Table 19.1). Overcrowded roads and trains occur because many people travel to work at the same time.

Figure 19.1 Weekday commuting to London

In the north east of England, the Tyne and Wear Integrated Transport Network was opened in 1984. It was built to improve public transport in an area where forty per cent of households were without cars. Look at Figure 19.2 which shows the network.

Figure 19.2 The area served by the Tyne and Wear Integrated Public Transport System.

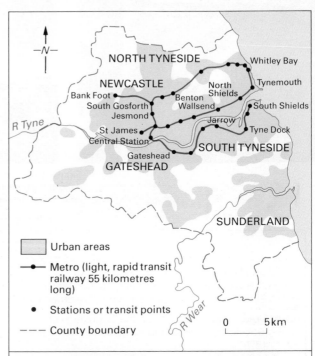

Metro-related travel time savings on public transport

Journeys between centres	Savings in minutes
Bankfoot – South Shields area	35
Regent Centre – Jarrow and South Shields area	20
Team Valley – North Shields area	20
Gateshead – Metro areas North of River Tyne	10–20
Sunderland – Metro areas North of River Tyne	8–10
Washington – Metro areas North of River Tyne	20–30
South Shields – West Newcastle	13–19
South Shields – Gateshead	15
South Shields – Sunderland	0
North Shields – Gateshead area	7–20

Light, rapid, metro trains carry commuters from housing estates to work on new industrial estates on the edge of the city (see Fig. 19.3). Passengers travel in to the centre of the city for work, shopping and entertainment. Buses are timed to fit in with the arrival and departure of trains. Tickets can be used on buses or trains throughout the network.

Within Western Europe many people migrate long distances to find work. They may remain away from home for weeks or months or even years. Many unskilled workers left Italy, Yugoslavia and Turkey in the 1960s and 1970s to seek work in richer countries like West Germany and Switzerland (see Fig. 19.4).

In Nigeria young men in their twenties helped build a new University at Ife in 1967. People now move to the University for study and to help run it.

Figure 19.3 The Tyne and Wear Metro System

Figure 19.4 Turkish migrant workers in West Germany

Figure 19.5 A migrant worker in Africa

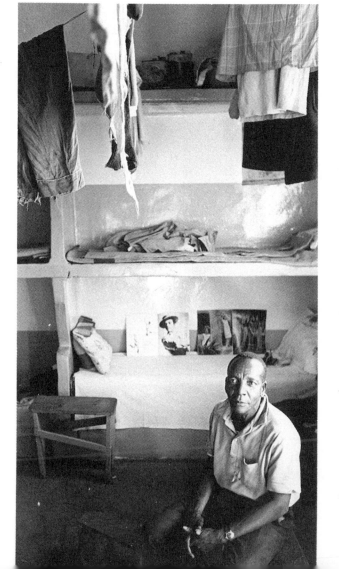

Black workers are employed in South African gold mines. Many men migrate for a year from neighbouring states like Botswana. While away from home they live in hostel accommodation. In 1982–83 1.4 million black migrants worked in the gold mines. Up to two million non-whites have migrated to squatter settlements on the outskirts of cities like Cape Town seeking work (see Fig. 19.5).

Traditionally pastoral farmers in countries like Norway, France and Nigeria have migrated seasonally with their livestock. This involves the movement to fresh pastures and is called **transhumance**.

Holidays

In the developed countries of the western world the number of people who travel on holiday has increased this century. They can do this because:

1 workers have paid holidays;
2 they can afford to travel;
3 they have time to travel;
4 they are interested in seeing or being in another place where the scenery, climate and way of life may be different;
5 there are good communications by road, rail, sea and air links. Increasingly people are travelling abroad.

Figure 19.8 European and North A

Figure 19.7 Package holiday make Majorca.

Figure 19.6
Day trippers at the seaside in Cornwall

ourists in the Caribbean

at Palma Airport,

In Britain there are three types of tourist; the foreign tourist; the day-tripper (see Fig. 19.6) and the tourist who spends some days or weeks in one resort (see Fig. 19.7).

Increasingly people from the developed world are holidaying further afield in developing countries. For example, British and American tourists visit Caribbean islands like Barbados to escape the winter cold (see Fig. 19.8). People travel for beach holidays to countries like the Gambia or to see historical monuments like the pyramids in Egypt. Some go on safari in Kenya or to make treks in the Himalayas. It is fast air travel, often in specially chartered 'planes, which allows more people to go further in a shorter time.

Travel can help people to learn more of the world and gain a better understanding of problems and issues facing us all. (The tourist industry is covered in the *Energy and Industry* book.)

Figure 19.9 A comparison between the relative linear and time distance by rail between Peterborough and London.

ACTIVITIES

1 a Trace the true distance diagram showing British Rail stations between London and Peterborough from Figure 19.9. Beside it, using the time distances from Table 19.1 and the given time scale, plot the journey times along the same route.

 b Which stations have been brought relatively closer to London? What effect may this have on commuting?

Table 19.1 Shortest weekday rail journey time between Peterborough and London, Kings' Cross

| From | Linear distance to London, King's Cross | | Time distance to London, Kings' Cross |
	Miles	Kilometres	Minutes
Peterborough	76.25	122	57
Huntingdon	58.75	94	47
Hitchin	32	51.2	29
Stevenage	27.5	44	26
Welwyn Garden City	20.25	32.4	25
Hatfield	17.25	27.6	22
Potters Bar	12.25	19.6	16

Source: British Rail Timetable

ACTIVITIES

2 a Look at Figure 19.2 of the Tyne and Wear Integrated Transport System. In what way has it helped someone living in Bank Foot and working in South Shields?

b What are the benefits to the traveller of co-ordinating bus and train services and fares?

c Look at Figure 19.10. When are the peak travel times during a weekday? Who uses the network then? When are peak times on Saturday? Who uses the network then?

3 a On a copy of the map of Europe (one is in the *Teachers' Book*) use the data in Table 19.2 to draw arrows to plot the flow of migrant workers to the Federal Republic of Germany (West Germany) in 1982 from other European countries. Make a scale so you can draw arrows of different thickness according to the number of migrants.

b Study the figures for 1974. Describe the trend in migrants from 1974 to 1982. Try to account for it.

4 Study the population pyramids of immigrant workers to France in 1974 and 1980 shown in Figure 19.11.

a What are the two age groups most immigrants fit into? Why do you think this is so?

b Why do you think there are so few migrants aged over forty?

c Describe and try to account for the major differences between the two pyramids.

d Construct a bar chart to show the information given in Table 19.3.

e Under the title 'Immigration to France' use the work you have completed to write a few summary points.

Figure 19.10 Metro boardings weekdays and Saturdays May 1984 on the Tyne and Wear Integrated Transport System.

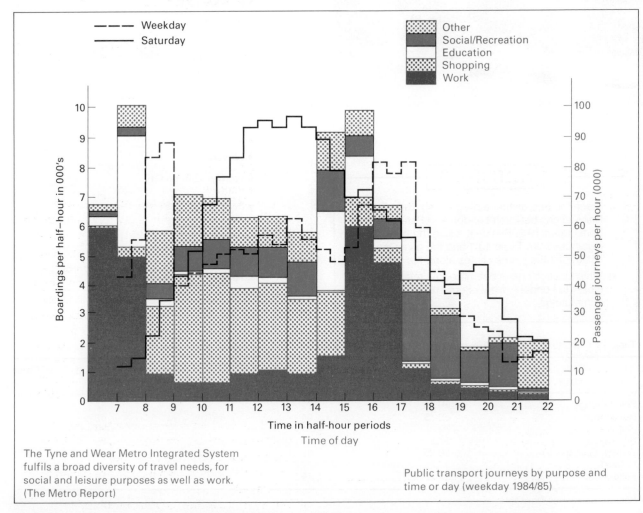

The Tyne and Wear Metro Integrated System fulfils a broad diversity of travel needs, for social and leisure purposes as well as work. (The Metro Report)

Public transport journeys by purpose and time or day (weekday 1984/85)

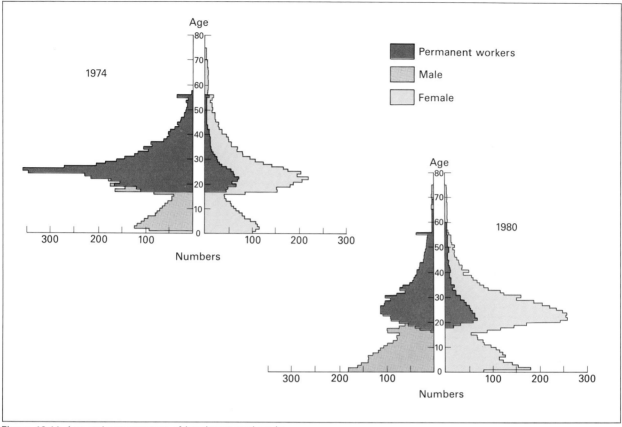

Figure 19.11 Age and sex structure of immigrant workers in France 1974 and 1980.

Table 19.2 Migrant workers to West Germany

Country of origin	1974	1982
Austria	7400	5600
Greece	1800	300
Spain	1200	700
Switzerland	–	500
Turkey	6100	400
Yugoslavia	7700	3300

Table 19.3 Employment of immigrant workers in France, 1979

	% of total employed in each job type
Manufacturing (especially cars)	9.9
Building and Construction	28.0
Commerce (especially catering)	6.0
Transport	6.3
Services (especially medical)	5.7

Coursework ideas

1 Find out where members of your class were born. Show this information as a flow map using arrows to link place of birth to the town or city where you live. Try to explain your map. Find out the reasons why your classmates' families moved. Was it just the result of parents changing jobs?

2 Carry out a survey of migration to Britain. Make up a questionnaire to see what number or percentage of pupils at your school have parents or grandparents who have moved to Britain. Try to discover why they moved. Question at least thirty people so you have a statistically reliable sample.

3 a Conduct a survey to find out where members of your form went for their summer holiday. Did most holiday near to home?

   b Map the location of where your classmates holidayed.

   c What factors influenced the location?

4 a Carry out an investigation to see if peak times for traffic flows reflect the journey-to-work periods. For ten minute periods at 8.30 am; 10.30 am; 12.30 pm; 2.30 pm; 4.30 pm and 6.30 pm on a weekday and on a Saturday or Sunday, count the number and types of vehicles and their direction of travel along a main road between the town centre and your area. Record your information on a tally sheet. Graph the results.

b Make enquiries of local schools and firms of their hours of opening and closing.
  i Devise diagrams to show your results.
  ii State your conclusions.

Figure 19.12 Nights stayed by overseas and British tourists in the UK in 1988

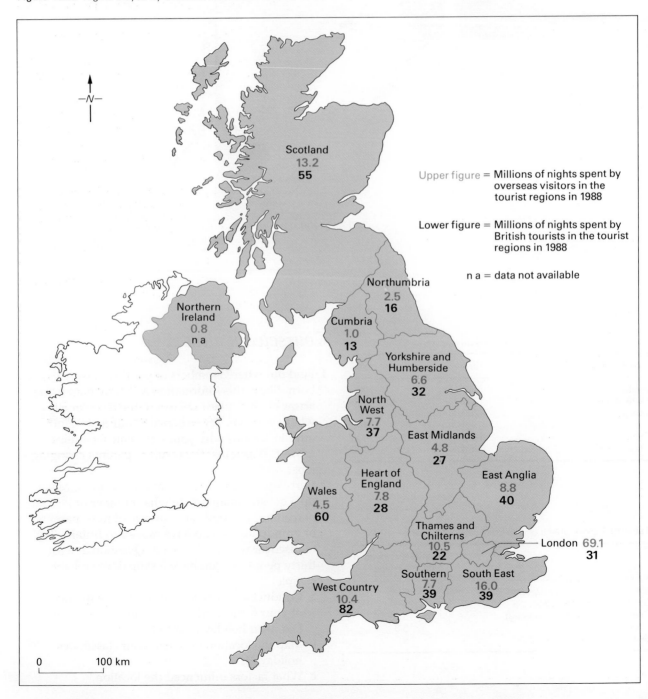

Upper figure = Millions of nights spent by overseas visitors in the tourist regions in 1988

Lower figure = Millions of nights spent by British tourists in the tourist regions in 1988

n a = data not available

Scotland 13.2 55

Northumbria 2.5 16

Northern Ireland 0.8 n a

Cumbria 1.0 13

Yorkshire and Humberside 6.6 32

North West 7.7 37

East Midlands 4.8 27

East Anglia 8.8 40

Wales 4.5 60

Heart of England 7.8 28

Thames and Chilterns 10.5 22

London 69.1 31

West Country 10.4 82

Southern 7.7 39

South East 16.0 39

0   100 km

Table 19.4 European visits to and from the United Kingdom in 1978

Country	Number of visits to the UK in thousands	Number of visits to Europe from the UK in thousands
Belgium Luxembourg	740	520
France	1435	2686
West Germany	1507	794
Italy	358	831
Netherlands	1003	571
Denmark	246	106
Republic of Ireland	873	1477
Greece	100	471
Yugoslavia	60	192
Spain	256	2322
Portugal	43	211
Austria	95	199
Switzerland	316	255
Norway	238	101
Sweden	398	73
Finland	55	27
Eastern Europe	83	147

Figure 19.13 Tourists in Italy 1979–1981.

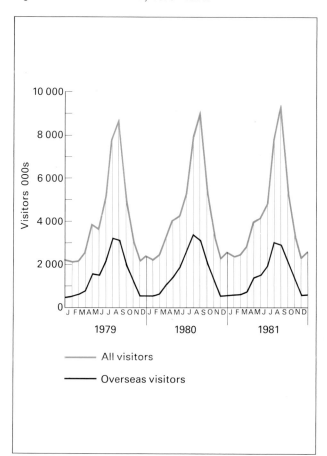

All visitors

Overseas visitors

ACTIVITIES

5 a Trace the tourist regions in Britain from Figure 19.12.
   i Shade in one colour the region which accommodated overnight the most overseas visitors in 1988.
   ii Shade in another colour the region which accommodated overnight most British tourists in 1988.
   iii Give the map a title. Make a key.
  b Try to find out reasons for the differences the map reveals.

6 a Using the information in Table 19.4 draw two maps of Europe.
    On one map show the number and countries of origin of tourists to the United Kingdom in 1988
    On the other map show the number and countries of destination of tourists from the United Kingdom in 1988. (Flow lines with the width proportional to the number of holidaymakers will show the information clearly.)
  b Is the number of holidaymakers related to the distance they travel?

7 a From Figure 19.13 describe the peak holiday month in Italy. Use a travel brochure advertising holidays in Italy to find out the location of resorts and what attracts tourists at that time.
  b Make up a poster advertising the attractions of one Italian holiday resort.

8 a Use an atlas map of world air routes to show on a world outline map routes from:
    i London to Barbados;
    ii London to Nairobi, Kenya;
    iii London to the Gambia;
    iv London to Cairo, Egypt;
    v London to Bankok, Thailand.
  b Try to find out from airlines such as British Airways the flight times of these journeys from London.
  c Explain how air travel has reduced the friction of distance between London and the places listed in a.

Index

LONGMAN
CO-ORDINATED
GEOGRAPHY

Series editor - Simon Ross

Settlement

Simon Ross

Acknowledgements

Office of Population Censuses and Surveys, Wychavon District Council, Jim Wallace, ASDA, John Tinworth, Cambridgeshire Community Council, Mark Vigor of Cambridgeshire Country Council, Development Commission for Rural England, Justin Berkovi, Cornerstone Estate Agents, Watford Borough Council, Tim and Hilary Kirtley, Northamptonshire County Council, Department of Environment, Department of Transport, John Cuffley of Telford Development Corporation, Pam Bradburn, Pamela Gibbons and Robin Palmer of OXFAM, C Mallinder of Rotherham Metropolitan Borough Council, HR Read of Cambrian News, Aberystwyth Tourist Information Office, MR Mitchell of Port of Felixstowe Authority, Singapore Port Authority, Council for the Protection of Rural England, Council for Small Industries in Rural Areas, Action with Communities in Rural England, The Unit for Retail Planning Information, SOS Childrens Villages, Centre for World Development Education Centre for West African Studies, Birmingham, Christian Aid.

Finally, I should like to thank my wife Nikki for putting up with undecorated rooms, uncleaned windows and endless 'phone calls. Also, thanks to Poppy and Leo for showing such interest in my work.

We are grateful to the following for permission to reproduce photographs and other copyright material:

The Cambrian News (Aberystwyth) Ltd for the adapted article 'Yachtsman warns of marina dangers' from *Cambrian News* 14.10.83. (c) Cambrian News Aberystwyth; Newspaper Publishing Plc for extract from the article 'Flaws that impair private bill system' by Richard North from *The Independent* 28.1.88; the Author, Christopher Reed for his article 'The City in a cloud of despair' from p16 *The Guardian* 20.1.84; Times Newspapers Ltd for article 'Queuing up for the retail revolution' by Terence Bendixan from *The Times* 13.1.87 (c) Times Newspapers Ltd 1987.

Aberdeen District Council, City Planning Dept., page 11 *above*; Aerofilms Limited, pages 3 *below*, 7 *below*, 47 *below*, 73; Edward Arnold (1984) 'The British Isles' by Young and Lowry, page 9; ASDA, 18 *right*; BBC Publications, 'BBC Geography Today, London's Docklands', page 48; Brackley and Towcester Advertiser, page 4; C S Caldicott, page 60; Cambridgeshire Community Council, Sketchworth Parish Council and Ellesmere Management Committee, page 26 *above*; Cambridgeshire County Council, Structure Plan Group, page 28; J Allan Cash, pages 3 *above*, 7 *above*, 40, 70 *centre*; Ceredigion District Council, page 74 below; Economic Development and Planning Dept. Middlesbrough Borough Council, page 50 (2); East Anglian Examinations Board — 'based on map from 16+ examination in Geography 1002', 1983, page 7; Paul Forster, pages 68, 69; the *Geographical Magazine*, Sept. 1986, page 59 above; *The Guardian*, page 43 left; Halifax Building Society page 33; Heinemann Educational Books Ltd (1978) 'Comprehensive Geography of West Africa'

by R K Udo, page 4; Hutchison Library, pages 63 *above*, 65; International Defence Aid Fund, page 23; London City Airport, page 47 *above right*; London Docklands Development Corporation, page 47 *above left*; Macmillan, London and Basingstoke (1974), 'Recreation and Environment' by P Toyne, page 15 *below*; Metrocentre, Tyneside, page 19; Moray House College Education, Edinburgh 'Migration and Urbanisation in West Aftrica' by J A Hocking and N R Thomson, page 59 *above*; Netherlands Board of Tourism, page 76; Network, page 18 (2) (Photo: Barry Lewis), 63 *below* (Photo: Barry Lewis); Ordnance Survey extract reproduced from 1:50,000 Landranger map with the permission of the Controller of the Majesty's Stationery office © Crown Copyright. Oxford University Press (1983), 'Development in the Third World' by M Morrish, page 58; Oxfam, page 61 *above* (Photo: Alison Speedie), 61, *below*, (photo: Deb Gibbs); Panos pictures, pages 62, 66, 67 *below*, all photos by Sean Sprague; Reflex, page 43 *right* (photo: Piers Cavendish); Rex Features, page 79; Rural Development Commission, page 25 *above*; The *Sunday Times*, page 45 *below right*, page 46 *above*; Telford Development Corporation, pages 54, 55, 56, 57; Times Newspapers Limited/Peter Sullivan, page 45 *below right*; Tropix Photo Library, pages 66–7 (photo; J Wickins); Warrington — Runcorn Development Corporation, page 57 *below*; Worcester City Council, page 36; Worcester G.A (1974) 'Droitwich: Salt Town Expanded' in 'Worcester and its Region' by R F Baker, page 10 *above*; World Bank (1981) 'Tackling Poverty in Rural Mexico', page 32; all other photographs were supplied by the author. We have, unfortunately, been unable to trace the copyright owner of the Photograph, page 25 *below* and would appreciate any information which would enable us to do so.

Settlement

Simon Ross

Contents

To the boys of Berkhamsted School

1 Settlement site

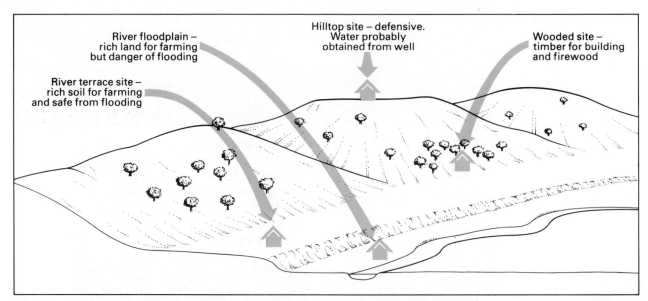

River floodplain –
rich land for farming
but danger of flooding

Hilltop site – defensive.
Water probably
obtained from well

Wooded site –
timber for building
and firewood

River terrace site –
rich soil for farming
and safe from flooding

Figure 1.1 Common sites for settlements

The **site** of a settlement is the position on the ground where it is built. When our ancestors decided where to build their settlements, they considered a number of factors:

1 Can the settlement be defended easily?
2 Is there plenty of water available?
3 How easy is it to get to other places?
4 Is the land safe – will it be flooded?
5 Can the land nearby be used for farming?
6 Is there fuel (wood) available?
7 Are there building materials nearby?

Study Figure 1.1. Here you can see some of the most common sites for settlements.

Today, it is interesting to try and work out the original reasons for the siting of a settlement. Study the site of Worcester on the map extract (see back cover). Notice how the city is near to water (River Severn) but raised up above it on a river terrace so that it is safe from flooding. It was the only place for kilometres around where the River Severn could be bridged. The land around is fertile – ideal for farming.

Sometimes we can use the name of the settlement to try and understand its siting. The '–cester' in Worcester is Roman and means 'castle or fort'. This tells us that Worcester was also a defensive site.

Try to find out about the reasons for the siting of settlements near your home. Table 1.1 may help you.

Table 1.1

Placename element	Meaning
AVON/AFON	STREAM OR RIVER
CET	WOOD
BRE	HILL
CESTER/CHESTER/CASTER	FORT/CASTLE
PORT	HARBOUR/GATE
BURY/BOROUGH	FORTIFIED PLACE
BRIDGE	BRIDGE
FORD	SHALLOW WATER CROSSING
FIELD	CLEARING IN THE WOOD
LEY	CLEARING
HOLT/WOLD/WEALD	WOOD
HURST	WOODED HEIGHT
FEN	WET PLACE
EY/EA/EG	ISLAND SURROUNDED BY MARSH

ACTIVITIES

1 Study the photographs of settlements in Figure 1.2. For each one, give the likely reasons for its siting. Remember that a settlement may have more than one reason!

2 Locate the village of Salwarpe on the map extract (see back cover) at grid reference 875620. Suggest some advantages of its site.

Figure 1.2
Variety of settlement sites

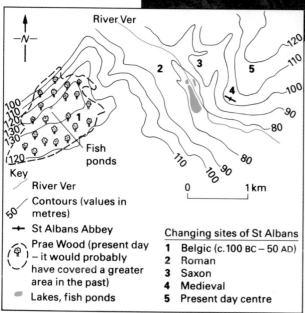

Figure 1.3 The changing sites of St Albans

Key

~ River Ver

Contours (values in metres)

+ St Albans Abbey

Prae Wood (present day – it would probably have covered a greater area in the past)

Lakes, fish ponds

Changing sites of St Albans

1 Belgic (c.100 BC – 50 AD)
2 Roman
3 Saxon
4 Medieval
5 Present day centre

River Ver

Fish ponds

0   1 km

ACTIVITIES

3 Study Figure 1.3 which shows the changing sites of St Albans.

a Four sites are near to the River Ver. Why?

b What were the advantages of the first site in Prae Wood?

c A number of sites had fishponds nearby. Why do you think these were important?

d Why do you think the Roman town grew up at Site 2?

e Present day St Albans extends inland from the Abbey. The land close to the river has not been built upon. Why do you think this is so?

ACTIVITIES

4 Study Fig. 1.4. It shows the village of Idanre in rural Nigeria. Old Idanre is the original site and New Idanre the modern site. Although most people live in New Idanre, a few people still live in the old village.

a Give reasons for the site of Old Idanre.

b Why do you think New Idanre is a preferred site today?

5 Figure 1.5 shows an area of unsettled land. You are the leader of a group of families who have been thrown out of your village. You are looking for a new site.

On a copy of Figure 1.5, mark on the site of your new village. Give reasons for your choice. It would be interesting to compare your choice with other members of the class.

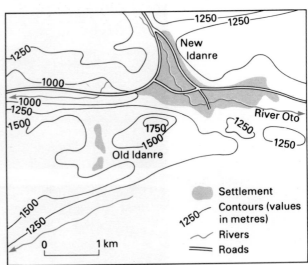

Figure 1.4 Sites of old and new Idanre, Nigeria

Figure 1.5 Siting a new village

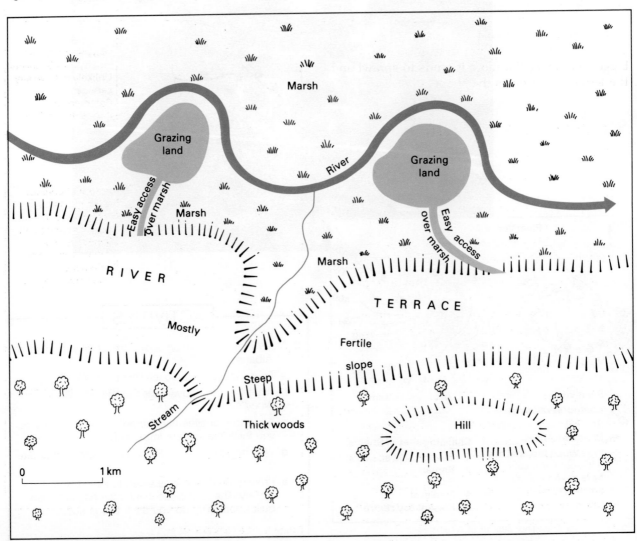

2 Settlement shape and size

Settlements come in many shapes and sizes. Some are only tiny villages whilst, at the other extreme, massive cities can be tens of kilometres across. One village may have all its houses very close to one another perhaps centred on a church, whilst another may be very spread out with no obvious central point.

This section will define and clarify the different **types** of settlement.

Settlement shape

Although every settlement is unique there are some common shapes which can be recognised (see Figure 2.1). The shapes in Figure 2.1 are most easily identifiable when looking at villages. It is possible to see these shapes in small towns but the bigger the town, the more it tends to sprawl and the less clear becomes the shape.

Settlement size

Words like 'town' and 'village' are used everyday but have you tried to think what is the difference between a town and a village? Interestingly, people living near to London talk about 'going into town' as meaning going into London − hardly a 'town' by most peoples standards! So, it is important to try and have clear definitions.

1 **Hamlet** − a small cluster of houses in the countryside usually with no services.
2 **Village** − a collection of houses and other buildings. There will usually be one or more shops, churches, a primary school, a garage and possibly small places of work. Population will be from about 100 to a few thousand. Villages are essentially rural in character.
3 **Town** − a settlement with facilities to live and work in a self-contained way. More shops including supermarkets and more specialised shops. More places of work, recreation facilities (e.g. cinema, swimming pool), schools (including secondary schools). A town will have an influence over the surrounding countryside − people will come into the town for certain shopping items or to go to the cinema.

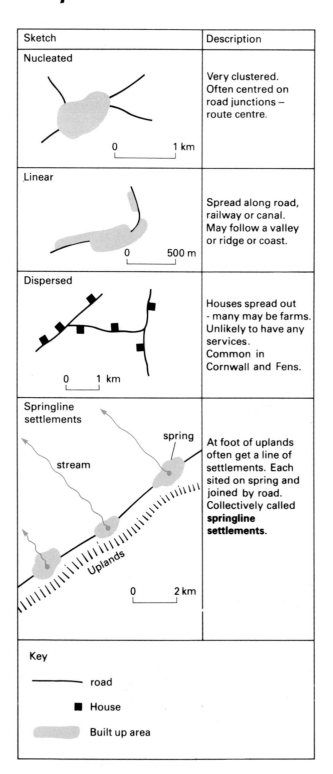

Sketch	Description
Nucleated	Very clustered. Often centred on road junctions − route centre.
Linear	Spread along road, railway or canal. May follow a valley or ridge or coast.
Dispersed	Houses spread out - many may be farms. Unlikely to have any services. Common in Cornwall and Fens.
Springline settlements	At foot of uplands often get a line of settlements. Each sited on spring and joined by road. Collectively called **springline settlements**.

Key

——— road

■ House

▬ Built up area

Figure 2.1 Settlement shapes

4 **City** − traditionally, a city contained a cathedral. This meant that very small towns like Wells and Ely were in fact 'cities'. Nowadays most people think of cities as having a population of about 100 000+. A city will have a large shopping and office area generally in the centre − such an area is called the **Central Business District** (CBD). There will be department stores and a much greater number and range of all facilities − sports centres, a university, factories, and so on.

5 **Conurbation** − this is a massive urban area with over 1 million people. It may have grown up as one huge area like London or may have involved the joining together of several nearby towns and cities, like Greater Manchester. Figure 2.2 shows the seven conurbations in Britain.

Figure 2.2 Conurbations

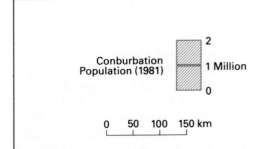

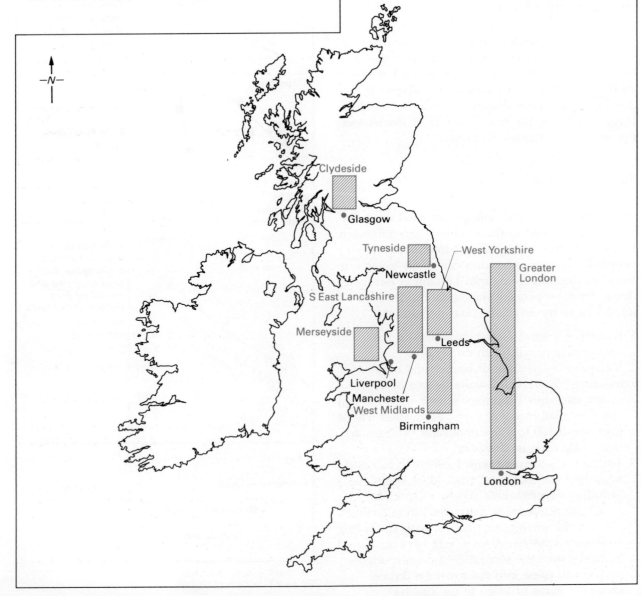

Special towns and cities

Some towns have grown up to have special characteristics and functions.

There are **resort** towns at the coast or high up in the mountains for skiers to use in the winter. There are **spa** towns like Cheltenham and Bath famed for their healing 'spa' water. Many small towns can be called **market** towns as they have important markets where local produce is bought or sold.

Commuter or **dormitory** settlements contain a high proportion of people who live but do not work there. Many small towns around the major conurbations are commuter settlements. **New towns** are newly planned, modern settlements like Milton Keynes aimed at providing homes and employment for people who would otherwise have to live in the conurbations.

Figure 2.4

Figure 2.3 A settlement in the UK

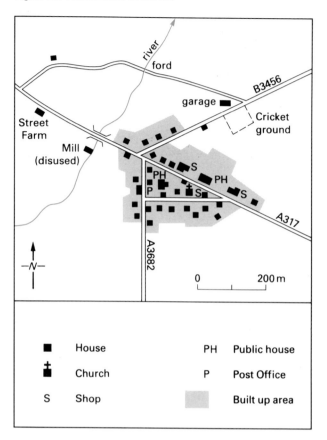

■	House	PH	Public house
✝	Church	P	Post Office
S	Shop		Built up area

ACTIVITIES

1 Study the settlement in Figure 2.3.

   a Approximately how large is the settlement (m²)?

   b What facilities are available in the settlement?

   c What type of shape is the settlement (see Figure 2.1 for help)?

   d Where and why would you expect to find the oldest and the newest buildings?

   e Is the settlement a hamlet, a village, or a town? Explain your decision.

2 Turn to the OS map extract of Worcester (see back cover).

   a Locate Grimley (8360). Make a simple sketch of this settlement. Comment on its shape and size.

   b Repeat **a** for the settlement of Newland Common (9060).

   c Worcester is a city yet Droitwich is only a town. Use evidence from the map to compare the two settlements. (Hint: look at the size, types of facility, etc.)

3 Study the photographs of settlements in Figure 2.4. For each one, describe its shape and size. Try to use the correct terms.

3 Settlement growth and decline

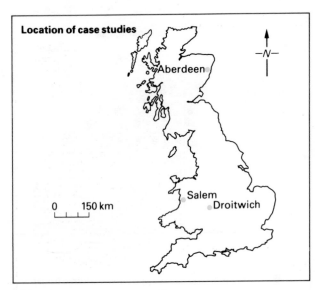

Figure 3.1 Location of case studies

Settlements rarely remain stagnant — without change. People, shops and industry come and go causing growth in some settlements but decline in others.

In Britain the trend has been for villages to decline as their occupants have migrated to towns and cities. It is not quite as simple as this for those who can afford the cost of travelling to work often choose to live in more attractive villages on the outskirts of towns.

The most likely reason why a settlement attracts people is because it offers better opportunities. These could include jobs, better schools and colleges, a wider range of shops, more entertainment, and so on. A settlement may decline if a major employer closes, if it appears less attractive than other nearby settlements, or if there has been a natural disaster of some kind.

To study settlement change we will look at three case studies in Britain (Figure 3.1). Droitwich, the town on the map extract at the back of the book, will serve to show the gradual historical growth common to most towns in Britain. Aberdeen is a town which witnessed sudden growth in the 1970s with the exploitation of North Sea oil but has more recently declined slightly. Finally, the little village of Salem in Wales is an example of a mining village which has declined greatly.

Droitwich

A settlement has existed at Droitwich since prehistoric times due to the deposits of salt which was a very important commodity at the time. The Romans named the settlement Salinae (the place of salt) and built a fort there.

The salt trade flourished through the Middle Ages with many roads being built to transport the salt around the country. The Droitwich canal was built in 1771 to bring in coal which was used as fuel in the salt industry. Early development of the town was along the Salwarpe valley where most of the salt was found.

In the mid-nineteenth century salt was discovered to have medicinal powers and Droitwich grew as a Spa town. Many hotels were built to cater for patients suffering from all sorts of ills. Droitwich then grew up on the higher ground to the south. Growth also occurred west towards the railway station.

Salt is no longer worked at Droitwich and the brine (Salt) baths are now only used for recreational purposes. Droitwich today is a pleasant market town of some 20 000. It has been designated an **expanding town** and has recently absorbed overspill from the Birmingham area (Figure 3.2).

Figure 3.2 Droitwich: population growth 1951—1986

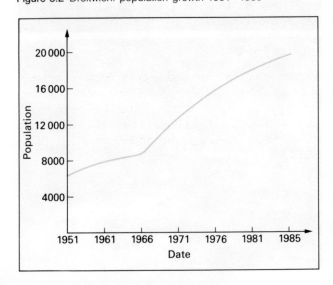

Aberdeen

Aberdeen is located on the north-east coast of Scotland. Since the 1960s there has been tremendous change in the city − Figure 3.3 shows why − OIL!

Before 1960, Aberdeen was a declining manufacturing city of some 200 000 people. Traditional industries such as fishing, textiles and food processing were closing down and unemployment rates were very high.

However, in the 1960s oil was discovered in the North Sea and Aberdeen's location suddenly became ideal as a centre for oil exploration. Oil companies established offices and many new jobs were created serving the rigs and the men. Aberdeen soon became the administrative HQ for the oil industry and virtually overnight it became something of a boom town.

The results of this good fortune have been varied:

1 the population has increased by 6000 since 1976;
2 employment has risen from 94 700 in 1971 to 145 900 in 1984;
3 since 1972, 395 000 square metres of new office space has been built;
4 new industrial estates have grown up mostly used by oil companies and oil service companies;
5 several new shopping areas and superstores have been built to cater for the increase in population and increased prosperity;
6 13 000 new houses have been built since 1976;
7 Aberdeen airport is now the fastest growing international airport in the UK;
8 £25 million has been spent on modernising the harbour, now one of the best equipped of its kind in Europe;
9 house prices have risen dramatically.

Unfortunately there has recently been something of a turn-around in Aberdeen's prosperity. The fall in the price of oil in 1986 caused a slight decline − about 6000 jobs in the oil industry were lost in that year. The situation is expected to stabilise in the 1990s.

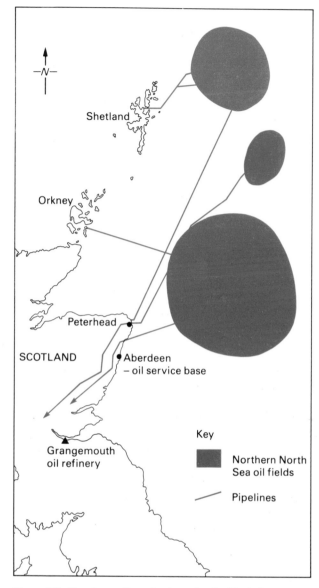

Figure 3.3 Aberdeen: oil service base for the North Sea

Salem

Today Salem is a tiny and rather derelict village set amidst hills some 12 km north-east of Aberystwyth in Wales. It grew up in the early nineteenth century as a mining village − lead mining took place in nearby valleys. All the mines have long since closed and the village has subsequently declined. Some houses have become derelict and the large chapel, itself an indication of earlier importance, now holds only one service a month. Only the English, buying cheap property for holiday use, keep the village going.

ACTIVITIES

1 Figure 3.4 maps the early growth of Droitwich.

  a Describe the changes that took place between 1831 and 1903.

  b Make a copy of Figure 3.5.

    Use the map extract (see back cover) to map the present day extent of Droitwich. Draw on bold arrows to show the main directions of growth — one towards the railway station and another southwards up the hill.

2 Study the information in Table 3.1.

  a Present the information as a series of bar charts or pie diagrams.

  b Describe and try to explain the trends in employment between 1976 and 1986.

  c Table 3.2 shows the helicopter passenger traffic using Aberdeen airport 1974–1986. Show this information as a line graph. Label the period of most rapid growth and the recent decline.

Table 3.1 Employment in Aberdeen district 1976–1986

Jobs by Sector for the District

	1976	1981	1984	1986
Armed Forces	350	410	450	–
Primary	5,590	13,190	22,460	22,100
Manufacturing	27,220	26,690	24,710	23,830
Construction	9,520	8,720	14,160	12,570
Services	71,210	83,100	84,120	86,120
Total for District	113,890	132,110	145,900	144,620

Source: City of Aberdeen District Council

Table 3.2 Helicopter passenger traffic at Aberdeen airport 1974–1986

Date	Helicopter passengers
1974	89,100
1975	141,200
1976	184,100
1977	225,300
1978	220,500
1979	222,600
1980	309,800
1981	379,100
1982	501,200
1983	644,300
1984	665,900
1985	608,300
1986	500,000

(Source: Aberdeen District Council)

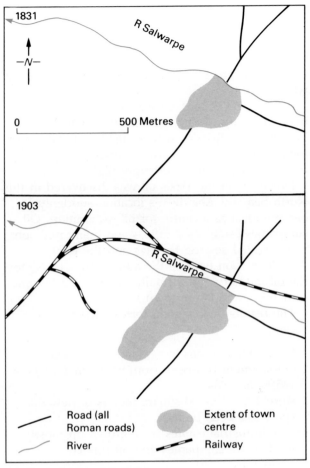

Figure 3.4 Growth of Droitwich

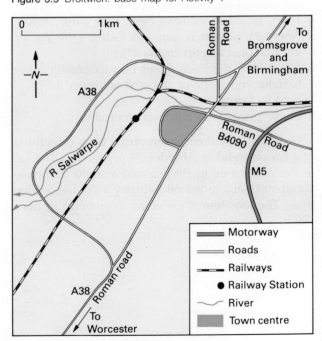

Figure 3.5 Droitwich: base map for Activity 1

Figure 3.6 Aberdeen

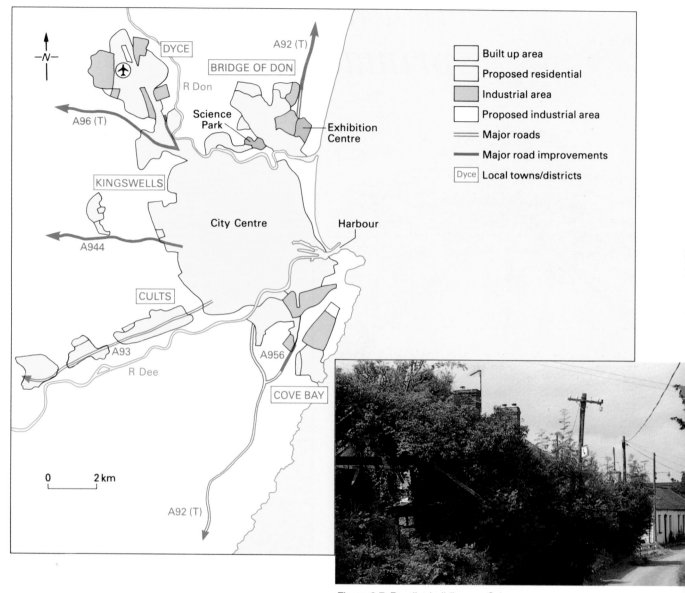

Legend:
- Built up area
- Proposed residential
- Industrial area
- Proposed industrial area
- Major roads
- Major road improvements
- Dyce Local towns/districts

Map labels: DYCE, BRIDGE OF DON, A92 (T), R Don, A96 (T), Science Park, Exhibition Centre, KINGSWELLS, City Centre, Harbour, A944, CULTS, A93, R Dee, A956, COVE BAY, A92 (T), 0 2 km, N

Figure 3.7 Derelict buildings – Salem

Figure 3.8 Chapel

ACTIVITIES

2 d Study Figure 3.6. What evidence is there on the map to suggest that Aberdeen is growing?

 e Although a great many people in Aberdeen have benefited from the recent growth, there have been some disadvantages. Try to think of some problems that would have arisen as the city grew and prospered. Are there any groups of people who might not be so pleased?

3 Figures 3.7 and 3.8 show the village of Salem – the photographs were taken in 1987. What evidence is there on the photographs that Salem used to be a more important and thriving village than it is today?

4 Urbanisation and primate cities

Figure 4.1 Reasons for moving to the city

Urbanisation can be defined as the growth of urban areas. In the early nineteenth century less than 10% of the world's population (about 100 million) lived in towns and cities. Now about 45% (about 2250 million) live in urban areas. By the year 2025 this is expected to rise to 60% (about 4920 million).

Until the 1940s urbanisation was most rapid in the developed world. Since then the most rapid growth has been in the Third World. As 2/3 of the world's population live in the Third World, this represents a huge number of people.

The growth of urban areas in the Third World results from:

1 Rapid population growth within cities;
2 Rapid in-migration from rural areas (see *Population* book).

It is the second factor that is most important. Figure 4.1 explains why people move to the cities.

The rapid growth of Third World cities is leading to great problems of overcrowding. There are not enough houses so that newcomers are forced to build their own; water and sanitation systems are unable to cope; and unemployment is becoming a serious problem. These problems will be studied in forthcoming chapters.

ACTIVITIES

1 Table 4.1. shows the trends in urbanisation in different parts of the world. Plot line graphs to show this information – one has already been done in Figure 4.2. Use a different colour for each region.

a Which parts of the world are most urbanised?

b Which parts of the world are least urbanised?

c Up until 1990, which part of the world is most urbanised?

d From 1990, which part of the world looks likely to be most urbanised?

e In which parts of the world has urbanisation slowed down since 1970?

f In which parts of the world has urbanisation been most rapid since 1970?

g Use your graph to predict the situation in 2010. Will the rank order be the same as in 2000? Which region(s) would be urbanising most rapidly? Would any be slowing down?

Table 4.1 Percentage of population living in urban areas

Region	1950	1960	1970	1980	1990	2000
Africa	15.7	18.8	22.5	27.0	32.7	39.1
S America	41.0	49.2	57.4	65.3	72.0	76.8
N America	63.9	69.9	73.8	73.9	74.3	74.9
E Asia	16.8	25.0	26.9	28.1	29.5	32.8
S Asia	16.1	18.4	21.3	25.4	30.2	36.5
Europe	56.3	60.9	66.7	70.2	72.8	75.1
Oceania	61.3	66.3	70.8	71.4	70.9	71.3
USSR	39.3	48.8	56.7	63.1	67.5	70.7

(Source: World Resources 1987)

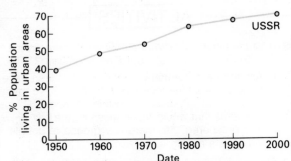

Figure 4.2 Growth of urbanisation in regions of the world

2 Table 4.2 gives population change in thirteen of the largest cities in the world. Show this information on a

world map using proportional bars. Locate each city and draw a bar to represent the estimated population for the year 2000 – depending on your map, a scale of 1 cm = 5 million might be suitable.

Now draw a horizontal line through the bar to show the 1985 population – you will have divided each bar into two. Shade the upper portion of each bar the same colour – this represents the increase in population between 1985 and 2000 and will show clearly which cities are growing most rapidly.

Give your map a suitable title and include the scale of your bars in a key. You might like to write the actual population figures alongside each bar.

Are the most rapidly growing cities in the developed or the developing world?

Table 4.2 Population in 13 world cities 1985/2000 (millions)

City	1985	2000
Tokyo/Yokohama	25.4	26.1
Mexico City	16.9	31.6
São Paulo	14.9	26.0
Seoul	13.6	18.7
Buenos Aires	10.7	14.0
Calcutta	10.4	19.7
Bombay	10.1	19.1
Rio de Janeiro	10.1	19.4
Los Angeles	9.6	15.0
London	9.4	12.7
Manila	8.4	12.7
Jakarta	8.1	16.9
Shanghai	6.6	19.2

(Source: United Nations Population Division 1987)

Primate cities

One common characteristic of urbanisation in the Third World is that often the vast majority of a country's urban population live in a single massive city. Such a city is called a **primate city** – a true primate city will have a population of more than twice that of the second largest city.

Primacy can be clearly seen in S America. The capital of Peru, Lima, is about eight times larger than the second city Arequipa; Santiago, capital of Chile, is seven times larger than the second city Valparaiso.

Primate cities tend to be located on the coast. They developed as ports during colonial times and expanded rapidly as people migrated from the countryside to work in newly established industries. The city holds a magnetic attraction and grows at the expense of other cities in the country.

Primacy can be a good thing. It concentrates resources (finance, trained labour, infrastructure, etc.) and enables economic progress to be made. However, some regard it as harmful. If resources are concentrated in a single centre, what happens to the rest of the country? Other towns and villages stagnate and decline. Also, what about shortages of houses and jobs for the thousands who move into the primate city?

Certainly primacy does lead to the development of a dynamic wealthy **core** area and a backward, economically stagnant **periphery**.

3 Table 4.3 shows the top ten largest cities in selected countries.

  a Which country has the most obvious primate city? What is the name of the city?

  b Are there any countries where primacy does not exist?

  c Is there a marked difference in primacy between developed countries and developing countries?

4 a Make a list of three advantages and disadvantages of the existence of primate cities.

  b You are the chief planning officer for a small Third World country which is developing rapidly. Already development is being concentrated in a single city. You have been asked by the Prime Minister whether the rapid growth of this single city should be allowed to continue. Write a paragraph giving your views – remember to give reasons for your decision.

Table 4.3 Populations of cities over 100 000 in rank order for several selected countries (thousands). (Mostly 1981.)

Rank	Algeria	Nigeria	S Africa	Zaire	Indonesia	Australia	Argentina
1	1740	1477	1726	2444	6503	3335	9927
2	543	847	1491	704	2028	2865	982
3	379	432	961	451	1462	1138	955
4	246	399	739	383	1379	969	597
5	224	282	585	339	1026	969	560
6	158	282	448	209	787	414	497
7	130	253	198	162	709	256	407
8	113	242	198	149	512	235	290
9	103	224	192	–	481	192	287
10	–	224	170	–	470	174	260

(Source: New Geographical Digest G Phillip)

5 Settlement functions

As you have already discovered in Chapter 2, there are a great many different types of settlement. Not only do they vary in their present day size but some are growing quicker than others. Some may be declining too.

In Chapter 2 we saw that different sized settlements had different types and amounts of facility on offer. These facilities are the settlement's **functions**. Generally, the larger the settlement, the greater the number and variety of functions offered. A settlement **hierarchy** exists with small villages at the bottom and large towns and cities at the top.

There are two broad types of function, **goods** and **services**. Goods are items such as bread, clothes and records, whereas services include schools, hospitals, sports centres, and so on.

It is possible to talk about **high** and **low order** goods and services. High order goods will be expensive items which are bought fairly infrequently, for example clothes and furniture. Low order goods are those items bought frequently, perhaps even daily, such as sweets, newspapers, and vegetables.

As low order goods are bought very regularly it does not need many shoppers to maintain the profitability of a shop. This is why villages tend to have grocers and bakers and why newsagents are very widespread in towns as they cater for a very localised population.

On the other hand, large department stores need a much larger number of potential customers as each customer will only buy goods very rarely. This explains why such stores are found in towns and cities where there is a large number of potential customers. It also explains why you don't tend to find branches of Marks & Spencer in villages!

The number of people required to make a business profitable is called the **threshold population**. The higher the order of good or service, the higher the required threshold population.

People travel to obtain goods and services. You may perhaps have travelled many miles to see a pop concert or a first division football match. It can be said that the pop group or football team exerts an influence over the surrounding area. This can be mapped — it is called the **sphere of influence**. The outer edge or **range** is the furthest distance people are prepared to travel. Again, the greater the order, the greater the sphere of influence.

ACTIVITIES

1 Pupils in a class decided to see whether there was any relationship between the number of functions and the size of settlement. To do this they each took a type of good or service and, using the 'Yellow Pages', discovered the total number offered in a selection of settlements. Their results are given in Table 5.1.

Table 5.1

Settlement	Population (1981)	No. of functions
Watford	74 462	96
Elstree	29 360 (1971)	6
Rickmansworth	18 319	43
Bushey	23 298	26
Abbots Langley	9 503	16
Aldenham	8 721	16
Chorleywood	8 471	12
Sarratt	2 598	2
Chipperfield	1 764	6

To see if there is any relationship a **scattergraph** needs to be drawn (Figure 5.1) plotting population (Y axis) against number of functions (X axis). Each settlement is plotted as a cross and, if a trend exists, a **best-fit line** can be drawn. If, as in Figure 5.1, most points are close to the best-fit line then we can say that there is a good relationship. With one set of values inceasing as the other set increases, we can call the relationship **positive**. Thus, on Figure 5.1 there is a good positive relationship between settlement size and number of functions.

a Plot the information in Table 5.1 as a scattergraph. Give it a title and remember to label the axes. Draw on a best-fit line (your teacher will help you if necessary). Describe any relationship that exists.

b There may be one or two points well away from the best-fit line. These are called **residuals** (see Figure 5.1). Label any residuals on your graph. Discuss some possible reasons why they are residuals. (Hint: a small settlement with many functions might be wealthy or it might be a very long way from another settlement. A large settlement with relatively few functions may be a declining settlement or perhaps have many competing settlements nearby)

You can, of course, do a similar activity for the settlements in your own local area.

ACTIVITIES

2 A group of pupils in Great Malvern, Worcestershire
   wanted to find out the spheres of influence of
   certain functions in their town. To do this they visited
   offices and shops to find the range of 4 functions —
   a local coach company, a carpet shop, a grocery
   shop and the circulation of the local newspaper.
   Their results are shown in Figure 5.2.

   a Which function has the largest sphere of
      influence?

   b Is the function identified in a low or high order?

   c Which function has the smallest sphere of
      influence?

   d Is the function identified in c low or high order?

   e What is the maximum range of the local
      newspaper and which is the furthest town served
      by the newspaper?

   f Why do you think the local newspaper circulation
      does not extend as far as Hereford?

   g Do the physical barriers of the River Severn and
      the Malvern Hills affect the shapes of the spheres
      of influence?

3 Figure 5.3 shows spheres of influence for leisure
   activities on offer in Torbay.

   a Which activity is the lowest order and which the
      highest?

   b Do you think there is a zoo in Exeter or Plymouth?
      Explain your answer.

   c Do you think there is a cinema in Exeter? Explain
      your answer.

   d There is a theatre in Exeter. How, therefore, can
      you account for the Torbay sphere extending
      beyond Exeter?

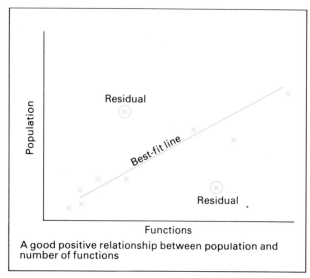

A good positive relationship between population and
number of functions

Figure 5.1

Figure 5.2 Spheres of influence around Great Malvern

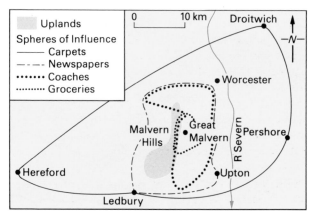

Figure 5.3 Spheres of influence around Torbay, Devon

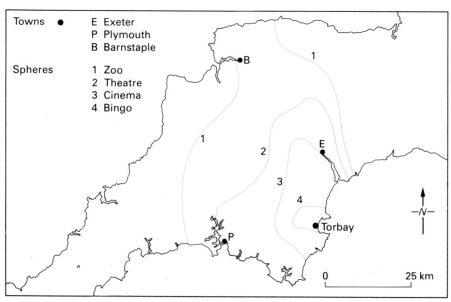

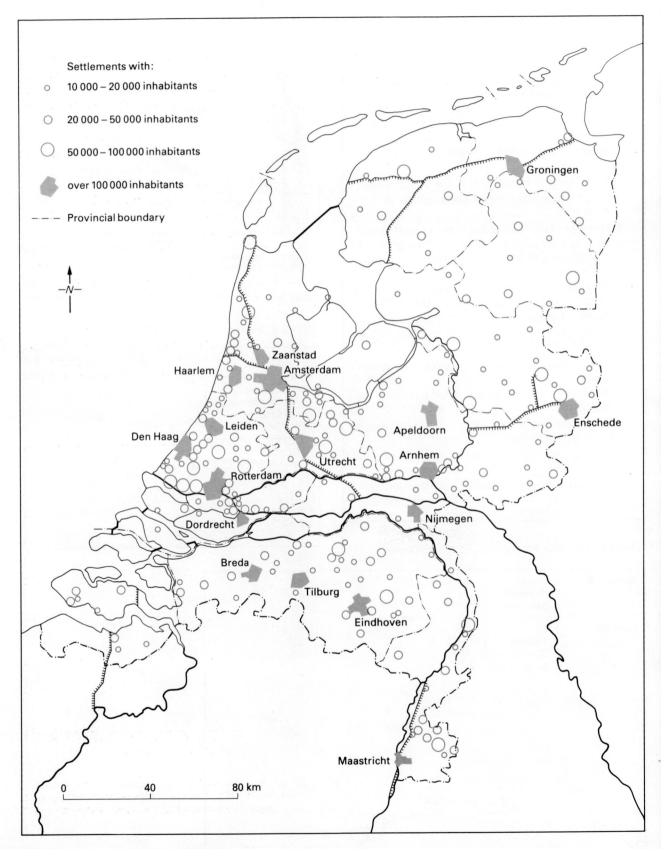

Figure 5.4 Settlement size, The Netherlands

ACTIVITIES

4 Make a study of settlement patterns and hierarchies in the Netherlands. Figure 5.4 plots the location of different sized setlements in the Netherlands. We can call the smallest settlements (10 000–20 000 inhabitants) 1st order settlements. The next size up (20 000–50 000 inhabitants) will be 2nd order, and so on.

a In pairs, carefully count the number of each order of settlement. Combine your results and take an average. Show your results in a table.

Now produce a graph with 'Number of Settlements' on the Y axis and 'Order' on the X axis. Once you have plotted the 4 points, draw a **best fit line** to show the trend.

i As order increases, does the number of settlements increase or decrease?

ii Try to explain your answer

b Trace a grid of squares 0.5 cm × 0.5 cm. It must be large enough to be laid over Figure 5.4. Now lay your grid over Figure 5.4 and count the number of 1st order settlements (10 000–20 000 inhabitants) in each square. Record the number in pencil in the squares.

Devise a colour key where the higher the number of settlements the darker the colour. Now shade each square the appropriate colour.

i Is there an even spread of 1st order settlements in the Netherlands?

ii With the help of your atlas, discuss with a friend some possible reasons for the spread of 1st order settlements. Try to explain why there may be many in one part of the country but few in another part.

Coursework ideas

1 Plotting spheres of influence for a range of different functions can make a very successful project. It is, however, important to construct a good questionnaire to ensure that the right sort of data are obtained.

Once spheres have been drawn it is interesting to consider their size and shape. Transport routes, physical barriers, and competition can all be influential and should be superimposed on any map drawn. It is important to try to explain the spheres and to look for geographical relationships.

2 You could look at settlement hierarchies in your home area. Try to order the settlements using population data and number of functions. You will need to carry out fieldwork to discover the number and types of function in each settlement. Your teacher will show you how to produce a **functional index** for each settlement. They can then be plotted on semi-log graph paper and different orders can be suggested (see Figure 5.5).

Figure 5.5 Discovering settlement orders

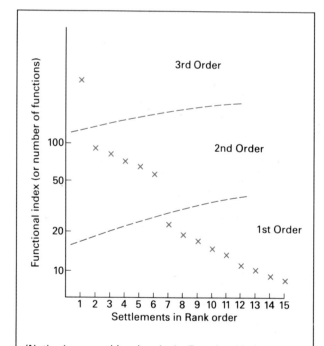

(Notice how a sudden drop in the Functional Index can be used to indicate the change from one 'order' to another.

In this way, it is possible to 'order' a group of settlements)

6 Shopping

Types of shop

Figure 6.1 shows the shops in a typical town high street. Notice that there are a range of different sized shops and that a variety of items are for sale. You will recognise the names of some shops such as Next as these are commonly found in most towns. These shops are called **chain stores**. The other shops are owned by individuals rather than large companies.

The shops in Figure 6.1 are **comparison** shops — people looking to buy a video recorder will compare prices in different shops and buy the best value item. Such shops selling, for example, furniture, carpets, and clothes are **high order** shops. They have a high turnover and are able to afford the high rates charged in town centres. Other shops are thought of as local shops — these include newsagents and greengrocers. They are **convenience** shops selling small items fairly frequently to local residents. Turnover is much less than the high street stores and so they tend to be found outside the town centre where rates are cheaper. They are **low order** shops.

Figure 6.1 A town centre street

Recent trends in shopping

Traditionally, shops located in town centres as these were seen to be the most accessible places. Recently, however, there has been a trend to locate shops outside town centres where there is more space for parking and where the rates are cheaper. As most families have a car nowadays, these shops can be easily reached.

Shops have grown in size too. We are now used to **supermarkets** where all food items can be bought under the same roof making shopping much easier. Supermarkets have massive turnovers of food so enabling food to be bought in bulk at cheaper rates. This food can then be sold more cheaply to customers undercutting smaller traders.

Figure 6.2

Massive **hypermarkets** (Figure 6.2) sell all sorts of different items under a single roof including food, clothes, records, gardening equipment, and so on. Clearly such huge shops require vast amounts of land both for the building and for car parking and deliveries. They need to be located in accessible places which is why they tend to be located near to motorways or major roads.

Large shopping centres have been developed in out-of-town locations. Such centres offer a range of different shops mostly chain stores all located together under one roof. Brent Cross near the start of the M1 in North London was the first such centre built in 1976. Recently a huge centre called MetroCentre opened on Tyneside (Figure 6.3).

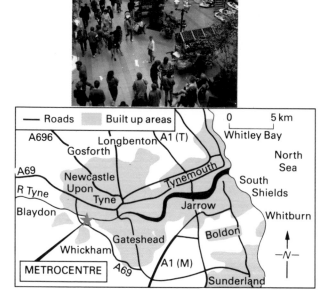

Figure 6.3 Tyneside

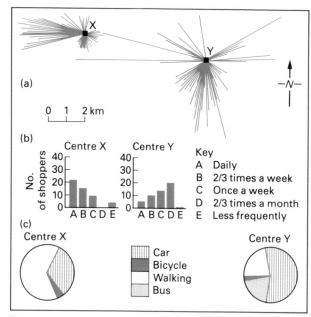

Figure 6.4

Metrocentre — facts

* The biggest drive-in-centre in Britain
* 7600 free car parking spaces
* 140 shops including Marks & Spencer and House of Fraser
* 1.3 million potential customers within 30 minutes travel time
* 6000 new jobs created
* £250 million invested by 1990
* 10 screen cinema and indoor 'fun palace'

Comments by John Hall, the man behind the scheme:

'Today we are seeing fundamental changes in society. People are buying different goods from those they used to and they are buying them in a different way. They do it now by motor car and the city centre does not cater for the motor car. There isn't the space.

'The MetroCentre is not just about shopping. It is about how people spend their leisure time. What I have tried to do is bring the outdoors indoors, to create the sort of atmosphere you get in a square in a town in Portugal. Here you can sit in the garden court, you can stroll, you can have something to eat, ...

(Information from Terence Bendixson 'Queuing up for the retail revolution' in *The Times* 13/1/87).

ACTIVITIES

1 a Give **three** examples each of **comparison** and **convenience** shops.

   b Imagine that one of the shops in Figure 6.1 became vacant. Give an example of a likely shop that would take over. What type of shop would you not expect to find?

2 Figure 6.2 shows a newly built ASDA store in Nuneaton.

   a How do you think most people travel to the store?

   b Do you think the photograph was taken on a Tuesday morning, a Thursday evening, a Saturday afternoon?

   c What factors do you think ASDA considered when locating this store?

3 Make a copy of the map in Figure 6.3 that shows the location of the MetroCentre in Newcastle. Imagine that you are a business tycoon wanting to locate a similar centre on Tyneside.

   a Show the location of your centre on your map.

   b Write a newspaper advertisement for your centre. Remember that you will have to try to attract people away from MetroCentre. Your centre will have to offer something special.

4 Figure 6.4 shows the results of a shopping survey. Fifty shoppers at two different shopping centres were asked where they had come from (Figure 6.4a), how often they shopped at the centre (Figure 6.4b), and how they had travelled (Figure 6.4c).

   a Study Figure 6.4a.

      i Which centre X or Y represents the town centre and which represents a local centre? Explain your answer.

ACTIVITIES

ii For centre X, from which direction did people travel furthest?

iii For centre Y, from which direction did most people travel?

iv Describe how **road networks, physical barriers** (like rivers) and **alternative centres** may influence the patterns shown in figure 6.4a.

b Study Figure 6.4b.

i Describe using actual figures the frequency of shopping visits to the two centres. Try to account for your observations.

c Study Figure 6.4c.

For each centre approximate the percentage of shoppers who travelled by **car** and by **walking**. Comment on your results. Are they as you had expected?

d At which of the centres X or Y would you expect that most shoppers had bought food, sweets and newspapers? Why?

5 Turn to the map extract at the back of the book.

a Locate grid square 8657. Draw an enlargement of this square and mark on the roads, railway and canal. Also mark on the location of three recently built retail outlets:

i 865574 Harris Queensway (furniture)

ii 863570 St Oswalds Park (Texas, MFI, Allied Carpets)

iii 866572 J Sainsbury plc (food)

b What do you think attracted these firms to the Blackpole district?

Coursework ideas

A shopping survey is a very good and popular choice for a project. It is relatively simple to carry out and it is quite easy to obtain a considerable amount of information. Here are three possible themes which you may care to explore:

1 Carry out a shopping survey questionnaire (Table 6.1) in one or more centres. It is a good idea to choose centres of different sizes so that you can compare and look for **hierarchies** (p. 14). Data can be presented in a similar way to Figure 6.4. Remember to try and *explain* your results as well as describing them.

2 Carry out a survey of different types of shop in your town. Obtain a base map which shows the outline of each separate shop and use Table 6.2 to colour each shop according to its type. Describe the pattern. Do shops of the same type locate close together? Do certain shop types

occur on the outskirts or in the centre? A statistical technique called **nearest neighbour analysis** could be used — ask your teacher for guidance here.

Table 6.1 Questionnaire for Shopping Survey

1 Goods shopped for	2 Origin of shoppers	(list below villages or towns, or roads)
Groceries:		
Chemist:		
Post Office:		
Pet supplies:		
Electrical:		
Clothes/shoes:		
Household goods:		
Fast food:		
DIY:		
Furniture:		
Newspapers/sweets/cigarettes:		
Laundrette:		
Financial:		
Others:		

3 Frequency of visits	4 Mode of transport
Daily:	Walk:
2/3 times week:	Car:
Once a week:	Bus:
2/3 times month:	Cycle:
Less often:	Motorbike:
	Lorry
	Train:

Table 6.2 Shop types

1 Convenience shops — e.g. baker, butcher, newsagent, chemist. (yellow)
2 General shoppers goods — e.g. shoes, clothes, hardware. (red)
3 Specialist shops — e.g. jewellery, antiques, books, computers. (orange)
4 Autosales — e.g. garages, spares. (purple)
5 Recreation — e.g. cafes, pubs, cinema. (green)
6 Legal/finance — e.g. banks, estate agents. (blue)
7 General services — e.g. laundrette, hairdresser, travel agent. (brown)
8 Chain stores — e.g. Marks & Spencer, Boots (black)
9 Vacant properties — (blank)

3 Study change in shops. A **GOAD** map of your town will show the types of shop a few years ago. It is interesting to compare this with the present day. What types of shop have gone? Which types have replaced them? Where have most shops closed — in the centre or on the outskirts?

There may be a local proposal for shopping change such as a new centre or new supermarket. This would make a good study.

7 Urban land use

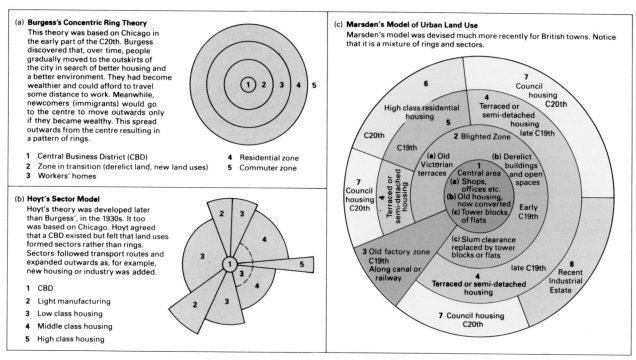

(a) Burgess's Concentric Ring Theory
This theory was based on Chicago in the early part of the C20th. Burgess discovered that, over time, people gradually moved to the outskirts of the city in search of better housing and a better environment. They had become wealthier and could afford to travel some distance to work. Meanwhile, newcomers (immigrants) would go to the centre to move outwards only if they became wealthy. This spread outwards from the centre resulting in a pattern of rings.

1 Central Business District (CBD)
2 Zone in transition (derelict land, new land uses)
3 Workers' homes
4 Residential zone
5 Commuter zone

(b) Hoyt's Sector Model
Hoyt's theory was developed later than Burgess', in the 1930s. It too was based on Chicago. Hoyt agreed that a CBD existed but felt that land uses formed sectors rather than rings. Sectors followed transport routes and expanded outwards as, for example, new housing or industry was added.

1 CBD
2 Light manufacturing
3 Low class housing
4 Middle class housing
5 High class housing

(c) Marsden's Model of Urban Land Use
Marsden's model was devised much more recently for British towns. Notice that it is a mixture of rings and sectors.

Figure 7.1 Theoretical patterns of urban land use

There are many different types of land use in settlements − industry, housing, shops and open spaces just to mention four. These different land uses tend to cluster occurring as **zones** within the settlement. Figure 7.1 shows some theoretical patterns of urban land use. It is interesting to consider the reasons why certain land uses are found in certain areas.

Older industrial areas are often located alongside railways and canals − they would have relied on these forms of transport. Rivers were also early sites for industry with their water powering factory machinery. More modern industry, usually in the form of **industrial estates**, tends to be on the outskirts where there is plentiful and relatively cheap land. Such industries rely on road transport.

The oldest housing will tend to be in the centre of settlements although much may have been demolished in the process of urban planning. As a settlement grows, new housing is built on the edges where there is space. This means that many settlements today show a pattern of gradually newer housing outwards from the centre. Sometimes, when space is created in the centre perhaps after an industry has been demolished, housing will replace it − this is called **infill**.

The main shopping and business area tends to be in the centre − the point, in theory at least, of maximum accessibility. It is called the **Central Business District** (CBD) and will typically contain large chain stores, estate agents, banks, and offices. Other shopping areas will be found either along main roads into the settlement or in 'parades' within housing estates.

Open spaces (woods, recreation grounds, picnic spots, etc.) are usually fairly evenly spread across the settlement generally close to housing areas. They may occupy river floodplains, canalsides or hilltops.

Today there are restrictions on building. All new developments and even conversions have to be granted **planning permission** from the local authority. Before permission is granted there is frequently a lengthy review of the likely consequences of development − not all people might be in favour! For example, a new shopping arcade may bring jobs and more customer choice, but it may cause other businesses, faced with competition, to suffer and it may also lead to greater traffic congestion.

ACTIVITIES

1 The following questions refer to Figure 7.2 and the map extract at the back of the book.

  a What else apart from shops are in the 'main shopping centre' to the east of the River Severn?

  b Why do you think another 'main shopping area' exists in the St Johns area of Worcester?

  c How many 'other shopping areas' are there? Describe their distribution.

  d Use the map extract to name the areas of public space marked on Figure 7.2. Why do you think one area has been kept as an open space and has not been developed in any other way? (Hint: think about the river!)

  e Use the map extract to identify some of the 'other uses' on Figure 7.2.

  f Use the map extract to approximate the size of the built-up city of Worcester – remember that

each grid square is 1 km². Now work out the area available for expansion. Express this as a proportion of the present area of the city.

  g Locate the areas of industry on Figure 7.2. Describe and account for its location. Use the map extract to refer to specific districts of Worcester.

  h Imagine that you needed 1 km² of land to locate a modern factory making component parts for artificial limbs. You have been given planning permission to locate anywhere within an 'expansion area'. Describe your preferred location giving reasons for your decision.

  i To what extent does the pattern of land use in Worcester resemble any of the theoretical patterns in Figure 7.1?

(See Chapter 10 for an activity examining the pattern of housing in Worcester)

Figure 7.2 Worcester: land use

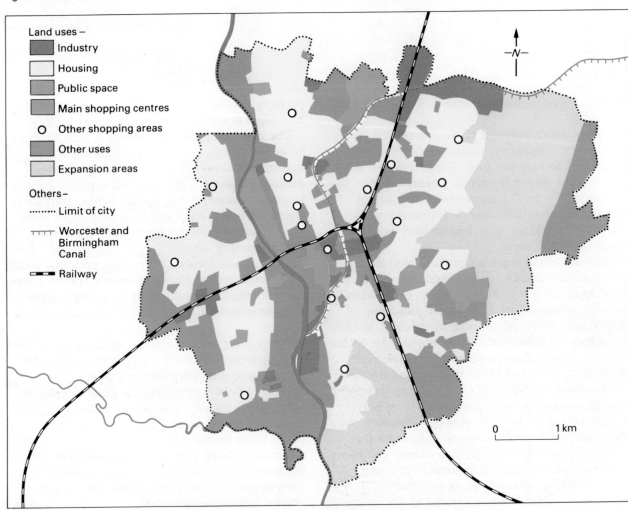

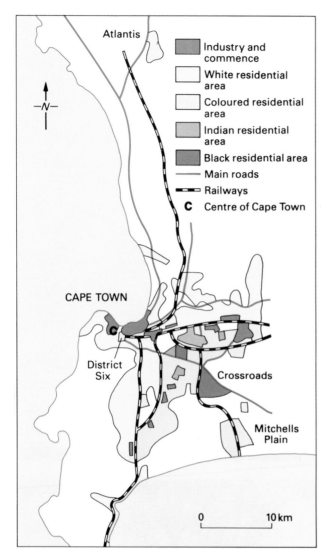

Figure 7.3 Urban land uses in Cape Town

Legend:
- Industry and commence
- White residential area
- Coloured residential area
- Indian residential area
- Black residential area
- Main roads
- Railways
- **C** Centre of Cape Town

Atlantis

CAPE TOWN

District Six

Crossroads

Mitchells Plain

0   10 km

Figure 7.4 District Six, Cape Town

ACTIVITIES

2 Figure 7.3 shows the pattern of land use in Cape Town, South Africa. Notice that there are distinctly separate housing areas for different races. This is largely due to the Group Areas Act of 1950 and the policy called **Apartheid** (racial segregation). Different 'townships' were built for different racial groups. They were fenced off and badly serviced. Shanties grew up on their outskirts as people continued to flock in from the countryside. One area. District Six (Figure 7.4), used to be a thriving coloured community. In the mid-1970s it was declared 'white' and was bulldozed with the previous occupants being moved to distant townships such as Atlantis and Mitchells Plain.

  a How far have the coloureds from District Six been moved to the townships of Atlantis and Mitchells Plain?

  b List in rank order (in terms of size) the different residential areas.

  c Is there any pattern in the location of the four different residential areas?

  d Comment on the location of areas of industry and commerce.

Coursework ideas

The plotting of land uses in a settlement be it a town or a village is relatively straightforward.

First you need to obtain a base map onto which you record the land uses. The scale of the map depends on how large an area you intend to use — your teacher will help you here.

Next you need to decide on what categories to use. Throughout this book you will find many maps to give you some ideas. It is entirely up to you what to do but remember that ideally you should have somewhere between 4 and 10 different categories.

You then need to do the leg-work collecting your data. Record the different land uses as you walk along roads in your designated area.

Back in the classroom you can produce a neatened version of your map.

Is there any pattern of land use (Figure 7.1)? Can you explain the location of each type of land use? Are there any areas due to be developed and for what purpose?

8 *Village life and rural issues in Britain*

Figure 8.1 Village shop closure

The problems of villages

In 1981 only 10% of the population of Britain lived outside towns and cities — in 1881, the figure was 33%. Although many villages are happy, thriving communities, some do have to face problems:

1 Steady decline in traditional rural employment particularly on farms as machines have taken over.
2 Reduction in population has led to the closure of shops (Figure 8.1).
3 Reduction in transport services. Railway stations are rare in villages and bus services are generally infrequent.
4 Only larger villages have a secondary school and many primary schools are being closed down.
5 Transport routes have remained unimproved — windy narrow roads make villages difficult to reach (Figure 8.2). Journeys are long and expensive.
6 Limited access to recreational and leisure facilities. This particularly affects the young.

Many young people, in the face of such problems, move out of villages and head for the towns. This movement leads to a further decline in services as the village population drops. The elderly residents left behind, who have probably lived in the same village all their lives, do not wish to move. They have to suffer the disappearance of shops, reduction in bus services, and general depression of a declining settlement.

Recently, the village population has shown a slight increase particularly those villages near to towns and cities. Urbanites, fed up with the noise and congestion of city life, move out to the countryside to live in more pleasant surroundings. This influx of 'townies' causes a series of other problems:

1 As demand for property increases so house prices rise. This makes it difficult for young villagers to buy a house in their own village.
2 Newcomers commute to work and are unlikely to use local shops or bus services.
3 Newcomers often upset the local population by suggesting changes in village life. They may also complain about typical country characteristics — the smell of pigs, early morning cockerels, etc.

Figure 8.2 Remote village

Solutions to the problems of villages

The Government has established a body to co-ordinate policy for the rural areas. The Development Commission for Rural England received over £22 million from the Government in

1986/7 which was then made available to rural areas for a variety of projects.

The Commission has identified those areas in greatest need. These **Rural Development Areas** are shown on Figure 8.3. It is these areas that receive the greatest help in the form of grants. Some examples of the work of the Commission are described below.

Figure 8.3 Rural Development Areas

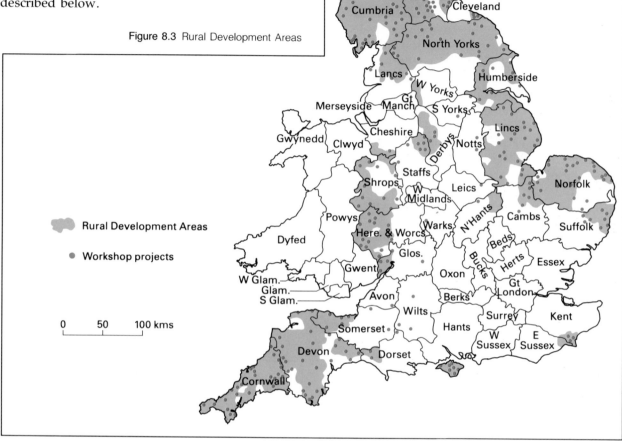

Figure 8.4

1 Industry. Grants have been used for constructing workshops. These have led to the creation of over 10 000 job opportunities and are used for small businesses in particular. Grants are available for the conversion and renovation of old buildings (Figure 8.4). Also, the Commission employs experts to give advice to small firms wishing to set up in rural areas.

2 Services. The Commission has been keen to promote community projects in order to maintain and improve services such as shops and schools.

3 Transport. Grants have been used to start up many small community transport schemes using buses and minibuses.

As you can see, the emphasis is on self-help and on community schemes.

Village study: Stetchworth, Cambridgeshire

Stetchworth is a village with a population of just over 600. It is located 5 km south of Newmarket in East Cambridgeshire – locate Newmarket in your atlas.

Figure 8.5 shows the layout and gives some details about the village. Stetchworth originated as an 'estate' village – the estate is to the north of the village. The owner of the estate would have housed his workers in the village and employed them on his land. He would have been responsible for the upkeep of the village. The estate workers' houses line the main village street.

Nowadays the estate has much less influence on the village. Few villagers are employed and many of the houses have been sold off. Nevertheless, some of the older inhabitants of the village still expect the estate to sort out village problems, such as overgrown verges.

Stetchworth and nearby villages were studied by the Cambridgeshire Community Council (all counties have such an organisation). They discovered that the villages were losing facilities and were stagnating – something was needed to invigorate them.

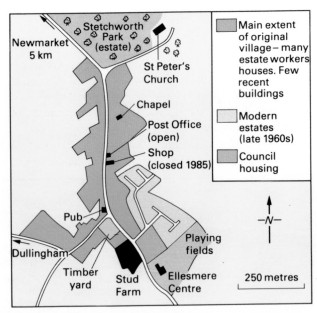

Figure 8.5 Stetchworth, Cambridgeshire

In 1984 the Ellesmere Centre was opened (Figure 8.6) in Stetchworth to serve the villages in the area. The centre, attached to the old village hall, has been designed as a multi-purpose centre where rooms can be used for a range of different activities. Figure 8.7 is a possible plan of the uses of the Centre.

Figure 8.6 Ellesmere Centre

ACTIVITIES

1 Study Figure 8.3.
   a Which parts of England have been designated as **Rural Development Areas**?
   b Which three counties have the greatest number of workshop projects?
   c Why do you think the Isle of Wight is a **Rural Development Area** with many workshop projects?

2 Study Figure 8.5.
   a Make a list of the facilities available in the village in 1988.
   b Try to account for the location of the Ellesmere Centre.
   c Whereabouts in the village are the council houses? Why do you think they were usually built on the edges of villages?
   d Make a list of the possible employees in the village.
   e List the uses being made of the centre (Figure 8.7). For each use, say which group of people (e.g. young, toddlers, young mothers, elderly, etc.) would benefit most. Suggest some other uses that could be made of the centre giving reasons for your choices.

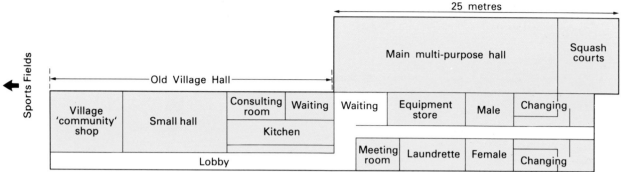

Figure 8.7 An example of the uses made of the Ellesmere Centre

---

ACTIVITIES

3 Table 8.1 gives the destination of workers from Stetchworth and Figure 8.8 shows the location of the destinations mentioned.

Table 8.1 Destination of workers in more than 8 hours per week paid employment (mini-census 1978)

Destination	No. of workers
Burrough Green	2
Dullingham	2
Stetchworth	53
Burwell	4
Cambridge	53
Newmarket	74
Haverhill	11
Rampton	1
Bury St Edmunds	5
London	2
No fixed place	31
Other	4

a Make a tracing of Figure 8.8 and use proportional lines (where the thickness of the line is drawn in proportion to the number of workers — e.g. 1cm = 10 workers) to show the flow of workers from Stetchworth.

b Briefly describe your map.

c To what extent would you say that Stetchworth is a commuter settlement? Describe some of the effects that such people might have on the village.

d What sort of jobs would those people who work in Stetchworth hold?

e In the census, 31 workers gave as their destination 'no fixed place'. What is meant by this?

f Put yourself in the position of an elderly resident of Stetchworth. You have lived in the village all your life. Describe how you would feel about:
  i The newcomers to the village who commute to work in Cambridge or Newmarket.
  ii The newly built Ellesmere Centre.

g Put yourself in the position of a newcomer to the village.
  i Why have you moved to Stetchworth?
  ii What do you think about the Ellesmere Centre?

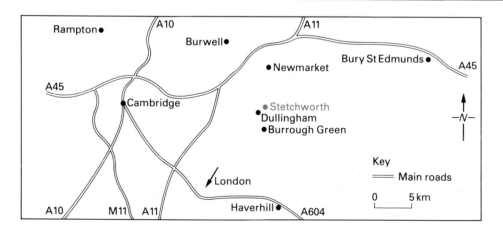

Figure 8.8 Destination of workers from Stetchworth (use with Activity 3 a)

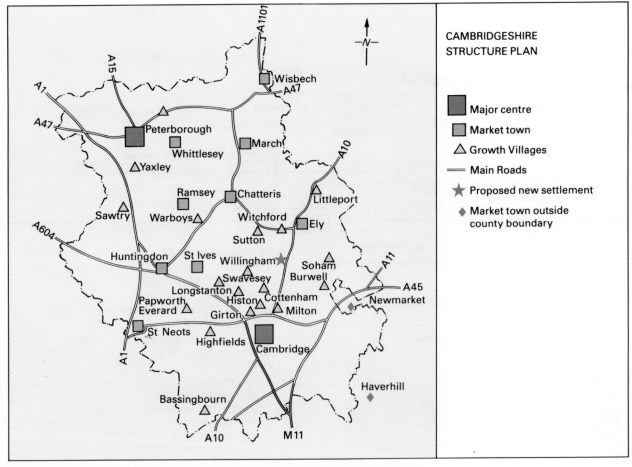

Figure 8.9 Cambridgeshire Structure Plan, 1987

ACTIVITIES

4 Figure 8.9 shows the Structure Plan for Cambridgeshire 1987. A Structure Plan details the policies of a County Council — it has to be approved by the Department of Environment.

Cambrigeshire is traditionally a rural agricultural county. Cambridge has recently become an important centre for hi-tech industry and many people have moved into the area. Also, with the M11 giving fast transport to London, south Cambridgeshire villages have witnessed the influx of many commuters. In contrast, the Fens remain very rural with villages generally showing no growth.

The Structure Plan 1987 identifies a number of villages where growth will be encouraged by less severe planning restrictions on development.

a How many villages have been identified as growth villages?

b Why do you think there are so few growth villages south of Cambridge?

c Why do you think there is such a concentration of growth villages in the Fens?

d Which villages do you think are intended to relieve the pressure on housing in Cambridge?

e Cambridgeshire County Council expects population to continue to rise in the Cambridge area. To accommodate these extra people a new settlement of between 2000 and 3000 dwellings could be established. Figure 8.9 shows the intended location of the new settlement. This site was chosen because it is close to the A10 and would hopefully spur development in this rural part of the county. The County Council will relax planning regulations allowing developers, like Wimpey and Barratt, to put up housing in the area. The Council will put in the necessary infrastructure (roads, mains supply, etc.).

Put yourself in the position of a public relations officer working for Cambridgeshire County Council. Your job is to produce a short press release about the proposed new settlement. The release would be made available to any company interested in developing the area. You should include a map and written account outlining the advantages of the location for development.

Coursework ideas

1 Carry out a survey of a village near to you to discover its character and whether it suffers from any of the problems discussed in this chapter. Figure 8.10 details a village survey. You may wish to change it slightly to take account of local aspects.

Figure 8.10

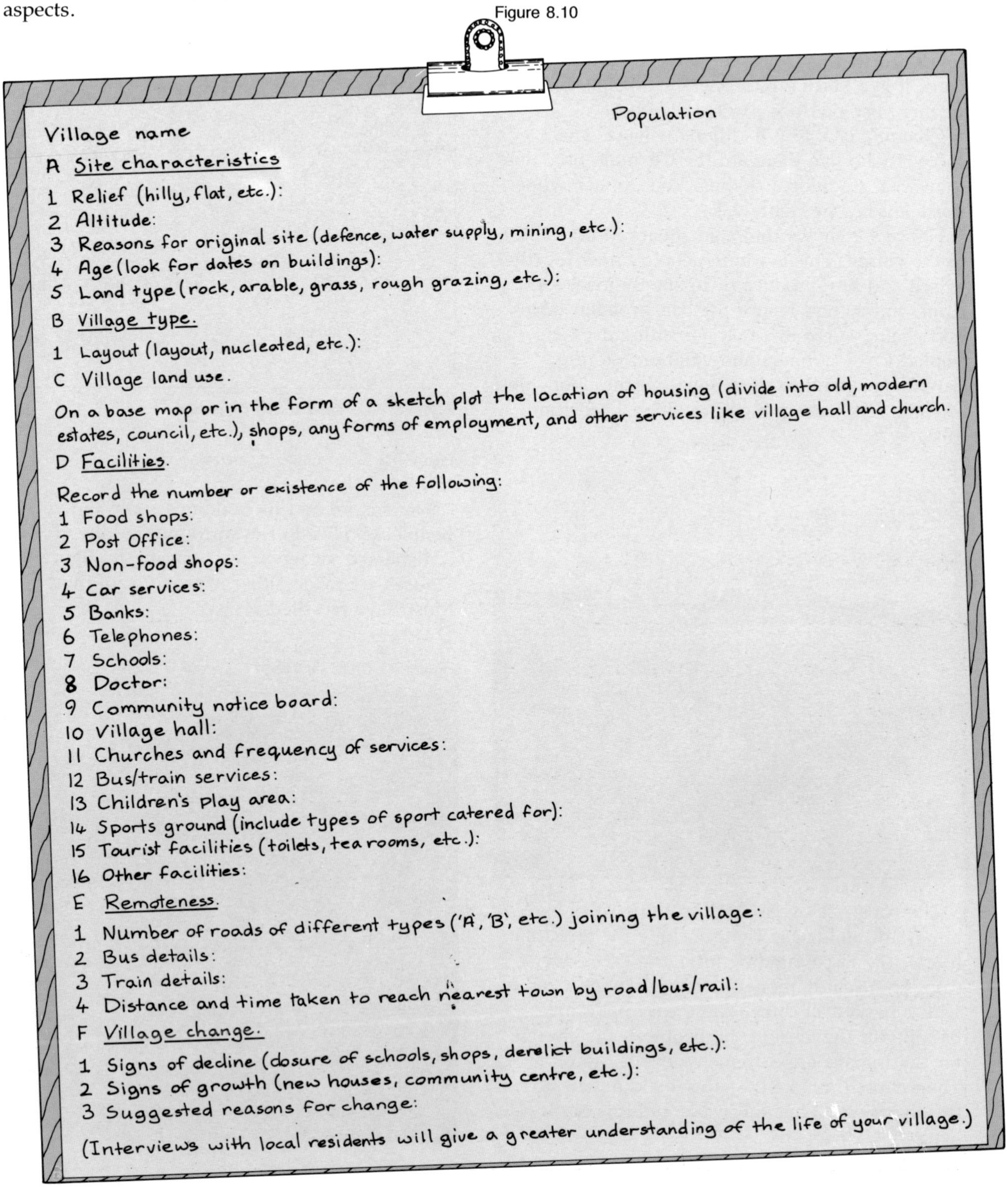

Village name   **Population**

A Site characteristics

1 Relief (hilly, flat, etc.):
2 Altitude:
3 Reasons for original site (defence, water supply, mining, etc.):
4 Age (look for dates on buildings):
5 Land type (rock, arable, grass, rough grazing, etc.):

B Village type.

1 Layout (layout, nucleated, etc.):

C Village land use.

On a base map or in the form of a sketch plot the location of housing (divide into old, modern estates, council, etc.), shops, any forms of employment, and other services like village hall and church.

D Facilities.

Record the number or existence of the following:
1 Food shops:
2 Post Office:
3 Non-food shops:
4 Car services:
5 Banks:
6 Telephones:
7 Schools:
8 Doctor:
9 Community notice board:
10 Village hall:
11 Churches and frequency of services:
12 Bus/train services:
13 Children's play area:
14 Sports ground (include types of sport catered for):
15 Tourist facilities (toilets, tea rooms, etc.):
16 Other facilities:

E Remoteness.

1 Number of roads of different types ('A', 'B', etc.) joining the village:
2 Bus details:
3 Train details:
4 Distance and time taken to reach nearest town by road/bus/rail:

F Village change.

1 Signs of decline (closure of schools, shops, derelict buildings, etc.):
2 Signs of growth (new houses, community centre, etc.):
3 Suggested reasons for change:

(Interviews with local residents will give a greater understanding of the life of your village.)

9 Village life in the third world

Village life in Chamula, Mexico

Chamula is an Indian village located in the mountains about 10 km from the southern Mexico town of San Cristobal (see Figure 9.1). In many ways it is a fairly typical Mexican village, being neither very rich nor particularly poor.

Chamula is a very traditional village. The Indians who live there are the Chamula tribe and they wear traditional clothes. You can see what these are like in Figure 9.2.

Figure 9.2 shows the main square in the centre of the village. This is where people gather together to sell and buy produce or to simply meet for a drink and a chat. Notice the bare wooden tables and chairs where the men are sitting drinking bottled Coke. Women and children, often bare-footed, tend to sit or squat on the floor. Men are still dominant in many traditional Mexican villages.

Figure 9.1 Location of San Christobal

Figure 9.2 Main square in Chamula

The square and its buildings serve the village people who live in the surrounding hills. It is quite a **dispersed** village as you can see in Figure 9.3. Houses are made either of wood or of mud bricks. Most have red tiled roofs.

Figure 9.3 Chamula village

The square is overlooked on three sides by important buildings. These include the *presidente's* house (the large building in Figure 9.2), several cafés and a fairly modern-looking school together with a basketball court. The most important building in the lives of the people is the church, for the Indians are extremely devout catholics. Their church, unlike the other buildings nearby, is freshly painted blue and white and it is beautifully kept.

The village has electricity, although not every house is connected. Mains water serves only those buildings close to the central square. Other buildings share a communal tap – on average, one tap serves four houses. There is no mains sewerage.

Each family grows staple food crops such as the maize (sweetcorn) shown in Figure 9.3, sugar cane and vegetables. Pigs, goats, sheep and chickens are also kept and allowed to forage around for food. Excess produce is sold to other villagers in the square. Women and children may make craft items such as woven wristbands which they sell to tourists.

Chamula is served by Volkswagen taxi vans which take villagers to San Cristobal to market their goods. The standard fare is 700 pesos (about 15p in 1988) one way. The taxis, which in Britain would never pass an MOT, are a lifeline for the villagers as they do not own cars. Only the most wealthy might be able to afford a truck.

ACTIVITIES

1 Locate San Cristobal in your atlas.
  a How far is it from the Guatemalan border and from Mexico City?
  b What is the altitude of the town? Compare this to that of your own town or village.
  c Try to discover what the climate and natural vegetation is like in the San Cristobal area.

2 Make a copy of the sketch map of the village in Figure 9.4. Use Figure 9.2 to help you add a label describing what happens in the main square.

3 Study Figure 9.3.
  a Why can the village be described as **dispersed**?
  b What evidence is there that the village is served by electricity?
  c What is piled up in front of the house in the foreground? What do you think it will be used for?
  d Are there any animals in the photograph? If so, what are they and what would they be used for?
  e How many storeys has each house? How does this compare to most houses in Britain?
  f Would you say that the houses and plots of land were generally well or poorly looked after? Explain your answer.

Figure 9.4 Layout of central Chamula

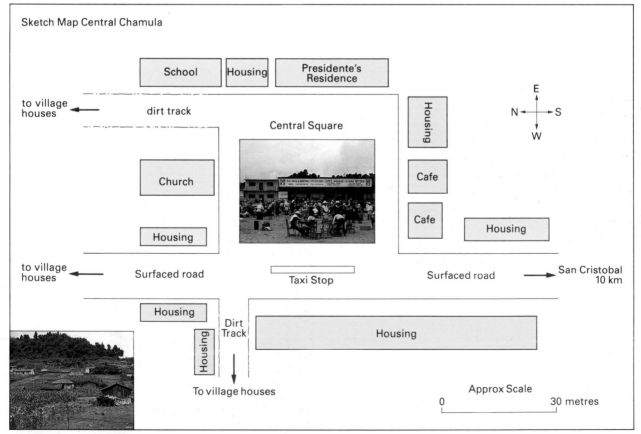

Rural improvements in Mexico

Although the Mexican Government had invested some money in the countryside in the 1960s, most remote areas like the San Cristobal district still lacked piped water, electricity, and other services.

In 1973 a programme called PIDER ('Integrated Programme for Rural Development') was launched (See Figure 9.5). Nearly $500 million is made available annually to improve life in rural areas.

Some of this money comes from the World Bank and the United Nations. Finance and expert help is available to improve farming, build roads, and establish services such as schools and health centres. Loans are available to farmers so that they can invest in fertilisers and machinery. So far, the programme has been quite successful. However, as Mexico is severely in debt, future financing of the programme is in doubt.

Figure 9.5 Mexico: Rural Development Areas

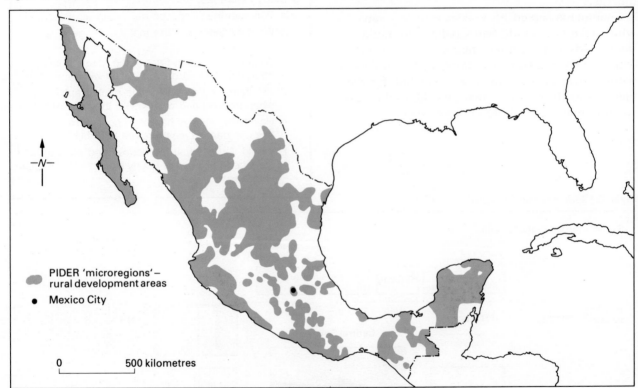

10 *Housing*

There are many different types of house that people live in. These include detached, semi-detached (one wall shared), terraces and flats. Can you think of any others?

The size and type of house varies within a town. Usually the centre of town has the older houses — these would have been the first houses built as the town developed.

Close to the town centre there might be early industrial workers' houses — these are often terraced 'back-to-back' houses. In many towns such housing was of poor quality with cramped conditions, little gardens, and often outside toilets. These houses have either been demolished or have been greatly improved (**renovated**).

The further out, the newer the houses become. This is because as a town grows, the newest housing tends to be added on to the outskirts — the **suburbs**. Here land is relatively cheap and plenty of space is available. As a result, houses tend to be larger and have bigger gardens.

It is important to realise that house prices vary. Flats are usually cheapest whereas detached houses are the most expensive. The price of houses also varies geographically (see Figure 10.1) with houses in southern Britain costing much more than similar properties in the north. This is because there is great competition for land in the south — this makes land expensive. Also, lots of people wish to live in the south. These two factors combine to push up the price of housing.

Figure 10.1

Semi-detached 3rd quarter 1987
Highest price county = Greater London

- █ £55 000+
- ▓ £35 000 – £54 999
- ░ £30 000 – £34 999
- ☐ less than £30 000
- * insufficient samples

Study Figure 10.2. It shows a number of different types of house in Watford, Hertfordshire. Notice where each type of house is located in the town.

Figure 10.2 Typical house types in Watford (Each map square = one hectare)

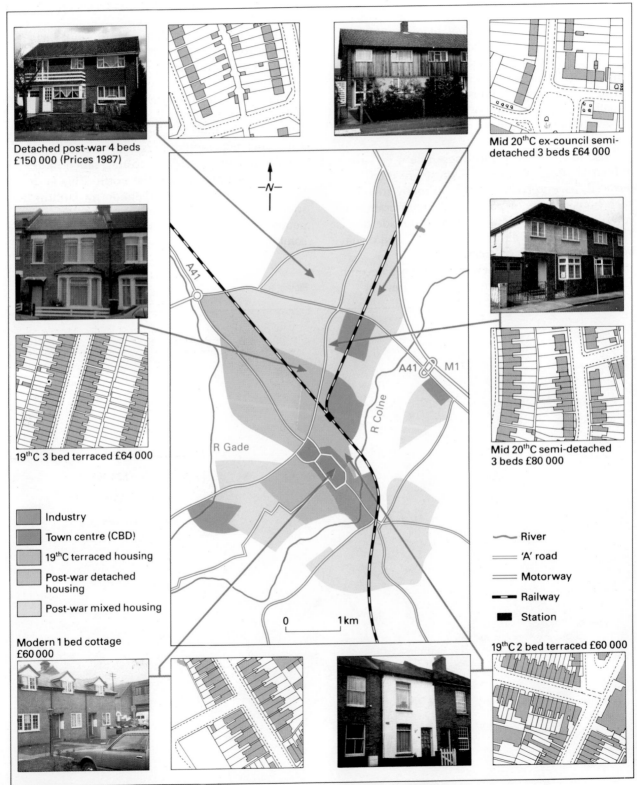

Detached post-war 4 beds £150 000 (Prices 1987)

Mid 20thC ex-council semi-detached 3 beds £64 000

19thC 3 bed terraced £64 000

Mid 20thC semi-detached 3 beds £80 000

Industry
Town centre (CBD)
19thC terraced housing
Post-war detached housing
Post-war mixed housing

River
'A' road
Motorway
Railway
Station

Modern 1 bed cottage £60 000

19thC 2 bed terraced £60 000

ACTIVITIES

1 Study Figure 10.1 and, with the aid of an atlas, attempt the following questions:

   a Which region of the UK has the highest house prices?

   What is the average price of a semi-detached house in this region? Why are prices so high in this region?

   b Which parts of the UK have the lowest house prices?

   Use an atlas to help you list some of the possible reasons for the prices being relatively low. (Hint: think about physical geography, accessibility, industrial decline, etc.)

   c On an outline map of Britain represent the information in Table 10.1 as proportional bars. Make up a scale, something like 1 cm = £10 000, and draw a bar the appropriate height at the location of each town. Next to each bar write the name of the town.

   Describe the overall pattern shown by your map.

Table 10.1 Average, semi-detached house prices 3rd quarter 1987

Town/City	House price(£)
Carlisle	33 050
Middlesbrough	27 900
Doncaster	22 850
Hull	29 750
Liverpool	31 850
Manchester	29 900
Derby	26 000
Hereford	37 450
Birmingham	32 900
Great Yarmouth	36 500
Cambridge	59 850
Swansea	33 100
Bristol	47 050
Exeter	46 850
Plymouth	39 100
Reading	74 800
London	105 950
Belfast	29 550
Glasgow	38 500
Aberdeen	40 750

ACTIVITIES

2 Study Figure 10.2.

   a Make a copy of Table 10.2 and complete it as follows:

   i work out the density of housing for each house type by counting the number of houses in each hectare. Count in 'whole' houses — if only part of a house is in your hectare then include it in your total only if more than half is showing;

   ii use the scale to work out the distance of each house type from the town centre (take this as being the location of the modern 1 bed cottage on Figure 10.2).

   iii fill in the average prices of the houses.

   b Draw a graph to see if there is a relationship between distance from the centre (X axis) and density of house (Y axis). Once you have plotted all the points attempt to draw a best-fit line — ask your teacher if you need help. Describe the relationship.

   c Draw a similar type of graph plotting distance from the centre (X axis) and average house price (Y axis). Add a best-fit line and describe the relationship.

Table 10.2

House type	Density (houses per ha)	Distance from centre (km)	Av price (£)
Modern 1 bed cottage			
19C 2 bed			
19C 3 bed			
20C semi-detached			
20C ex-council			
Detached			

Figure 10.3 Worcester: age of housing

ACTIVITIES

3 Study Figure 10.3 and the map extract at the back of the book.
  a What age of housing is most extensive in Worcester?
  b What age of housing is least extensive in Worcester?
  c What age of housing occupies land on the outskirts of Worcester? Why is this?
  d Is the oldest housing found in the centre of the city? Try to account for your answer.
  e What age of housing borders the railways and the main roads leading out of Worcester?
  f What age of housing is found at the following places:
    i Dines Green (8255)
    ii Astwood (8657)
    iii Rainbow Hill (8556)
    iv St Johns (8354)?
  g Write a few summary sentences describing the pattern of housing in Worcester. Does housing get more modern away from the city centre?

4 Make a study of **street patterns** in Worcester. Locate the area in the centre of the city between the racecourse and the A38 (approx. 845555). For comparison, locate the Dines Green estate centred on 827550.
    Make a comparison of street patterns in these two areas. Comment on the density of streets and on the pattern (grid square, crescentic, etc.?). Use sketches to illustrate your answer. Think carefully about the likely reasons for the contrasts you observe.

Coursework ideas

It is easy to carry out a similar activity to Activity 2 using your own town. Maps of a scale 1:2500 or 1:1250 are readily available from Borough or District Councils—these can be used for discovering densities. Local estate agents will supply you with information on house prices as will your local paper. Fieldwork will enable you to identify house types.

11 Recreation

Figure 11.1 In-town recreational facilities

There are two broad types of recreation, **formal** and **informal**. Formal recreation involves joining clubs and paying fees whereas informal recreation tends to involve individuals or families walking or picnicking. It is important to cater for both forms of recreation in towns and cities.

Recreational facilities

There are many different forms of recreational facility each of which has special requirements in order that they may be successful. For example, people picnicking may require a few picnic tables, toilets and waste bins. A tennis club would need changing facilities, a car park and well maintained courts. Planning for both these examples requires considerable thought if people are to be satisfied. Also, transport and accessibility need to be carefully considered. Congestion needs to be kept to a minimum and plenty of car parking space made available.

The following list describes some of the facilities that may (or should?) exist in a town as identified by Worcester City Council:

1 Children's play spaces — ideally no more than five minutes walk from home and in areas away from heavy traffic.

2 Kickabout areas — generally grass areas with perimeter walls or fences ideally no more than ½ mile from home.

3 Open spaces — grassy areas perhaps landscaped that may include children's area and kickabout space. Ideally everybody should be within ½ mile of such an area.

4 Large open spaces — this includes woodland, water areas, parks, commons, etc. Mainly for informal recreation.

5 Sports facilities — includes playing fields, sports centres, tennis courts, golf courses, etc. Mainly for formal activities.

6 Social facilities — includes museums, churches, pubs, theatres and cinemas. Colleges and schools are often used for evening courses.

7 Footpaths — footpaths, bridleways (intended for horses), towpaths and cycle ways are heavily used and need considerable upkeep.

8 Allotments — vegetable plots have become greatly reduced in total area over the years due to the spread of housing. The result is that there is frequently a long waiting list to obtain an allotment. Plots need sheds and car parks.

The photographs in Figure 11.1 illustrate some of the facilities described above.

ACTIVITIES

1 Make two lists, one of formal and the other of informal, recreational activities that are available in your home town.

2 Study the map extract at the back of the book.

  a What recreational activities might be based on or by the River Severn and the Worcester – Birmingham Canal? Separate them into formal and informal.

  b How many golf courses are there in or close to the city?

  c Locate square 8754. Describe how you would develop the resources in this square for recreation.

  d Identify areas on the extract that could be described as 'large open spaces'.

3 Not all recreational activities go well together. Power boating and swimming, for example, conflict with each other.

  Consider the activities that may be based on or by water. Devise a number of groups of activities that would **not** conflict with each other.

4 On a map of your home town locate the recreational activities available. To do this, work as a class locating as many different facilities (open spaces, play areas, sports grounds, etc.) as possible. You can then produce your own map using colours and symbols if you wish to group together similar activities (see text).

  Produce a report about the provision of recreational facilities in your town. Comment on the map. Are the facilities spread across the town or are they concentrated in one part? Is there a good range of activities available or are certain groups of people poorly catered for? How would you improve recreation in your town?

5 The information in Figure 11.2 describes a controversial proposal for a new leisure complex on the edge of Watford. Many people welcome the proposal as Watford only has one small cinema (another was recently closed), the nearest ice rink is in London and the nearest bowling alley, at Harrow, is extremely busy.

  Nevertheless, not all local residents near to the proposed development are happy. Some are concerned about increased traffic, noise and possible vandalism. Imagine that you are a resident and that you have been given the job of writing to the local newspaper to publicise the fears of the local community.

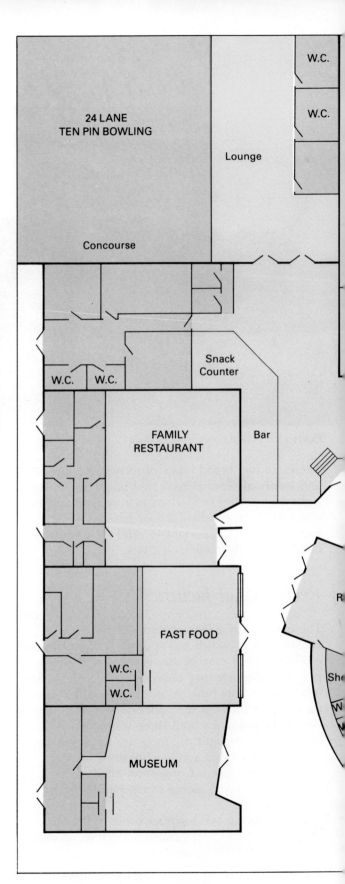

Figure 11.2

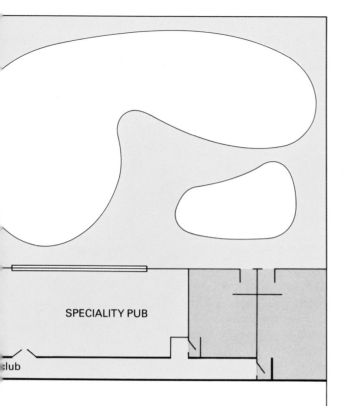

SPECIALITY PUB

club

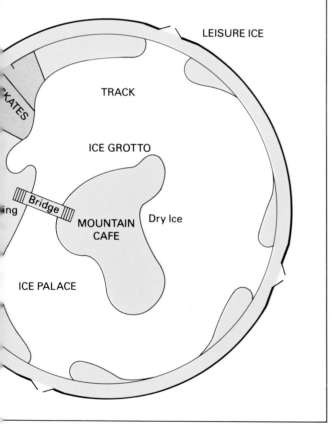

LEISURE ICE

TRACK

KATES

ICE GROTTO

Bridge

ing

MOUNTAIN CAFE

Dry Ice

ICE PALACE

Coursework ideas

1 Carry out a survey of recreational activities undertaken by your class over a period of a fortnight. Discover what activities are in greatest demand and where the activities are based. How often are the activities carried out? Which activities are well/poorly catered for?

  You could extend the study to cover other age groups. A questionnaire could be devised to discover recreational habits in your town. From your results you could suggest types of recreation that need better facilities or areas of your town that need improved provision.

2 Discover whether there is a hierarchy of recreational activities. To do this you could identify a variety of different activities and, by means of questionnaire interviews, discover how far people travel to partake in the activities. Desire lines could then be drawn (see Figure 11.3) and the resultant patterns analysed. The greater the distance travelled, the higher the 'order' of the activity.

3 There may well be a controversial proposal for a town near you like the one described in Activity 5. You could investigate the reasons behind the proposal and identify the advantages and disadvantages, perhaps using a questionnaire.

  There may be a derelict area or unused plot of land in a town near you. How might such an area be developed for recreation?

Figure 11.3 Desire lines for three recreational activities in a town

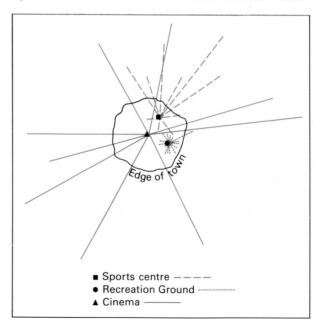

■ Sports centre ----
● Recreation Ground ············
▲ Cinema ——

12 Movement in towns and cities

Figure 12.1 Traffic jam

You have already learnt that many towns developed as 'route centres' (p. 5). At the very centre of these towns was a major road junction, often a crossroads. This junction would be the most accessible point in the town for people travelling to the town from all directions. It became the **town centre** and, being accessible, attracted shops and other businesses.

As towns have grown up these central points have become congested. People trying to pass through the town *en route* to other places or simply wanting to park and shop causes massive traffic jams (Figure 12.1). As car ownership increased rapidly in the 1960s it became necessary to manage traffic.

Today all towns and cities have traffic management plans in order to cope with both cars and pedestrians.

Figure 12.2 shows the 'Borough of Watford Traffic Arrangements 1987'. As you can see it is quite a complicated map. Look at the key to identify the varous forms of management.

In Watford, traffic used to use the High Street but as this got more congested planners decided to construct an **inner ring road** — identify the High Street and the ring road on Figure 12.2. Shoppers and delivery lorries have access to car parks and shops by taking one of the service roads leading off the ring road. The town centre (in this case the High Street) has been largely pedestrianised so that shoppers do not have to worry about the dangers of traffic.

Some towns have **ring roads** that actually skirt the whole town keeping traffic out of the urban area altogether. Study Figure 12.3. It shows the location of the small town of Brackley. Notice that it is on the main road between Northampton and Oxford. For many years Brackley has been a 'bottle-neck' as through traffic combined with local traffic to cause huge jams. Heavy lorries caused pollution, noise and were a danger to pedestrians.

In 1987 the Brackley by-pass was opened (see Figure 12.3) at a total cost of about £9 million. The by-pass involves 3 km of dual carriageway, a number of roundabouts and several new bridges. The local paper reported that 'local people and traders have branded the by-pass as a huge success'. Figure 12.4 shows the town's main street before and after the opening of the by-pass.

Recently a number of towns and cities have introduced **park and ride** schemes. Shoppers park outside the town centre in large car parks and regular buses with cheap fares ferry people to and from the shops.

(See *Population* book for studies of commuting.)

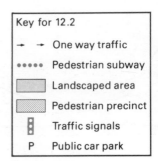

Key for 12.2

→ → One way traffic

••••• Pedestrian subway

Landscaped area

Pedestrian precinct

Traffic signals

P Public car park

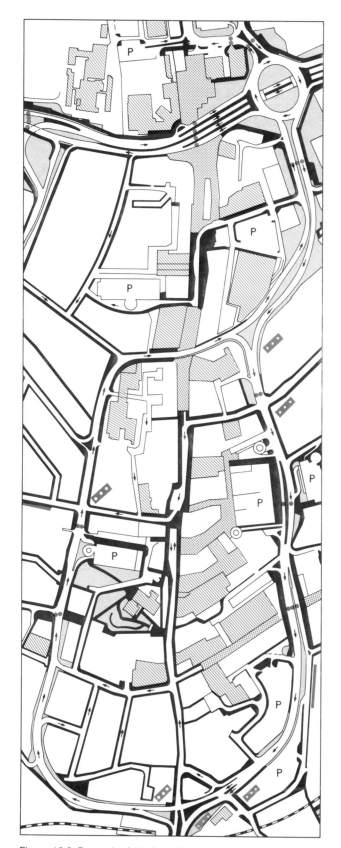

Figure 12.2 Borough of Watford, Town Hall, Watford

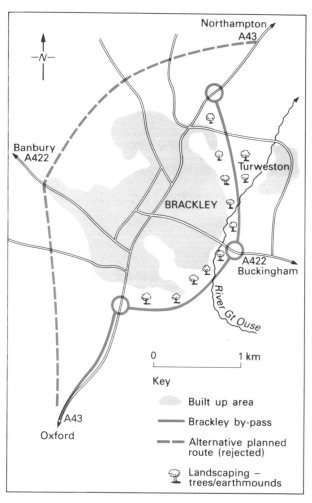

Figure 12.3 Brackley by-pass

Figure 12.4 Brackley before the by-pass

and afterwards

ACTIVITIES

1 Study Figure 12.2.

   a Make a list of the different forms of traffic management in Watford town centre.

   b How many car parks are there? What do you notice about their location? Try to explain your answer.

   c Some people in Watford think that the car parking charges are too high. They often try to park elsewhere in side streets. What problems might this lead to?

2 Make a case study of the Brackley by-pass.

   a Copy Figure 12.3 and use a different colour to show the route of the by-pass. Use an atlas to work out the distances from Brackley to Oxford, Banbury, Buckingham and Northampton. Show these on your map.

   b Study Figure 12.4. What effect does the by-pass appear to have made to the traffic in Brackley?

   c Figure 12.5 shows some opinions of local people about the newly completed by-pass. After reading through the accounts write a few sentences describing **your** reaction if you were:

      i the owner of a small supermarket on the main street;

      ii the owner of a fish and chip shop on the main street;

      iii a family with young children living near the town centre;

      iv a pensioner who shops daily on the main street;

      v the owner of a cottage close to the new road.

   d Local people are now hoping that the by-pass will be extended to link up with the Banbury road. Suggest a possible route on your map.

3 Study the Worcester map extract at the back of the book. A new road has recently been opened between 836523 and 853516.

   a Which two roads are joined by the new road?

   b How many kms long is the new road?

   c Is it single or dual carriageway?

   d Look carefully at Worcester city centre. Before the new road was built, where would you expect major traffic congestion to have occurred?

   e Use an atlas to list the towns which can now be reached without having to pass through the centre of Worcester.

   f What do you think was the potential natural hazard that needed to be borne in mind in the planning and building of the new road? Suggest measures that might have been used to protect the road.

Coursework ideas

1 Study the traffic management in your local town. Local councils will have maps which you can use. It is interesting to identify changes by looking at historic maps. Can you suggest any additional management schemes? Is there a local traffic problem (danger, parking problems, congestion) that needs studying and solving?

2 Studying traffic flow in order to identify peak periods or congestion 'black spots' makes a good project. Carry out a census on the relevant roads at different times of day and different days of the week. Count the different vehicles for a ten minute period. Traffic flows can be shown as lines whose thickness is proportional to the number of vehicles. Different colours can be used to show different types of vehicle.

3 Is there a town near you that needs a by-pass? Suggest possible routes. Devise a questionnaire to find the views of local people.

Figure 12.5

Ed Barge (Deputy Chairman of Brackley and District Chamber of Trade): *'The quieter town centre would encourage more pedestrian traffic in the town largely to the benefit of local businesses. From an environmental point of view it's been wonderful.'*

Derek Newman (Oliver and Newman): *'It's been such an improvement, it's unbelievable. I'm sure that 99% of Brackley people are very pleased they can walk across the High Street feeling safe without the lorries screaming down.'*

Sheila Davies (Tiffany's Restaurant): *'It can do nothing but good. Brackley is a lovely town — now it could become a magnificent town.'*

Hilary Kirting (mother of four): *'At last it is safe for me to let my children make their own way to school and run errands to the shops for me. Before the new road I had to accompany them everywhere.'*

Lesley Stubbings (resident of Turweston): *'The access to my home is right by the bypass roundabout which is dreadful with so many cars tearing down the hill to join the bypass. I worry dreadfully about my children.'*

Dr John Holman (resident of Turweston): *'This road has completely changed our lives. The noise wakes us up at night...and our views are totally ruined by a wide open road and an enormous footpath bridge.'*

13 *Inner cities*

Figure 13.1

Figure 13.2 Typical inner city problems

You will probably have heard about the 'inner city problem' (Figure 13.1). But do you really know what the 'problem' is? In this section you will discover something of the character of inner city areas — it is for you to decide whether or not there is a problem!

Something like 4 million people in Britain live in the inner city — this is about 7% of the population. The areas where these people live and their quality of life can generally be described as poor. Unemployment rates are high; there is a large proportion of public housing often in bad condition; there are few open spaces for recreation; a high proportion of minority groups lives there; a high proportion of the residents is elderly; the environment is generally gloomy and depressive with much derelict land. The photograph in Figure 13.2 illustrates these characteristics.

The inner city has become a weighty political issue in recent times. Even Prince Charles has become involved in publicising the plight of the inner city dwellers. The Government has identified those areas in greatest need and, through a variety of organisations, makes available grants and other incentives for improving these areas. Figure 13.3 describes some of these Government measures.

In March 1988 the Government pledged a further £3 billion to be spent on the inner city over the next four years. A large proportion of this money will be spent on training schemes and on providing help for small firms wishing to locate in the inner city.

Figure 13.3 Inner City Aid

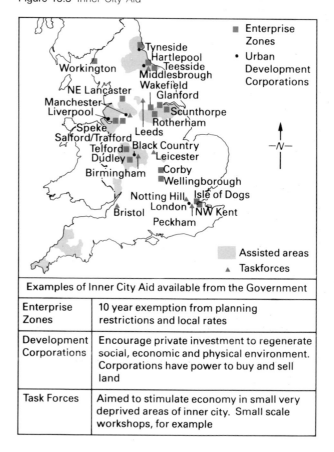

Examples of Inner City Aid available from the Government	
Enterprise Zones	10 year exemption from planning restrictions and local rates
Development Corporations	Encourage private investment to regenerate social, economic and physical environment. Corporations have power to buy and sell land
Task Forces	Aimed to stimulate economy in small very deprived areas of inner city. Small scale workshops, for example

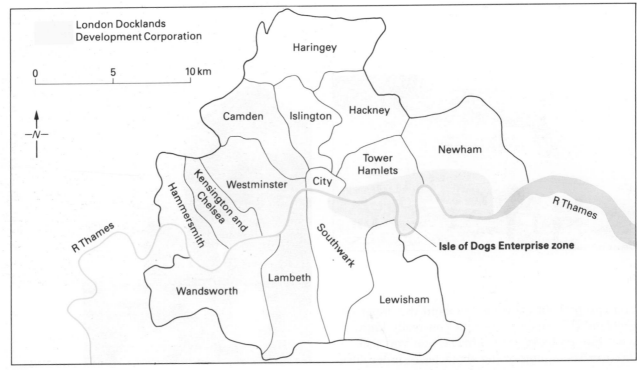

Figure 13.4 Greater London boroughs (all Inner London boroughs named)

London's inner city

1 Housing in Southwark

Some of the worst housing areas in inner London are located in the borough of Southwark (Figure 13.4). This area contained slum housing (now regarded as desirable properties when renovated!) which in the 1960s was mostly demolished and replaced with blocks of flats (Figure 13.5). These flats, with their high level walkways, were considered to be just what the residents wanted. In actual fact they have turned out to be a disaster.

There is a lack of privacy with long dimly lit corridors, graffiti covered and littered with human waste, providing the only means of access to people's homes. In between the blocks of flats are open plots of land – land owned by everybody yet nobody. A lot of young unemployed drug pushers haunt the area carrying out burglaries in order to finance their habit. When asked, a number of residents gave this as the major problem in the flats. Fires are common (Figure 13.6) and most of the shops

situated within the flats have been boarded-up (Figure 13.7).

The council spends a lot of money on these housing estates – painting corridors and fitting special re-inforced security doors. Yet despite this, the graffiti and vandalism continues (Figure 13.8). Some of these estates are considered 'no-go' areas by postmen and milkmen (Figure 13.9).

Figure 13.5

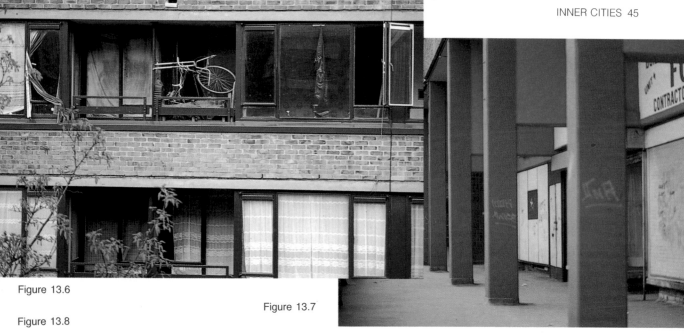

Figure 13.6

Figure 13.7

Figure 13.8

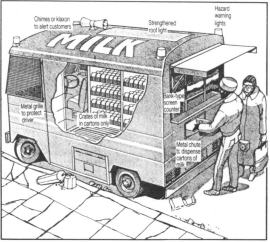

Milk wars: new plan to deliver to no-go estates

One pint or two? Metal grilles, steel plates and reinforced glass would protect the new-model milkman from his customers

Inner cities get some bottle

Milkmen have not set foot on the North Peckham estate for three years. The Express Dairy was the last to withdraw, after a series of incidents in which roundsmen were threatened, beaten up and robbed.

Now Southwark council has agreed to give the Co-operative Wholesale Society a £5,000 grant for a feasibility study on re-starting milk deliveries.

Figure 13.9

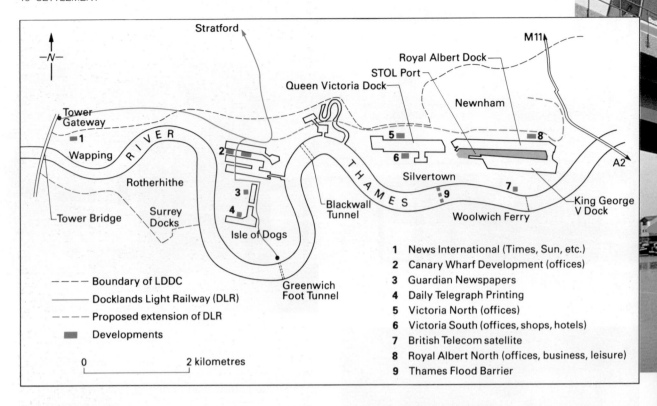

Figure 13.10 London docklands

Figure 13.11 Hi-tech industry

Map labels:

Stratford

M11

Royal Albert Dock
STOL Port
Queen Victoria Dock
Newnham

Tower Gateway
1
RIVER
Wapping
5
8
2
THAMES
6
A2
Rotherhithe
Silvertown
7
Tower Bridge
Surrey Docks
3
9
King George V Dock
4
Blackwall Tunnel
Woolwich Ferry
Isle of Dogs

Boundary of LDDC
Docklands Light Railway (DLR)
Proposed extension of DLR
Developments

0   2 kilometres

Greenwich Foot Tunnel

1 News International (Times, Sun, etc.)
2 Canary Wharf Development (offices)
3 Guardian Newspapers
4 Daily Telegraph Printing
5 Victoria North (offices)
6 Victoria South (offices, shops, hotels)
7 British Telecom satellite
8 Royal Albert North (offices, business, leisure)
9 Thames Flood Barrier

2 Re-development of the London docklands

London's dockland area, now under the planning control of the London Docklands Development Corporation (LDDC) (Figure 13.10), has seen a complete facelift since the early 1980s.

The docks of central London were gradually closed down during the 1950s and 1960s. This was mainly due to the increased size of ships, no longer able to easily navigate the meandering River Thames, and the opening of the large container port at Tilbury. By 1981 all the central London docks had closed and, as a result, the once thriving port now suffered serious unemployment. Much of the housing was of poor quality and in urgent need of modernisation. A lot of the land became derelict and the environment became depressive.

The LDDC was set up in 1981 to re-vitalise the docklands area. It received a large amount of money from central government and set about re-developing the area. The first job of the LDDC was to turn the derelict land into land fit for development − this involved considerable clearance and the building of sewers and transport connections. Grants and advertising campaigns attracted new industry, mainly hi-tech, to the area (Figure 13.11). Some of the improvements in docklands include:

Figure 13.12 Docklands Light Railway (DLR)

Figure 13.13 London City Airport

1 the Docklands Light Railway (Figure 13.12) linking docklands to the rest of the Underground;
2 London City Airport (Figure 13.13) provides links to Europe;
3 the creation of 33 000 jobs by 1991;
4 the building of 15 000 new houses by 1991.

There is no doubt that the LDDC has given this formerly depressed area a facelift (Figure 13.14). A visit to the area, (well recommended), shows great activity and there is certainly an air of prosperity and optimism.

However, there are critics. Research has shown that many of the new people moving in are simply commuters. The area has become fashionable amongst the 'young upwardly mobile' able to afford the high property prices beyond the reach of the older original dockland residents. Many of the new jobs, being hi-tech, are inappropriate to the manual labour skills of those dockers made redundant by the docks' closure. Residents feel that their home area has been taken over by outsiders with little interest in their needs. An alternative scheme for the development of docklands called the 'People's Plan' was drawn up by residents only to be politely turned down.

Figure 13.14

ACTIVITIES

1 Some people think that the type of housing design described in the section on Southwark is largely the **cause** of inner city problems such as crime and vandalism. Perhaps corridors should be blocked off and private access provided for households. Open spaces should be divided up into private gardens. Walkways, at present a means of quick escape for criminals, should be removed.

Imagine that you are a resident in one of the flats shown in Figure 13.5. Write a letter to the local council describing life in the flats and suggesting some measures which you feel the council should adopt in order to solve the problems you describe.

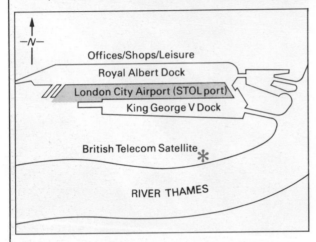

Figure 13.15 LDDC Plan

2 Figure 13.15 and 13.16 show two contrasting plans for part of the London docklands.

Hold a mock radio interview to debate the two opposing plans. To do this, you need to form three small groups of pupils to represent:

a the radio station;

b the LDDC.

c the group suggesting the People's Plan.

Each group should prepare their case. The radio station will need to think up some probing questions. The other two groups must be ready for awkward questions but should do all they can to promote their own plan. This preparation must be done away from the other groups.

d To hold the radio discussion you need:
  i the presenter of the radio show;
  ii one representative from the LDDC;
  iii one representative from the 'People's Plan'.
  The rest of the class can act as the 'studio audience'.

Put a time limit on the discussion, say ten minutes. Like in real radio broadcasting , this must be stuck to!

It would be interesting to take a vote following the discussion to see what the 'studio audience' felt.

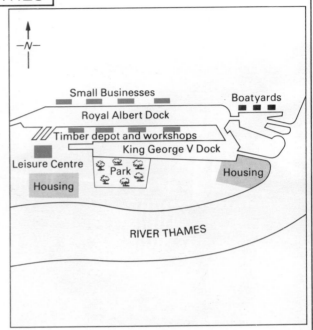

Figure 13.16 People's Plan

3 Figure 13.17 and 13.18 (p.50) contain details about the inner parts of Middlesbrough, in particular the area of St Hilda's. Study the information carefully and then produce a report/project to act as a case study of an inner city area.

You should try to discover to what extent Middlesbrough has an inner city problem. Also, using maps, try to identify exactly which wards make up the inner city. You will need to think very carefully and do not worry if you do not share the same opinion as others in your class.

For the maps, you could use the choropleth technique or proportional symbols such as bars and circles. By means of an example, suggested choropleth categories for **unemployment** are given below:

Unemployment (%)

41+	Black
36–40	Brown
31–35	Red
26–30	Orange
20–25	Yellow

Notice that **white** is not used – this is because some wards will be left blank on your map.

Ask your teacher for help in devising categories for the other indicies but have a go yourself first. Remember that the ideal number of categories is 4–8, that they should be equal in size and that they should not overlap.

Use sketches to illustrate your report.

Describe some of the ways in which the area in being improved.

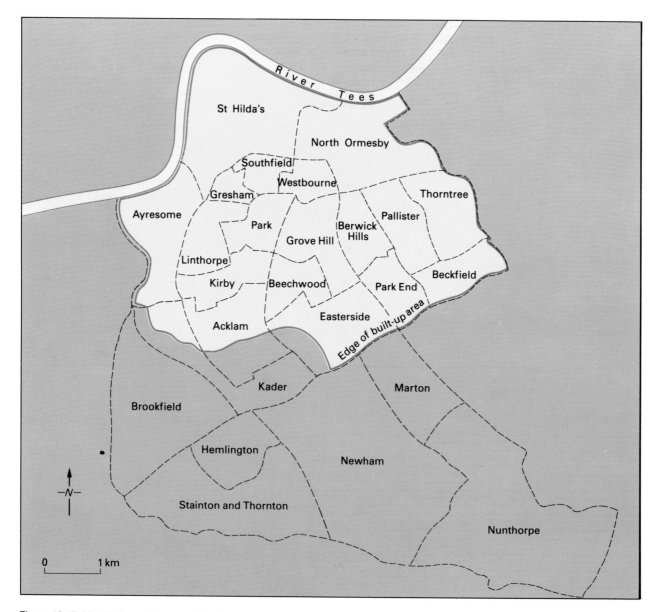

Figure 13.17 Middlesbrough Borough Wards

Coursework ideas

It is possible to obtain census information for wards in inner cities. These are best obtained from **county monitors** available from the Office of Population Censuses and Surveys. It is then possible to plot such information on maps (as in Activity 3). Such maps should be augmented with personal fieldwork where possible. A questionnaire could be used to discover if people were afraid to go out at night or whether or not they felt litter or graffiti to be a particular problem.

A study of environmental quality could be undertaken. Decide on a number of environmental aspects such as the state of roads and pavements; state of housing; amount of trees/open space; amount and type of litter; amount of graffiti; etc. Devise an index for each ranging from 0 (bad) to 5 (good). For each area or road tot up the score to give you an 'environmental index'. Values can then be compared. Such an exercise could be done for all parts of a settlement and not just restricted to the inner city.

(Teachers will find an excellent article about Glasgow's inner city in *GEO* Series 9 Issue 4 1986–87)

SOCIAL INDICATORS FOR SELECTED MIDDLESBROUGH WARDS
(per 1000 relevant population or households except*)

	`Unemployment (%)*	Pensioners living alone	Head of Household born in New Commonwealth Pakistan	Households without car	Households sharing/lacking bath	Households of density of more than one person per room
St. Hilda's	31.9	164	26	781	35	94
Southfield	39.5	160	158	733	82	96
Westbourne	44.4	122	243	601	98	99
North Ormesby	30.7	149	87	678	84	83
Gresham	29.2	127	107	630	69	71
Beechwood	28.7	185	8	694	8	67
Thorntree	39.5	122	4	766	12	83
Beckfield	31.9	168	3	632	2	50
Pallister	30.0	80	1	761	5	133
Park End	28.8	72	2	646	1	91
Ayresome	27.7	136	14	656	7	58
Stainton & Thornton	25.4	145	14	431	1	46
Berwick Hills	24.2	91	3	673	3	53
Easterside	23.5	144	5	623	1	56
Grove Hill	23.6	160	15	597	11	69
Hemlington	27.8	123	6	435	1	44
Middlesbrough	23.6	122	32	502	18	54

Notes: 1 All information is from the 1981 Census, except unemployment figures which are from the Cleveland County Planning Officer, for July 1986.

(Source: Middlesbrough Borough Council)

Figure 13.18 Middlesbrough Borough Wards

(a)

During the middle part of the last century the St Hilda's area formed the town of Middlesbrough itself – laid out on a planned grid iron pattern with the old Town Hall at the centre of the grid. The town developed and boomed on the basis of the creation of Middlesbrough Dock and the arrival of the railway, on coal, iron and steel. It was only as Middlesbrough grew to the south and a 'new' town centre developed, and later on as the heavy industry moved down river towards the new Tees Dock area (culminating in the closure of the Middlesbrough Dock), that the old Middlesbrough of St Hilda's declined due to lack of investment.

By the 1950s the area had become so run down that a major redevelopment was seen as the only solution. A series of flats and maisonettes was constructed. However, due to poor design, and continuing social and environmental problems which were not tackled in parallel with the redevelopment, the decline was not arrested. St Hilda's rapidly became a 'difficult to let' estate.

Against this background of physical and social decline, Middlesbrough Council began the major revitalisation of St Hilda's in 1979. The bulk of this regeneration has involved spending from the Housing Investment Programme. By a combination of improvements, conversions, refurbishment and redevelopment of the blocks of flats and maisonettes by traditional 2 storey houses, the community of St Hilda's being retained. This process should be complete in 1987.

Enterprise Centre

A former derelict biscuit factory was acquired and converted into an Enterprise Centre-a seed bed for new industrial ventures – as part of Middlesbrough Council's package of economic initiatives. It consists of 15 small units plus a communal workshop and equipment for hire, together with back up technical, management and advisory staff.

In its 2 years of operation the Centre has already proved its worth. Although some would-be entrepreneurs have been unsuccessful in developing ideas, 29 new businesses have been established and 15 of these have moved on to new premises.

The whole venture has been financed from the Inner Area Programme and has had the additional benefit of focusing attention on the St Hilda's area and bringing in new activity.

Yuill's Urban Development Grant

Middlesbrough Council has been keen to attract private housing development into St Hilda's to offer residents a wider choice, and to boost investment confidence in the area.

Early attemps foundered due to lack of viability. However with the announcement of Urban Development Grants, CM Yuill Ltd have put foward an imaginative scheme to develop 120 dwellings for sale. The scheme has won the support of Middlesbrough Council and has been submitted to the Department of the Environment for approval.

(Source: Middlesbrough Borough Council)

(b)

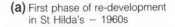

(a) First phase of re-development in St Hilda's – 1960s

(b) Second plase in the 1980s. Lower density housing has been much more successful

14 Green belts

After World War I, planners became increasingly concerned at the outward spread of towns and cities. They decided to create a ring around major cities where development could be controlled. These rings are called **green belts** (Figure 14.1).

There are three aims of green belts:

1 To stop further outward growth (**sprawl**) of urban areas;
2 To stop cities becoming joined together;
3 To preserve the character of historic towns/cities.

The first green belt, around London, dates back to 1947. Most others were established in the 1950s. Since then, new ones have been added and old ones enlarged. At present, about 11% of England is now 'green belt'.

Figure 14.2 is a photograph of part of the green belt around London. Notice that it is mostly countryside although there are roads, and houses. The important thing about the green belt is that there are very strict controls on any form of development be it housing or quarrying.

Recently there has been a lot of debate about the future of green belts. Some people say that the time has come to relax the restrictions and allow some development. Already the M25 has been built in the London green belt and there are plans to build 'new towns' and shopping and leisure centres. Clearly, a question mark hangs over the future of green belts.

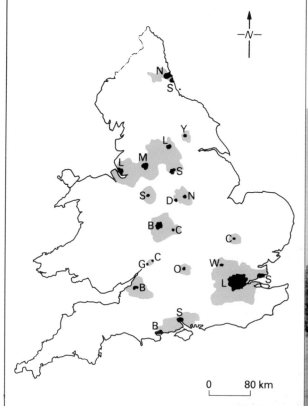

Figure 14.1

Green belts in England

Figure 14.2

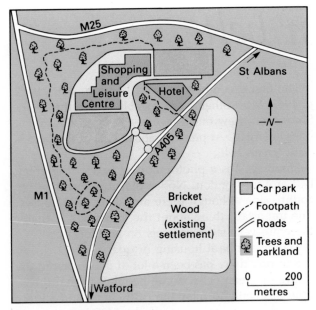

Figure 14.3 The Golden Triangle

One proposal involves an area called the 'golden triangle' near St Albans, Hertfordshire (see Figure 14.3). The plan is to build a shopping and leisure complex. There will be restaurants, a multi-screen cinema, children's play areas, parkland, shops and a large 250 room hotel. Local roads which are at present very noisy will be improved and built through cuttings to reduce the noise. Large car parks will also be built.

Figure 14.4 Pressures on the London green belt

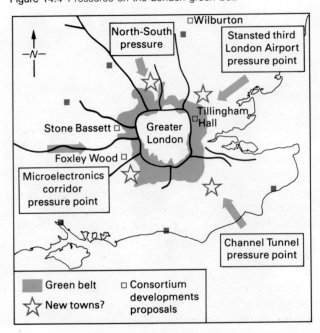

ACTIVITIES

1 Make a copy of the map of green belts, Figure 14.1. Use an atlas to discover and label the cities and towns. Their initial letters are on the map to help you.

2 Study Figure 14.1 or your own map drawn in Activity 1.
   a Give an example of two towns which may have become joined together if a green belt had not been there. Do you think it is important to stop cities becoming joined? Why?
   b Give three examples of 'historic towns' with green belts. Why do you think there are green belts around these towns?

3 Study Figure 14.3.
   a Why do you think the area is called the 'golden triangle'?
   b At present the area can be divided up as follows:
     i 63% farmland (not top quality);
     ii 10% woodland;
     iii 27% derelict land and spoil tips.
     Show these figures as a pie chart.
   c You are director of the company proposing to develop the 'golden triangle'. Use the pie chart, Figure 14.3, and the text above to write a few sentences describing the benefits of the scheme.
   d If you lived in nearby St Albans, would you be in favour of the scheme? Why?
   e Many residents of Bricket Wood (see Figure 14.3) are against the development. Why do you think this is so?

4 Study Figure 14.4. It shows some of the pressures on the green belt around London.
   a Make a copy of the map, use an atlas to name the towns and motorways. There is no scale on Figure 14.4 (it came from a newspaper!) – use your atlas to give **your** map an approximate scale.
   b Describe how the various pressures mentioned on the map increase the need for more housing in the London area.
   c Consortium Developments is a group of major builders. It has proposed several sites for housing development. Tillingham Hall was turned down by the Government in 1987 but the others are still being debated. There is no doubt that several thousand new homes are needed in the London area. Should they be built on areas of low quality land **in** the green belt, or on attractive land just **outside** the green belt? What do you think? Should the green belt be preserved at all costs?
   d Imagine you live in a small attractive village on the outskirts of a major town (if you do, this activity will be easier for you!). A developer plans to build a new housing estate of 200 houses on land beside the village. Would you approve of this development? How might it affect life in the village?

15 New towns

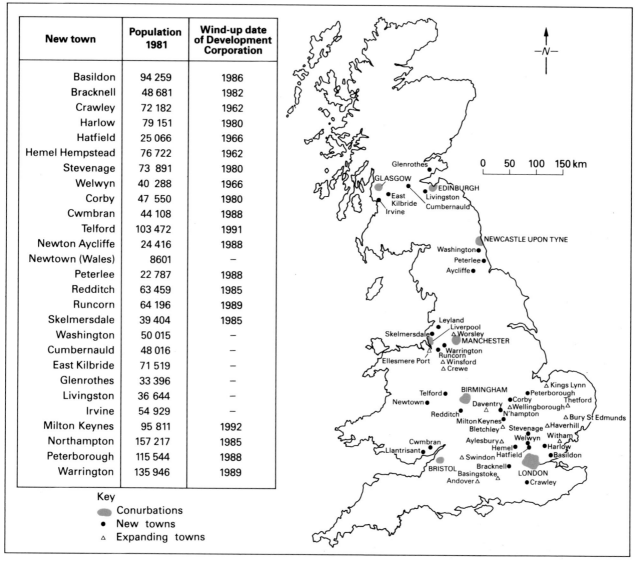

New town	Population 1981	Wind-up date of Development Corporation
Basildon	94 259	1986
Bracknell	48 681	1982
Crawley	72 182	1962
Harlow	79 151	1980
Hatfield	25 066	1966
Hemel Hempstead	76 722	1962
Stevenage	73 891	1980
Welwyn	40 288	1966
Corby	47 550	1980
Cwmbran	44 108	1988
Telford	103 472	1991
Newton Aycliffe	24 416	1988
Newtown (Wales)	8601	–
Peterlee	22 787	1988
Redditch	63 459	1985
Runcorn	64 196	1989
Skelmersdale	39 404	1985
Washington	50 015	–
Cumbernauld	48 016	–
East Kilbride	71 519	–
Glenrothes	33 396	–
Livingston	36 644	–
Irvine	54 929	–
Milton Keynes	95 811	1992
Northampton	157 217	1985
Peterborough	115 544	1988
Warrington	135 946	1989

Key
- Conurbations
- New towns
- Expanding towns

Figure 15.1 New and expanded towns in England

Towns and cities grew very rapidly in the nineteenth century as people moved in from the countryside. These settlements soon became overcrowded. Houses were cramped and in poor condition, there were few open spaces for children to play, and pollution led to smog and disease. In short, these towns and cities became unpleasant places to live.

Many people had thought up ideas for new, clean towns located in the countryside but it wasn't until the twentieth century that the idea of a 'new town' was put into practice.

In 1946 the Government decided to create several new towns around London to absorb some of the people living in poor conditions in the capital.

Over the next few years other new towns were built. Most were designed to absorb overspill from the major industrial cities such as Manchester. Some, however, were planned for special reasons – Corby, for example, was planned to be an iron and steel making town using local supplies of iron ore.

Look at Figure 15.1. It shows the location of the new towns. Notice how most of them are located on the outskirts of major cities.

Inside a new town: Telford

Telford is located in the county of Shropshire 12 miles east of Shrewsbury and about 35 miles north-west of Birmingham — look it up in your atlas. Telford was developed as a new town for two reasons:

1 It was predicted that there would be a large population increase in the West Midlands (centred on Birmingham) — these people would need to be accommodated as the West Midlands was already overcrowded.
2 There were thousands of hectares of derelict land in the Coalbrookdale-Ironbridge area due to the decline of mining in the East Shropshire coalfield.

In 1963 the Government decided to develop this derelict land as the site for a new town originally called Dawley. During the 1960s Government money improved the already existing towns such as Wellington and Oakengates and developed the areas in between to create the 30 square miles that today form the city of Telford (Figure 15.3).

It was originally planned that Telford would house 150 000 people — the already existing towns held 70 000. At present its population is just over 100 000 and it seems unlikely that it will reach the target. This is because population growth has slowed down since the 1960s.

Look at Figure 15.2. It will give you some idea as to what it is like to live in Telford. Let us consider a few aspects of the new town.

1 Housing
There are many different types of house for sale or for rent. There are lots of different styles to suit all tastes. Despite this, some people would say they lack character and all look 'the same'. This is probably simply because they are modern.

2 Shopping
Look at Figure 15.3. Notice that there is a single town centre and several smaller 'regional centres'. This means that for everyday items people can go to their local shops rather than having to travel into the town centre. Only when they require larger items do they need to go into the town centre.

3 Work
Jobs in the Telford area used to be in heavy industry but nowadays they are mainly concerned with high-technology — the jobs are better paid, more secure, and involve much better working conditions. 400 new firms providing 8000 jobs in, amongst other things, electronics, electrical engineering and robotics have set up in Telford.

4 Environment
The beginnings of the Industrial Revolution were at Ironbridge — here coal was first used to smelt iron. (If you get the chance, Ironbridge has an excellent museum which is well worth a visit.) Many of the old buildings in the Telford area have been preserved — they help to give the town some character.

There are large areas of parkland and woods as well as several recreation centres, an ice rink, golf courses and a dry ski slope. The River Severn to the south of Telford has been developed for a range of water sports.

Figure 15.2

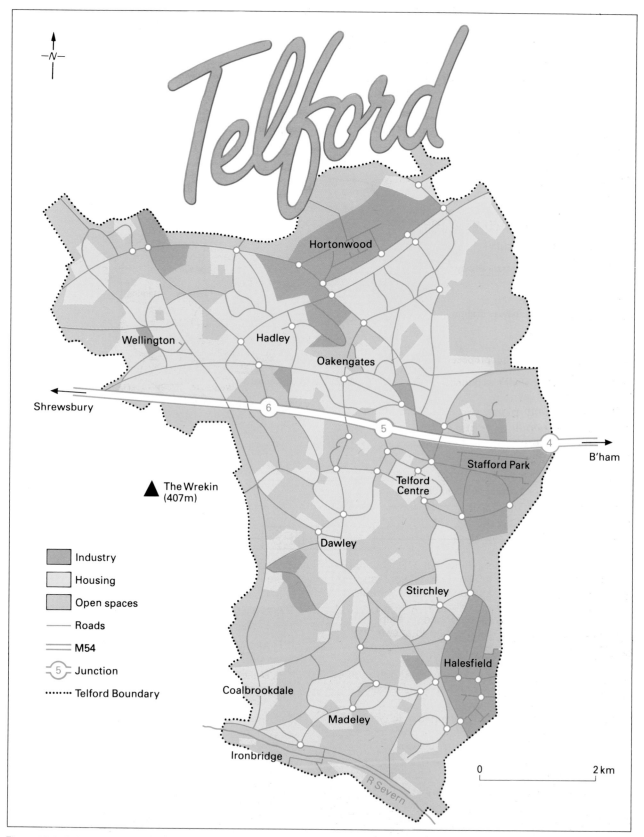

Figure 15.3 Telford

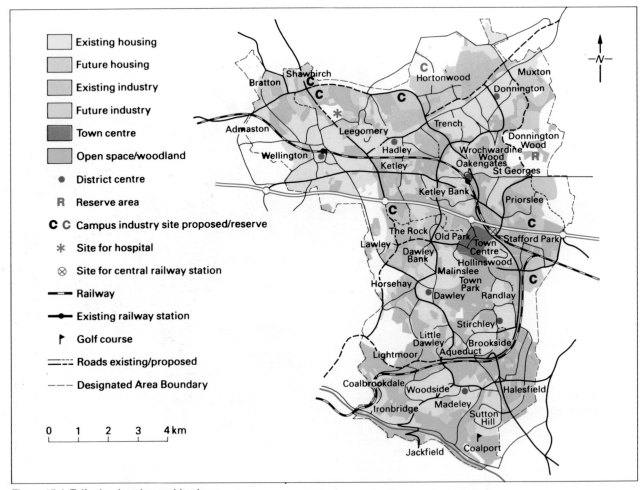

Figure 15.4 Telford — housing and land use

Legend:
- Existing housing
- Future housing
- Existing industry
- Future industry
- Town centre
- Open space/woodland
- ● District centre
- R Reserve area
- C C Campus industry site proposed/reserve
- ✳ Site for hospital
- ⊗ Site for central railway station
- ━━ Railway
- ━●━ Existing railway station
- ┏ Golf course
- ═╍═ Roads existing/proposed
- ─── Designated Area Boundary

0 1 2 3 4 km

ACTIVITIES

1 Study the map of Telford Figure 15.3. Notice that it uses different colours to show the different land uses. Locate the five original towns that were later joined to form Telford — Wellington, Dawley, Oakengates, Madeley, and Ironbridge. With reference to the key, use the map to answer the following questions:

a The present day housing areas are shown yellow. Why do you think the planners decided to scatter the housing estates rather than concentrating them in a single area?

b Do the main roads have a tendency to go right through the housing estates or do they tend to ring them? Try to explain your answer.

c Describe and try to explain the distribution of 'open space'. This is mostly woods and parkland.

d Describe the location of industry in Telford. Use the following list of factors to explain its location; accessibility; land prices; labour force; pollution. Refer to other factors too (see Figure 15.4).

2 Study the photographs in Figure 15.2 which show life in Teford.

a How does it differ from your town/city?

b Would you like to live somewhere like Telford? Explain your answer.

3 Figure 15.5 shows the layout of shops in the town centre.

a Make a list of the top five shops according to floor space. Write down what each one sells. Are the names familiar to you?

b Count up the number of chain stores — these are shops found in most large towns, e.g. *Chelsea Girl*, *Boots* and *Mothercare*. Compare your result with the total number of 'local' shops such as the *Wrekin Bookshop* and *Barber and Son*.

c What do you notice about the location of the chain stores in comparison to the 'local shops'? Try to explain your answer.

ACTIVITIES

d Locate Brodie House. Notice that some of the units are empty. Which of the following shop types might you expect to occupy the empty units: Marks & Spencer? Hutchins Jokes? Kirtley Beauty Salon and Sauna? Tesco? British Home Stores? Explain your answer.

e The town centre shopping centre was deliberately planned to have a number of smaller 'local shops'. Why?

f How many cars are catered for in the nearby car parks?

g Other than by car, how might people arrive at the shopping centre?

h Do you think Telford town centre is a good shopping centre? Which types of people do you think would like it? Can you think of any people who would dislike it?

4 The Telford Development Corporation, which has overseen the whole development of the new town, has been eager to get new industry to locate in the town. Telford is an **enterprise zone** (see *Energy and Industry* book) which means that there are certain incentives for new industries such as ten years free of rates.

Devise a short newspaper advert promoting Telford to modern industries. Figure 15.6 will give you some ideas about what to say and how to present your advert. You will see other examples in the Sunday papers and on T.V. Imagine that you were advertising space in Stafford Park. Try to think of all the advantages that Telford offers.

(Teachers may like to contact the Environmental Interpretation Centre, Stirchley, Telford for further details.)

Figure 15.6

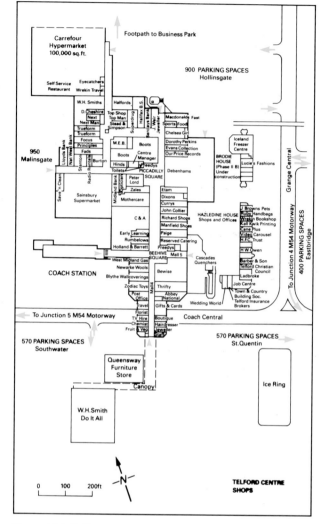

Figure 15.5 Telford centre shops

16 Housing

Figure 16.1 Third World city

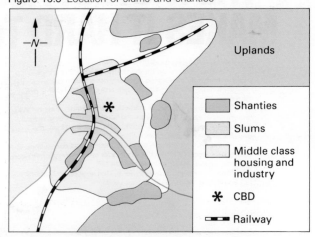

Figure 16.2 Shanty area

If you were to go to a city in the Third World your first view from the airport might look just like any city in Europe or USA with high rise flats and offices. It is not until you travel out of the city centre that you will see what the city is really like (Figures 16.1 and 16.2).

In most Third World cities the vast majority of the population (e.g. Bogota, Colombia 60%; Ibadan, Nigeria 75%) will live in DIY houses made from whatever people can obtain. Materials, which may be bought or simply scavenged from dumps, will include bricks, concrete, cardboard and corrugated iron.

Areas of these houses are called **shanty towns** (locally called **favelas** in Brazil, **barrios** in Mexico, **ranchos** in Venezuela, and **bustees** in India). Sometimes whole groups of people move at the same time causing a shanty to be created almost overnight – this is an **invasion**. It is often carefully planned sometimes by Government officials as shanties partly solve the housing crisis which will save the Government money!

The quality of life in shanties is generally poor. Conditions are overcrowded, there are poor services such as electricity and water, and there are few play areas, shops or forms of entertainment.

Slums, in the centre of cities, are perhaps even worse. They are usually sub-divided older houses with tremendous overcrowding and often dangerous living conditions. These areas are

usually lived in by newcomers to the city. They are prepared to live in such conditions just until they find a job. Then they will tend to move out to build their own house in a shanty.

Figure 16.3 shows the location of different types of housing in a typical Third World City. Compare it to the housing in British cities, Chapter 10.

Figure 16.3 Location of slums and shanties

	Uplands
	Shanties
	Slums
	Middle class housing and industry
*	CBD
	Railway

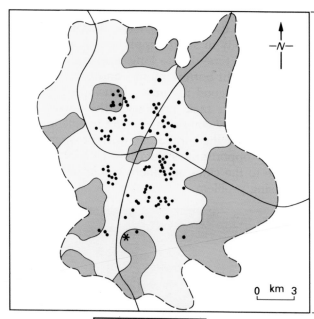

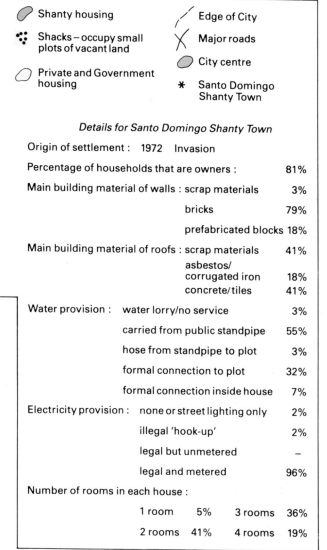

Figure 16.4 Mexico City: housing

Details for Santo Domingo Shanty Town

Origin of settlement : 1972 Invasion

Percentage of households that are owners : 81%

Main building material of walls : scrap materials 3%

bricks 79%

prefabricated blocks 18%

Main building material of roofs : scrap materials 41%

asbestos/corrugated iron 18%

concrete/tiles 41%

Water provision : water lorry/no service 3%

carried from public standpipe 55%

hose from standpipe to plot 3%

formal connection to plot 32%

formal connection inside house 7%

Electricity provision : none or street lighting only 2%

illegal 'hook-up' 2%

legal but unmetered –

legal and metered 96%

Number of rooms in each house :

1 room 5% 3 rooms 36%

2 rooms 41% 4 rooms 19%

ACTIVITIES

1 Study the information in Figure 16.4. It will enable you to make your own report of shanty housing in Mexico City. Your report should describe where shanty housing is found and why, and what life is like in a shanty town. Include maps and diagrams to illustrate your report.

2 Shanty towns are often located in the least desirable areas of the city. These will include alongside roads the railways, near docks, near rivers that may be mosquito infected or prone to flooding, and on steep slopes.

Figure 16.5 shows a shanty settlement in Kuala Lumpur (Malaysia) Describe its location and some of the likely problems to be faced by the residents

3 Different people see things in different ways. Put yourself in the position of the following people and describe your feelings about shanty towns from what you have learnt:

a a peasant man in his mid-20s moving to a shanty with his wife and three young children;

b an old farmer who has lived all his life in the countryside;

c a wealthy businessman living in a middle-class area of the city;

d a European visiting a Third World City for the first time;

e a teenage girl living with her family in a shanty town;

f the owner of poor quality land planning to rent it out at a high price to migrants;

g a Government housing official;

h a man who earns his living selling firewood as fuel to shanty dwellers.

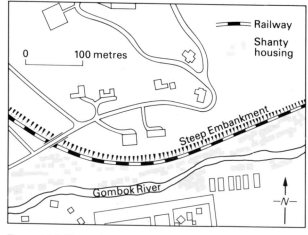

Figure 16.5 Shanty housing in part of Kuala Lumpur, Malaysia

Solving the housing problem

As you have already discovered, squatter
settlements and slums are generally unpleasant
places to live. There are two ways in which the
situation can be improved:

1 Alternative housing
This involves the removal, often by bulldozer, of
shacks and the rehousing of the occupants in
newly built flats. Whilst this sounds attractive it
rarely works well as authorities run short of money
for new buildings or rents are fixed too high for
ordinary people to afford.

2 Improvement and self-help
This is a far less dramatic approach and has been
recently adopted by many Third World countries.
It usually involves the following (see Figure 16.6):

1 retain all but the worst existing homes;
2 provide building materials for the residents to
   improve them;
3 give residents the rights to the land so that they
   are legal owners rather than illegal squatters,
4 install services like electricity, water, sanitation;
5 provide loans for people to improve their
   homes;
6 involve the people in the planning — let them
   decide what improvements to make.

Although this approach has been widely
successful, there are problems. Building materials
are generally in short supply; residents may have
little spare time due to long working hours; some
people might borrow money and end up in debt.

However good the urban housing schemes
might be, the main problem is the **cause** namely
the poor rural conditions forcing people to move to
the cities. If the rural areas were prosperous and
offered opportunities then fewer people would
move to the cities and there would be less of a
housing problem.

Figure 16.6 Self-help in Bolivia

Figure 16.7 Housing in Lusaka 1987

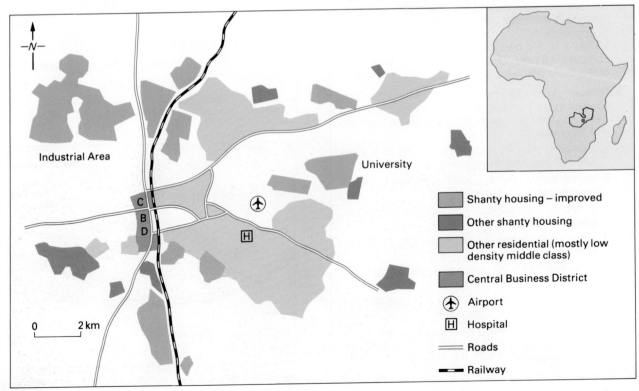

Self-help in Lusaka, Zambia

Lusaka is the capital of Zambia (see Figure 16.7, inset). It has a rapidly growing population – in 1964 it was 127 000 rising to over 600 000 by 1980. Over 40% of the people live in shanty housing made of scraps of cardboard and corrugated iron.

The Government turned to upgrading as the most appropriate solution gratefully accepting guidance and financial help from foreign sources such as Oxfam.

Figure 16.8 Typical shack in Lusaka

ACTIVITIES

4 Study Figure 16.7
  a What type of housing borders the railway in Lusaka? Why is this location generally thought to be unsuitable for housing?
  b Describe the location and extent of 'other shanty housing'.
  c What types of job do you think the shanty dwellers would hold?

5 Study Figure 16.8.
  a What materials have been used to build the house?
  b Do you think the house is safe to live in? (Hint: think about heavy rainfall, fire, etc.)
  c Explain why disease was a major problem in such areas prior to upgrading.
  d Make a simple sketch of the house adding labels to highlight its characteristics.

6 Study Figure 16.9.
  a What advantages does this house have in comparison to Figure 16.8? Make a sketch of the house and use labels to show its advantages.
  b How could this house be improved further?
  c Why do you think the government is improving the **infrastructure** (roads, schools, etc.) as well as the housing?
  d Describe some of the problems that might face the shanty dwellers as they try to improve their living conditions.

Figure 16.9 Modern housing unit in Lusaka

17 Services

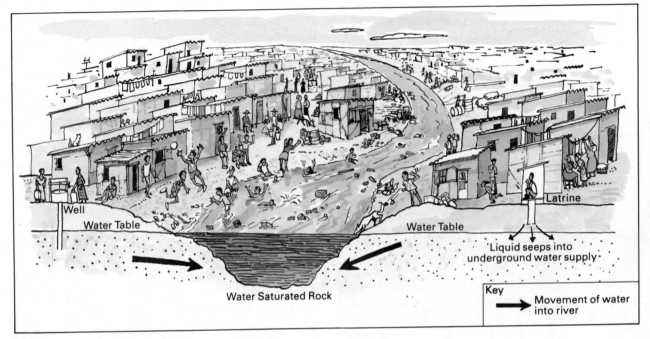

Figure 17.1 Water supply in shanties

The provision of services in many Third World cities is very patchy. The middle-class suburbs would be little different from Britain, however, the majority of the city's population would expect to suffer a great shortage of services such as water and electricity which we in Britain take for granted.

Water

Many urban dwellers obtain their water from rivers or shallow wells. These tend to be polluted (see Figure 17.1) causing disease. If too many people rely on a well or if the rains fail, the well will dry up.

Some people obtain piped water. We tend to think of this as being safe but, if untreated, it can be just as dangerous as river water. Very few of the urban poor have their own private tap. Generally there is one tap between several families with water from that tap being used for cooking, washing clothes, etc. Little wonder that queues exist and that the area around the tap soon becomes foul (Figure 17.2).

Water supplies frequently run dry and people

have to buy water from 'hawkers' who may charge £1 for a kerosene tin, which holds 15 litres of water. Although £1 may not sound much to us, it represents a great deal of money to the city poor.

Water is perhaps the most vital resource. Not only is it necessary for domestic use but it is needed in industry and for fire-fighting. Reservoirs and water treatment plants are essential so that adequate safe water can be provided.

Figure 17.2 Communal tap

Figure 17.3 Open drains in Calcutta

Sewage disposal

Private flush toilets are almost non-existant in the poorer parts of Third World cities. Apart from any other reason, there is a shortage of water with which to flush them!

Instead, people either go in the open or, most commonly, use pit latrines dug into the ground (Figure 17.1).

Most new houses are being fitted with proper flush toilets linked up to sewers. This is a very expensive operation and is, therefore, not very widespread particularly in the poorer cities.

It is important to realise that a number of city authorities consider shanties to be illegal — after all, shanty dwellers have not bought the land on which their houses have been built. Therefore, with no rates paid, why should the authorities provide for safe sewage disposal or indeed any other service such as electricity?

Waste disposal

Figure 17.3 shows open drains in Calcutta, the most common way of removing waste water. Clearly such drains are very unhygienic.

Furthermore, after heavy rain, they overflow whereas in the dry season they become stagnant and stinking.

Solid waste is generally dumped in fields or on waste ground (Figure 17.4). One such open dump in Mexico City takes over 20 minutes to cross — by helicopter! A lot of this waste may be inflammable and non-biodegradable (it doesn't rot away naturally). Fires are common on these dumps often spreading to engulf nearby shanty housing which also occupies such waste ground. Dumps become rat infested and dissolved waste may seep through soil and rock to pollute water supplies.

Some cities do have refuse collection systems but these tend to be irregular and only cater for the richer areas. Incinerators have been installed in some cities to burn solid waste — again, this is expensive to run and may cause pollution.

Electricity

Most houses in most large cities do have a supply of electricity. However, supply is generally unreliable — there may be several power cuts a day. Electricity is essential as open fires and gas lights can be dangerous in areas where houses are mainly wooden structures.

Smaller towns are less likely to have so many houses with electricity.

In some shanty areas it is common to see 'electric piracy' where people illegally tap electricity from someone else's supply. As this overloads the system, it can be very dangerous.

Figure 17.4 Rubbish tip

Roads

The following quote describes traffic in Lagos, Nigeria.

'At the best of times the traffic in Lagos gives the impression that the whole population has suddenly taken to the streets to escape some great disaster. The jams are horrendous. Every other driver appears to be under the impression that it is necessary to sound the horn permanently or the engine will stop'.

(Source: *Migration and Urbanization in W Africa*. Moray House College of Education. p. 80)

Third World cities suffer major traffic and road problems. Many cities, being very old, were not designed for cars and their narrow streets easily get clogged-up (Figure 17.5). Lack of money means that road improvements, as have happened in Britain, are rare and patchy. The result is that congestion is severe with traffic jams in Mexico City, for example, lasting all day.

Accident rates are high as cars mix with bicycles, rickshaws, and pedestrians. The rule of survival is stronger than the Highway Code! Pollution levels (lead from exhausts) are high also.

Roads tend to be pot-holed so reducing the life of a vehicle drastically. In heavy rain, the potholes become pools and the poor drainage leads to roads becoming awash and even turning into temporary rivers!

Many vehicles are old and, combined with the terrible state of the roads, breakdowns are commonplace. In many cases vehicles are in such a poor state that they will simply be abandoned where they break down. This, of course, makes congestion even worse.

In Lagos and in other cities expressways and ring roads are being built to try and reduce the problems. It is important to remember, however, that these countries are poor. They do not have vast amount of money to spend and it is often a case of choosing between providing safe water for the poor or better roads for those who can afford cars.

One interesting policy for reducing the number of private vehicles on the road is to allow cars with registration numbers ending in an even number access on certain days, and those ending in an odd number access on other days. The obvious way round this is to have two number plates!

ACTIVITIES

1 Study Figure 17.6.
  a What proportion of the population of Bichi obtain water from pipes?
  b Explain why this water is generally of a higher quality than water obtained from wells.
  c Work out the percentages of population obtaining water from *each* of the four sources. Present your answers in the form of a table.
  d Describe some of the problems associated with communal or shared wells and taps. Look back to Figure 17.2 for some help.
  e Why do you think the supply of safe water is generally top of the priority list for authorities trying to improve the life of the poor?
  f What proportion of the population of Bichi uses pit latrines?
  g Why do you think pit latrines were the main form of 'toilet'? 1980.

2 The following is a quote taken from the Daily Times of Nigeria in 1980.

'I knew this must be Ibadan (a large town in Nigeria) by its noise, dirt and stench. Refuse swirled in the streets. Makeshift incinerators emitted continuous thick black smoke...children play among rusting bins and abandoned junk. At Ojaoba market, garbage still pursued its course, forcing the closure of many sections of the road and turning short trips into long ones. Still amid the stench, hawkers and flies jostled for a place under the sun.'

(Source: *Migration and Urbanization in W Africa*, Moray House College of Education.)

Look back at Figure 17.4.
  a Try to identify as many types of rubbish as you can.
  b What is the main type of rubbish on the dump?
  c What do you think the people are doing on the rubbish dump?
  d Using the quote above and Figure 17.4 describe the dangers associated with such open dumps.

3 Read the newspaper article Figure 17.7 which describes the many problems in Mexico City (you have already learnt about housing in Mexico City in Chapter 16)
  a List the different problems.
  b Imagine that you were an official given the task of deciding what to do to improve life in Mexico City. You do not have a great deal of money to spend but obviously have to benefit as many people as possible.
    Write a report outlining your suggestions. You may like to devise a priority list using what you have learnt from this chapter.

Figure 17.5 Congested streets

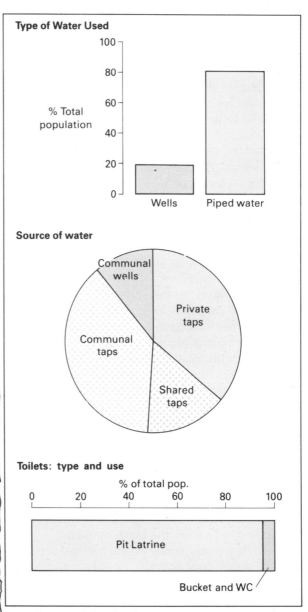

Figure 17.6 Details of water and sewage in Bichi, Nigeria

Figure 17.7 The *Guardian* 20 January 1984

The city in a cloud of despair

IT WOULD be difficult to imagine any new horrors to add to those already suffered by the inhabitants of Mexico City – the traffic, the dangerous driving, the poisonous air, the relentless din, the crime, the altitude, and the sheer size.

Experts have suggested that the toxic fumes rising constantly into the atmosphere, many from uncollected rubbish, could one day ignite. In 1981, a municipal rubbish dump caught fire. The flames and explosions continued for 10 days.

The city lies sprawled across its plateau a mile and a half up, surrounded by still higher mountains. The bowl acts as a trap for smoke and fumes, and an estimated 11,000 tons of pollutants compete with the oxygen daily. On most days a brownish–grey smog haze hangs over the city.

A city health officer has publicly reported that just breathing the air is equivalent to smoking 40 cigarettes a day.

The city is literally killing its inhabitants. Thousands die annually from diseases directly caused by or related to the contamination and pollution.

Pollutants come from the area's 130,000 industries, many belching uncontrolled smoke, and the 2.7 million vehicles that circulate in the metropolis.

The mountains of garbage from households alone accumulate at the rate of 800 tons a day, and are beyond the city's capacity for disposal.

The millions of poor who inhabit the slums which surround the city have no toilets or sewage systems, so they use any spare patch of ground they can find. Health authorities have suggested that 750 tons of human excrement are deposited this way daily. The waste matter dries up and disperses as dust, which is then blown through the city, causing serious illness.

Peasants from impoverished rural communities reach the city at the rate of about 1,000 a day. Their ramshackle slums have no infrastructure or services, and contain millions of people.

The worst are haphazard collections of hovels put together from corrugated tin, industrial cardboard, discarded plywood, and stolen building materials.

During daylight, the men and women who are able to tolerate hours of queuing and travelling in jammed and delapidated buses and (modern and efficient) electric trains to the city in search of work, have gone for the day.

These make up the final horror of Mexico City. Most drivers are single-occupant commuters, and 97 per cent of vehicles carry only a fifth of all travellers.

These commuters, Volkswagen "beetle" taxis, and broken-down buses emitting clouds of black diesel smoke, hurtle along main boulevards and choke side streets, sounding their horns incressantly, shouting insults, and playing music on their radios at ear-shattering volume. Driving is fast, chaotic, aggressive, and extremely dangerous. Triple parking is commonplace.

There are pleasant parts of Mexico City. Sumptuous public buildings of the Spanish colonial period stand as reminders that Mexico City was once regarded as the Paris of the Americas.

18 Employment

It is estimated that about 20% of all Third World urban dwellers are unemployed, although it is impossible to be precise as accurate statistics are unavailable.

Of the rest, the majority are involved in the **tertiary** sector. In many cities over 50% of all jobs are in this sector. Such jobs include clerks, porters, cleaners, taxi drivers, shop assistants, hospital and university workers. The high proportion employed in this sector is higher than in many developed world cities. The manufacturing (secondary) sector employs few people as such industry tends not to be very labour intensive.

People employed in the sorts of jobs mentioned above can be said to belong to the **formal** economy. They will have contracts, pay taxes and probably be part of a pension scheme.

There is an **informal** economy which can employ a large number of people. It involves irregular jobs often self-employed and with no contracts, pensions, etc. There are legal concerns such as shoe cleaning, car watching, and crafts, and illegal concerns like prostitution, theft, scavenging in rubbish tips (see Figure 17.4) and drug pushing.

In Mexico City the 'rat trap' is one such illegal 'job'. A rat is thrown into the open window of a car — the occupant panics and hurries out of the car — the rat owner then leaps into the car, drives it away and sells it on the black market!

Although we may condemn such illegal acts, the informal sector does provide the urban poor with enough money to get by. Some people would argue that there is no alternative in cities where there are more people than jobs.

Perhaps the major characteristic in Third World cities is **underemployment**. This is where too many people are employed on a particular job — for example, four people each changing a tyre at a garage rather than one person doing them all. Underemployment leads to short hours and low wages.

Figure 18.1 A variety of Third World jobs

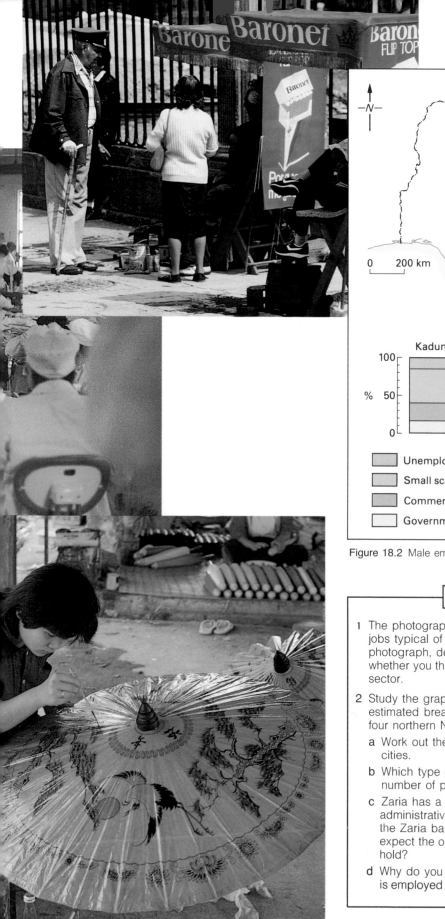

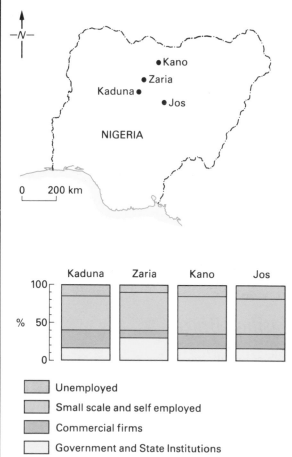

Figure 18.2 Male employment estimates for four Nigerian towns

ACTIVITIES

1 The photographs in Figure 18.1 show a variety of jobs typical of Third World cities. For each photograph, describe the job being done and state whether you think it belongs to the formal or informal sector.

2 Study the graphs in Figure 18.2 which show an estimated breakdown of adult male employment in four northern Nigerian towns.

a Work out the average unemployment rate for the cities.

b Which type of employment employs the greatest number of people?

c Zaria has a large university and many administrative buildings. How is this reflected in the Zaria bar? What sorts of jobs would you expect the ordinary employees of the university to hold?

d Why do you think a small percentage of people is employed in commercial firms?

19 New towns

New towns are not only found in countries of the developed world, they also exist in the developing world. There are generally two types of 'new town':

1 a town, much as in Britain, built to absorb overspill from a nearby major settlement;
2 a town or city built away from the economic core of a country with the aim of promoting growth in a hitherto less developed region. This type of settlement is called a **growth pole**.

In this section we will look at an example of a growth pole — the new capital city of Brazil, Brasilia.

Brasilia, Brazil

Much of Brazil's wealth has been concentrated on its south-eastern coastline. The stretch between Rio de Janeiro and São Paulo has for several centuries been Brazil's core economic region. People have flooded from the interior in the hope of finding better opportunities than are available to them in the countryside. The vast interior of the country saw little development as the coastal belt became steadily more and more crowded.

The idea of a large new city in Brazil is not new with the first recorded suggestion being made in 1789. However, it wasn't until 1957 that work first started on the construction of the 'new capital'. The then president set the deadline for completion as April 21st 1960!

Not surprisingly, the city was built in great haste — the president thought this to be a challenge for the newly developing Brazil to prove itself to the outside world.

40 000 peasants were brought in from the north-east to do the labouring but there were many early problems. It was 125 km away from the nearest railway and over 6000 km from the nearest paved road. Steel had to be transported some 1600 km to the site and timber, some 1200 km.

Nevertheless, as April 21st 1960 dawned, a new city stood on what had previously been rolling upland plains (Figure 19.1). The Presidential Palace was built along with Government offices, the Supreme Court, nearly 100 apartment blocks, and 700 houses with shops and schools. Figure 19.2

shows the layout of the planned city of Brasilia.

Perhaps most importantly of all, a main road was built linking the city with the coast to the south and the Amazon to the north.

Being a newly planned city, Brasilia should have been able to avoid the problems described in earlier chapters. Not so. The haste with which it was planned and built led to a range of problems:

1 a massive debt had to be re-paid, money borrowed to finance construction;
2 only four of the eleven ministry buildings were usable in 1960. Government activity moved slowly to Brasilia;
3 too few houses were built for low-income groups — by 1976, 600 000 people lived outside the city's limits in shanties;
4 shops were built facing service roads rather than the residential areas they were meant to serve;
5 jobs tended to be a long way from housing areas;

Figure 19.1

6 planned tree-lined streets are still awaiting trees!
7 a general shortage of bars and markets, important features of Brazilian life.

The main positive feature of the development of Brasilia has been the increased wealth brought to the interior of Brazil. Nearby towns and cities have witnessed rapid growth.

Brasilia is still being built – it is by no means complete. Many of the early 'teething' problems have been sorted out but already, sadly, Brasilia is characterised by the ills of city life that planners had intended to avoid (Figure 19.3).

Figure 19.2 Brasilia layout

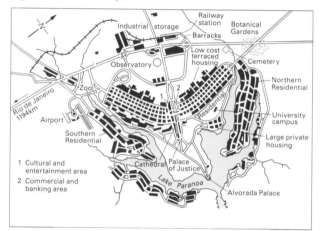

Figure 19.3

1 Draw a map to show the location of Brasilia. On an outline of Brazil mark on the location of the following:
   a cities – Brasilia, São Paulo, Rio de Janeiro, Belo Horizonte, Belem, Goiania, Manaus;
   b rivers – Amazon, Tocantins, São Francisco, Parana;
   c main roads – Brasilia to Belem; Brasilia to Rio de Janeiro; Brasilia to São Paulo via Goiania;
   d uplands – over 1000 m;
   e label the Amazon Basin and the Atlantic Ocean Add a scale and north point to your map.

2 Use your map and an atlas to discover:
   a the distance in km by road from Brasilia to Belem, Rio de Janeiro and São Paulo;
   b the climate, relief and vegetation of the Brasilia area. How might these aspects have affected the building of the city?

3 Study Figure 19.2. Notice how the city plan resembles the shape of an aeroplane with the administrative and cultural buildings forming the fuselage and the residential areas the wings!
   a What buildings are found in the centre ('fuselage') of Brasilia?
   b Compare the location of low income and high income housing.
   c Compare the road layout in the two types of housing area.
   d In which direction and how far would you have to travel from the cathedral to:
      i the airport
      ii the railway station?
   e As you can see, the city is built around a large lake. What are the advantages of such a situation?

4 Under a title 'The two sides of Brasilia' use Figures 19.1 and 19.3 to help you write a few sentences describing the characteristics of the city.

(An excellent source of information on new cities is *Geofile*, April 1988, No. 111.)

20 Rotherham — an industrial town

Rotherham is an urban area containing some 250 000 people. It is located in the county of South Yorkshire and holds a very central position in Britain. It is located a few miles to the north-east of Sheffield (Figure 20.1).

Rotherham dates back to Roman times although most of its growth has occurred since 1746 when local iron was first smelted. Rotherham soon became a thriving industrial centre based on coal mining and metal smelting. Although Rotherham prospered, industrialisation brought with it a number of problems — street after street of gloomy back-to-back terraced housing (Figure 20.2); pollution from coal waste; dereliction and spoil tips associated with mining; etc.

Problems of declining industries

Recently these problems have intensified. Many of the older coal mines have been forced to close as have several metal smelting works (see *Energy and Industry* book). In 1975 13 600 people were in coalmining and 14 500 in steel making — by 1985 these numbers had dropped to 7300 and 6100 respectively! Unemployment between the two dates soared from 6.7% to 22.1%. The effect has been particularly serious in Mexborough (a small town 10 km north-east of Rotherham) where, in 1971, 54.1% of working men were in coal mining.

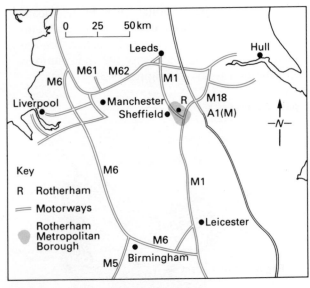

Figure 20.1 Location of Rotherham

As these two major industries have contracted there have been 'knock-on' effects forcing the closure of smaller related industries (Figure 20.3). Shops too have had to close as consumer spending power has been reduced.

Figure 20.3 Factory closure

Figure 20.2 Back-to-back housing

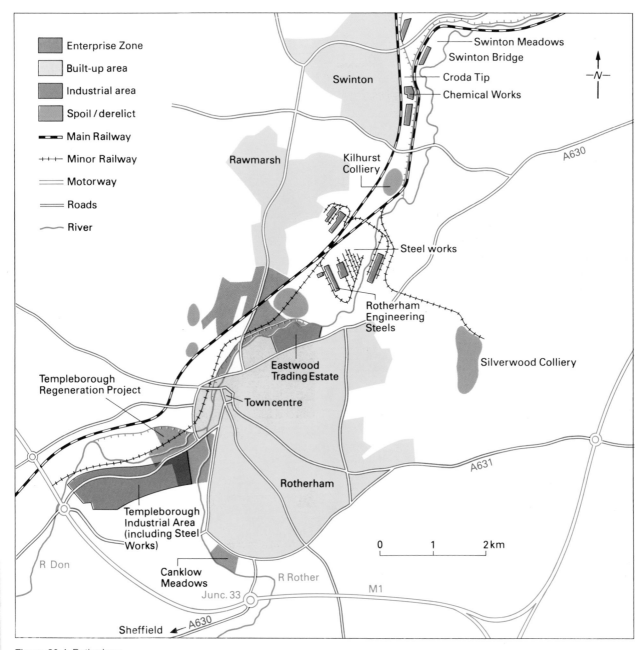

Figure 20.4 Rotherham

Recent improvements

Since 1974 over £10 million has been spent by
Rotherham Borough Council on improvements.
The money for these improvements has come from
the EEC and from the British Government. Part of
Rotherham has been designated an Enterprise
Zone (Figure 20.4). There are great attractions to
new industry moving into an EZ (see *Energy and
Industry* book).

Figure 20.5 Swinton Bridge Workshops

One area with particularly severe problems was Swinton (see Figure 20.4). An old industrial area, it exhibited all the typical features of decline — derelict land, abandoned buildings, poor quality housing and few areas of park and woodland.

It has now been designated an **Industrial Improvement Area**. This enables special financial incentives to be available for improvements. The following list describes some of the improvements:

1 former junior school converted into workshops (Figure 20.5) (cost £100 000);
2 new bridge built crossing canal and railway (cost £1.4 million);
3 improvements to roads and services (cost £400 000);
4 provision of serviced land ready for new industries.

ACTIVITIES

1 a Use the text and photographs (Figures 20.2–20.3) to describe the quality of life for many of the inhabitants of Rotherham during the height of industrial decline.
  b What do you understand by the expression 'knock-on effects'? Give an illustration of this.
  c Use the text and the before/after photographs in Figure 20.6 to describe some of the ways in which life in Rotherham is being improved. Illustrate your answer with sketches of the photographs.

Before

Figure 20.6 Bridge Street
After

21 Aberystwyth: a seaside resort

For the second part of last century and the first part of this, holidaymaking was centred on seaside resorts. When factories closed for a fortnight the masses flooded to their nearest resort. Coastal towns developed amenities for these holidaymakers – piers were built, parks laid out and many hotels constructed along the sea front.

Since the 1950s, however, there has been a change. Cheaper air transport opened up resorts abroad where sunshine could be guaranteed. It is now as cheap to go to Spain as it is to stay in Britain. In 1986 only some 30% of British holidaymakers stayed in Britain. This trend has led to the decline of many resorts with hotels and small businesses being forced to close. Some resorts have embarked on ambitious schemes to attract people – Brighton is now a major conference centre and Morecambe has its Western Theme Park which sells over 200 000 tickets a year.

Aberystwyth

The coastal resort of Aberystwyth (Figure 21.1) is almost in the very centre of the Welsh coastline. Look it up in your atlas. It is, in many ways, a typical small seaside resort although it began life as a port for exporting local lead, slate, timber and wool.

During the Victorian era Aberystwyth became a fashionable resort as the railway enabled large numbers of holidaymakers to reach the town. In 1896 the pier was built and around the turn of the century the resort grew considerably in order to cater for the tourists.

A major problem facing seaside resorts is the seasonal nature of trade. Aberystwyth is fortunate in having the University of Wales as the 2000 students buoy up the economy and enable hotels and gift/souvenir shops to remain open all the year round. Also, the Welsh Tourist Board advertises cheap off-season weekend breaks which are popular.

Figure 21.1 Aberystwyth (aerial photo)

Figure 21.2 Map of Aberystwyth GOAD

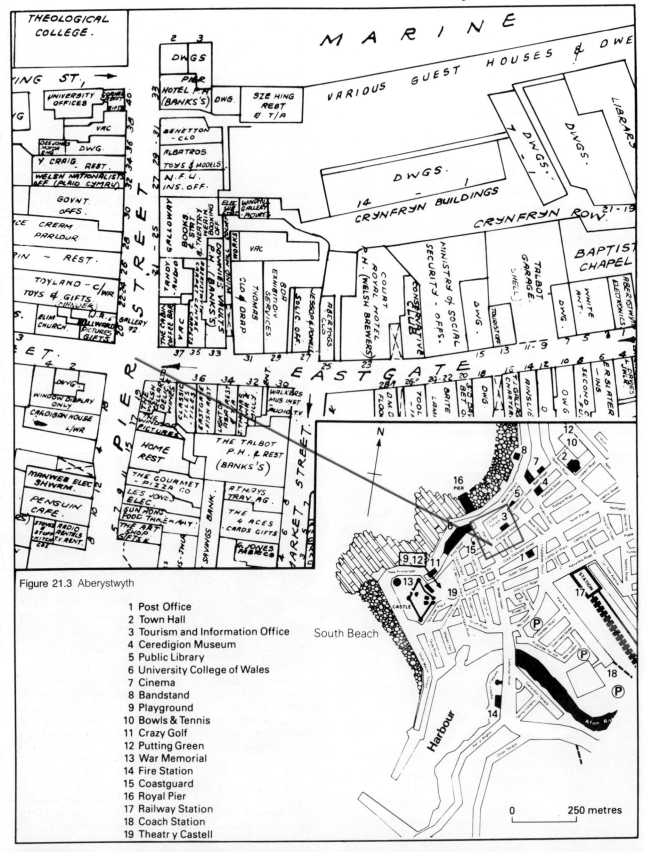

Figure 21.3 Aberystwyth

1 Post Office
2 Town Hall
3 Tourism and Information Office
4 Ceredigion Museum
5 Public Library
6 University College of Wales
7 Cinema
8 Bandstand
9 Playground
10 Bowls & Tennis
11 Crazy Golf
12 Putting Green
13 War Memorial
14 Fire Station
15 Coastguard
16 Royal Pier
17 Railway Station
18 Coach Station
19 Theatr y Castell

1 The majority of Aberystwyth's holidaymakers come from the West Midlands (Birmingham) and the Black Country (Dudley area). Use your atlas to explain why this is so.

2 Study Figure 21.3 which is a map of part of Aberystwyth.

   a The most popular stretch of coast is north of the pier. Why do you think this is so?

   b The photograph Figure 21.4 looks along the promenade — it was taken from the pier. Describe the scene.

   c Describe the route you would take to get from the railway station to the Tourist Information Office.

   d Make a list of the tourist facilities available in Aberystwyth. For each, say which groups of people (e.g. teenagers, elderly) it would appeal to.

3 Figure 21.2 is a GOAD map of part of Aberystwyth.

   a Make a tracing of the map and use colours to shade those buildings/land uses specifically catering for holidaymakers. Use different colours to show different uses, e.g. hotels, gift shops, etc.

   b Roughly what proportion of the total buildings have you shaded?

4 Locate the harbour on Figure 21.3. There has been a proposal to develop this into a marina with a hotel, new shops and flats, a swimming pool and a boating complex. The aim is to try to attract more tourists to Aberystwyth.

   a Imagine that you have been given the job of producing a plan for this development. Make a copy of the area around the harbour and show how you would develop the harbour. Include any developments that you think worthwhile — do not forget to plan car parks and possible road improvements.

   b Some local people are against the proposal. Read Figure 21.5 which is a local newspaper article concerning the proposal. Describe some of the fears of the locals.

5 Design a newspaper advert for holidaying in Aberystwyth. You should describe the facilities on offer (assume that the proposed marina has been built), include maps and illustrations and use your atlas to refer to the climate. Try to think of an eye-catching heading.

Figure 21.4 Aberystwyth sea front

Yachtsman warns of marina dangers

by Glenys Hammond

An Aberystwyth yachtsman with first hand experience of a marina in North Wales warned this week that they are not always what the developers promise.

Retired film cameraman, Ted Brown, remembers what happened when a scheme similar to the one proposed at Aberystwyth was undertaken at Port Dinorwic on the north Wales coast — a local community lost its harbour to outsiders.

"I'm not saying that will happen in Aberystwyth, but I am very much against private developers coming along and forcing ordinary people out", he said.

Mr. Brown lived in Port Dinorwic and had a boat in the harbour.

"Several people in the village had boats and we kept them in the old dock for next to nothing. We had it to ourselves and then came big plans for a massive marina.

There were to be three phases — houses, flats, marina dwellings, caravan park, the harbour would be dredged and all sorts of improvements were to be done to it. It was something like the scheme for Aberystwyth."

"Boat owners flocked down from Manchester and Liverpool. They came on Friday nights in their big cars, with their own food, petrol and diesel oil and contributed nothing to Port Dinorwic. Local people thought their properties would go up in price but not one shop opened in the village, some closed. Port Dinorwic is just as derelict now as it was.

"The marina is there and it all looks very nice. But the first thing that happened was that mooring charges went up sky high. The local chaps were forced out."

Figure 21.5 Cambrian News

22 *Amsterdam: capital city of the Netherlands*

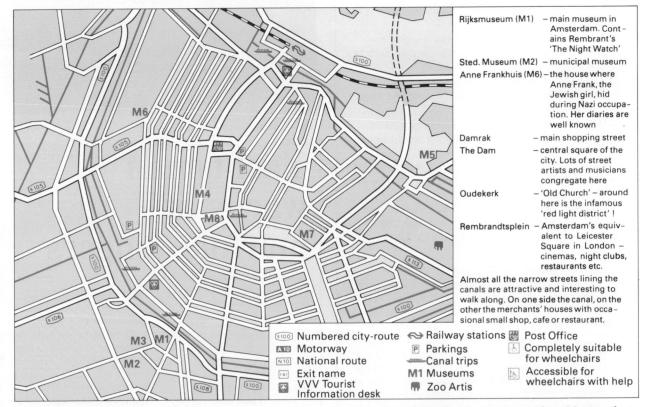

Rijksmuseum (M1) – main museum in Amsterdam. Cont- ains Rembrant's 'The Night Watch'

Sted. Museum (M2) – municipal museum

Anne Frankhuis (M6) – the house where Anne Frank, the Jewish girl, hid during Nazi occupa- tion. Her diaries are well known

Damrak – main shopping street

The Dam – central square of the city. Lots of street artists and musicians congregate here

Oudekerk – 'Old Church' – around here is the infamous 'red light district' !

Rembrandtsplein – Amsterdam's equiv- alent to Leicester Square in London – cinemas, night clubs, restaurants etc.

Almost all the narrow streets lining the canals are attractive and interesting to walk along. On one side the canal, on the other the merchants' houses with occa- sional small shop, cafe or restaurant.

S100 Numbered city-route	🚉 Railway stations	📮 Post Office
A10 Motorway	P Parkings	♿ Completely suitable for wheelchairs
N10 National route	Canal trips	
rai Exit name	M1 Museums	♿ Accessible for wheelchairs with help
VVV Tourist Information desk	🐘 Zoo Artis	

Figure 22.1 Amsterdam: tourist sites

Figure 22.2

Locate the Netherlands in your atlas. Notice that almost all of the country is at, or just below, sea level. This explains why the Netherlands and its neighbours are often called the 'Low Countries'!

There are several very large cities in the Netherlands. Locate Rotterdam, The Hague, Haarlem, and Amsterdam. Notice that Amsterdam is located on the shores of a large body of water called Ijsselmeer. This is gradually being reclaimed to form new areas of land called **Polders**.

Amsterdam is the capital of the Netherlands, although the seat of government is actually in the Hague. The Netherlands can be said to have two capitals!

Amsterdam is a delightful city. The centre is split up by several horseshoe-shaped canals (see Figure 22.1). Most of these were built in the seventeenth century to move goods from the docks to merchants houses which still line the banks of the canals (Figure 22.2).

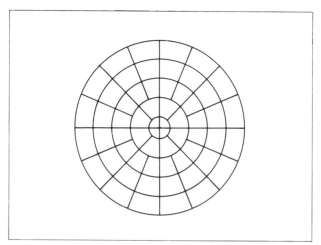

Figure 22.3

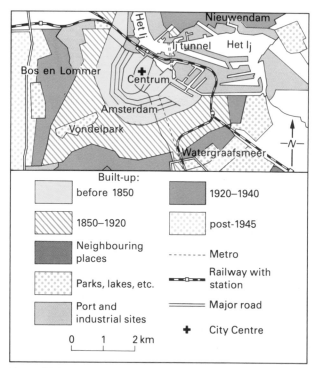

Figure 22.4 Land use in central Amsterdam

ACTIVITIES

1 Study Figures 22.1 and 22.2 and, in pairs or small groups, discuss the following issues concerning tourism in Amsterdam:

  a What are the attractions to tourists in Amsterdam?

  b What groups of people living and working in Amsterdam might be glad to have tourists visiting the city?

  c Why might some groups of people not welcome tourists? (Hint: think about the effect of tourism on the price of goods in shops, car parking, the quality of the environment, etc.)

  d Use Figure 22.1 to suggest a tourist route in Amsterdam to occupy a single day. Your planned route should be aimed at a typical family with teenage children like yourselves. You need to include a variety of attractions on your route to keep everybody happy. Intend that the family will walk between the attractions − this is perfectly possible as the centre of Amsterdam is quite compact.

    Beware of trying to cram too much into the day. Take into account walking time, stops for lunch and drinks, and remember that people will spend at least one hour looking round a museum.

    Produce a sketch map to show your route marking on it the main attractions.

    When all groups have completed this activity, read out the suggested itineries to the rest of the class.

2 Discover the pattern of land use in Amsterdam. Make a tracing of the grid in Figure 22.3. Place this on Figure 22.4 matching up the central +. For each segment of your grid, decide on the main land use. Use a number or letter for each type of land use and write this lightly a pencil in each segment.

    When complete, devise a series of colours for the different land uses. Now colour the grid accordingly.

    The final coloured grid will make it easier to discover whether any pattern exists. Look back to Chapter 2 to compare your pattern to the models of urban land use. Has Amsterdam got a concentric pattern or a sectoral one? Maybe it has a mixture of both.

Although Amsterdam developed into a major commercial centre from the seventeenth century, it had started as a small fishing port at the mouth of the River Amstel. A dam was built across the river to prevent flooding by the sea. Today, the dam gives its name to the central Dam Square.

Amsterdam is not a major industrial centre. It does have some industry, such as the Heineken brewery and some port related activities, but it is mainly an **administrative** city. There are many offices, financial institutions, and business headquarters. The many museums and markets, along with the beautiful 'old town' with canals and tall, narrow merchants houses, make Amsterdam a major tourist attraction.

Amsterdam does have its fair share of problems. The city centre, with its narrow streets and bridges, is totally unsuited to vehicles. Over 7000 premises are listed monuments − they need a lot of money to preserve them and cannot be altered. There is a housing shortage in Amsterdam particularly for the single poorly paid. There are few flats for rent and what there are tend to be expensive. It is estimated that between 20 000 and 70 000 squatters exist in the city. So, it is not only Third World cities that have shortages of housing!

23 Rotterdam Europoort: a port city

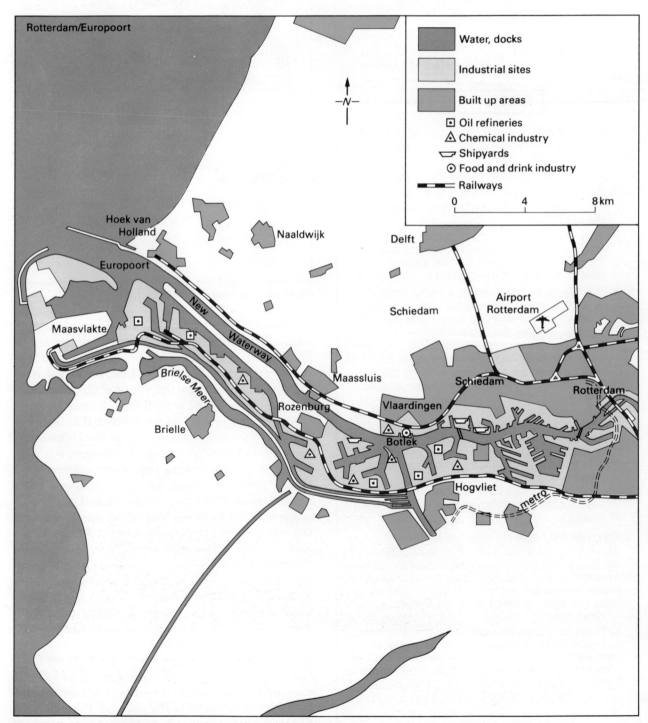

Figure 23.1 Rotterdam /Europoort

Locate Rotterdam in your atlas. Notice that, although it is near the mouth of the River Rhine, the river actually running through Rotterdam has a different name. What is it? Is it a separate river or simply another name for part of the Rhine?

Figure 23.1 shows the port of Rotterdam. Notice the docks near the centre of the city. There are many small basins where boats can berth to load and unload goods. This is the 'old port' of Rotterdam which grew rapidly after the opening of the New Waterway in 1872. This gave the port much better access to the sea. Rotterdam grew up as a transit port as it linked the North Sea with the major internal waterway, the River Rhine.

In the second world war, the docks were severely bombed. Following the war, development of the port spread westwards towards the coast. By 1955, the Botlek (see Figure 23.1) had become an important port zone with many flourishing industries.

Since the 1960s new basins were constructed on the coast itself to form Europoort – the new port of Rotterdam. Maasvlakte (see Figure 23.1) was in fact a platform of sand which has since been reclaimed. Europoort concentrates on the larger ships carrying ores, oil and containers (large metal boxes in which goods can be packed).

Unlike Amsterdam, Rotterdam is essentially an industrial city. It depends greatly on its function as a port with many industries being port-related. There are many repair yards and dry docks particularly in the 'old docks'. Oil refineries and chemical works are located nearby to make use of imported oil products, for example.

ACTIVITIES

1 Study Figure 23.1.
  a Why do you think the early port of Rotterdam was located inland from the coast?
  b How far is it by river from the centre of Rotterdam to the new port of Europoort?
  c Why is Europoort a better location for a modern port?
  d Which industries shown on the map depend on the port?
  e Describe and try to explain the location of 'industrial sites' in Rotterdam.
  f Recently there was a plan to build a huge modern iron and steel works at Maasvlakte. It has, however, been adandoned.
    i Why was Maasvlakte chosen as a good site?
    ii Suggest reasons why the plan has been turned down.

2 Developing a port is often controversial. Figure 23.2 describes a scheme to enlarge the small port of Falmouth in Cornwall. Read through the article and answer the following questions:
  a Why would Falmouth be a suitable site for such a development?
  b What are the objections to the development?
  c If you had to decide whether or not to develop the port, what would you do and why?

Figure 23.2 'Port in a storm' containers at Falmouth

If the new port is built, it will stretch about 20 metres beyond the port's eastern quay and occupy about 75 acres of the mouth of the River Fal, in one of the loveliest drowned river valley estuaries in the country.

It will bring 263 jobs within the terminal and the developers say that many more will be created outside. The town has an unemployment rate of 25 per cent.

Michael Tragett, managing director of the Falmouth Container Terminal Company, said yesterday that the new port should create little or no road traffic.

Most of the containers will arrive weekly in a large ship from the Far East, and be transhipped to smaller vessels for destinations in Europe. A rail link is also planned.

Conservationists have complained that the new terminal will damage marine wildlife, but failed to halt the Bill's initial successful passage through the Houses of Parliament.

Permission already exists to build the new port, but runs out this year.

The Bill's promoters believe that they can start building within the present time limit, but are asking to have the time limit extended, as a precaution.

Yesterday, Mr Tragett said that although he did not have complete funding for the new terminal "I know where it is going to come from".

Mr Tragett added that Peter de Savary, a shareholder in his company who has his own plans for port facilities in Falmouth, had agreed to co-operate fully with the new scheme. However, Mr de Savary is known to be sceptical about the viability of the new container port.

Photographs by Herbie Knott.

Index

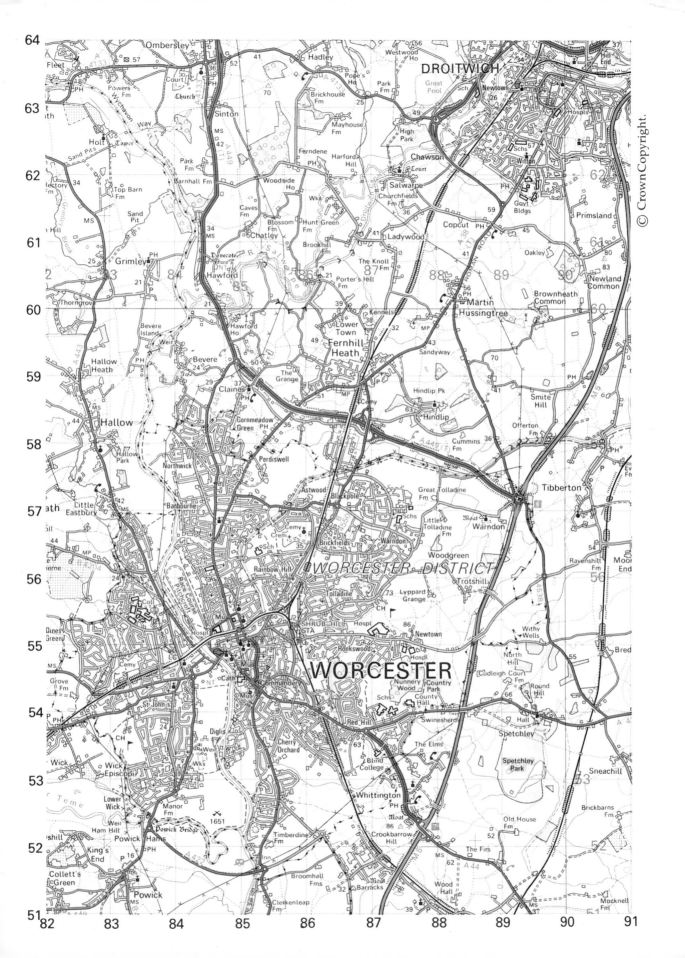

© Crown Copyright.

LONGMAN
CO-ORDINATED
GEOGRAPHY

Series editor - Simon Ross

Agriculture and Rural Issues

Kerstin Jarman & Anne Sutcliffe

We are grateful to the following for permission to reproduce photographs: Aerofilms, page 17; Broads Authority, page 53; J. Allan Cash, pages 3, 55; Farmers Weekly, pages 13 (photo: Peter Adams), 24; Intermediate Technology, page 40; Frank Lane Picture Agency, page 21 (photo: W. Wisniewski), 29; Tony Morrison, page 8; National Remote Sensing Centre, page 19; Oxfam, pages 36–37 (photo: Ced Hesse), 41 above, 42 (photo: Peter Stalk); Panos Pictures, pages 39 (photo: Mark Edwards), 41 below (photo: Paul Harrison); David Pratt, page 28; Reflex, page 54 (photo: Bellavia-Rea); Frank Spooner, page 22–23; Tropix Photo Library, pages 2–3, 63 (photo: John Schmid); A.C Waltham, page 44 above; Reproduced from the Ordnance Survey 1:50000 Wensleydale & Wharfdale Sheet 98, with the permission of the Controller of Her Majesty's Stationery Office, Crown copyright reserved. We are unable to trace the copyright holders of the following and would be grateful for any information that would enable us to do so, page 45.

Cover: Images Colour Library

Acknowledgements

Konrad Bailey
John Bedford
Thomas Bibby
Frank Brennand
Graham Deane
C.G. Bloomfield
Finally, Andrew for his own brand of humour and support

We are grateful to the following for permission to reproduce copyright material: Moray House College 'Land and Water Resources in West Africa – A Teacher's Guide, by J A Hocking and N R Thomson, figs. 8.1, 11.3 and 12.8; John Murray (Publishers) Limited 'Agricultural Change in France and EEC' by H Winchester and B Ilbery figs. 17.4 and 17.7 (17.7 based on data from 'A Rural Policy for the EEC' by H D Clout (Methuen 1984); Thomas Nelson and Sons Limited 'Land Resources and Development' by Robert Prosser, fig 20.1; *New Scientist*, fig 8.4; *Sunday Times Magazine*/Line and Line for the original drawing, 1.5.88 (c) Times Newspapers Limited; (c) Times Newpapers Limited/Peter Brookes 28.11.86.

We are unable to trace the copyright holder of the following and would be grateful for any information that would enable us to do so, fig 19.3

Agriculture and rural issues

Kerstin Jarman and Anne Sutcliffe

Contents

1 Rural areas

Rural is a word used to describe the countryside. It includes activities and land uses such as farming and perhaps small scale industry. **Urban** is the opposite: a word which describes town or city environments.

Traditionally rural populations are mainly involved in agriculture as in the developing countries today. In the developed world, however, improvements in transport and the rise in car ownership have meant some people prefer to live in pleasant rural areas and commute to nearby urban areas for work. The urban population in 1981 was approximately 49 million in the UK whereas the rural population was approximately 6 million. This gives only a general idea and disguises the fact that some rural areas like Norfolk have increased their population by as much as 10% since 1970 due to migration from cities like London. (See Figure 1.1.) (See also *Population*.)

Contrasts in rural areas

Study the photographs in Figures 1.2 and 1.3 carefully. Compare how people have made use of the land. These photographs help illustrate the large differences between rural areas in developing and developed countries. In both cases the rural area is being used as a valuable **resource**. The countryside may need protection to make sure it is not over **exploited** by human activities.

In Britain, which is densely populated with a small land area, rural areas are used for farming, forestry, leisure, mining, energy and wildlife. This may mean **demand** is greater than supply. In the developing world, population pressure may mean that good agricultural land is in short supply. Misuse can lead to erosion. In India most people live in rural areas (525 million compared to 159 million urban dwellers) and try to earn their living from the land. Rapid growth of the rural population in the last twenty years has meant increasing shortages of farmland.

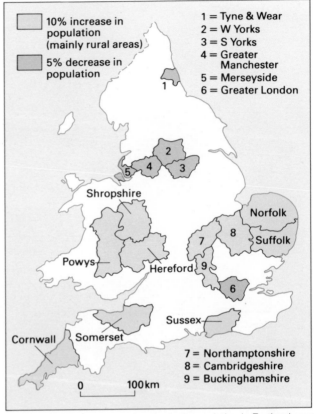

10% increase in population (mainly rural areas)

5% decrease in population

1 = Tyne & Wear
2 = W Yorks
3 = S Yorks
4 = Greater Manchester
5 = Merseyside
6 = Greater London

Shropshire

Norfolk

Powys

Hereford

Suffolk

7 = Northamptonshire
8 = Cambridgeshire
9 = Buckinghamshire

Sussex

Cornwall   Somerset

0   100 km

Figure 1.1 Changes in rural and urban population in England and Wales 1971–81 (census data)

Figure 1.2 Rural scene in the developing world

Figure 1.3 Village scene in the developed world

ACTIVITIES

1 What is meant by the terms rural and urban?

  a Study Figure 1.1 and name four counties in Britain where population has increased by 10%.

  b What is common to the four counties which have a decrease of population of 5% or more?

2 Suggest reasons why people may be prepared to live in more rural areas despite having to travel to urban centres for work.

3 **a** Using Figures 1.2 and 1.3, draw a table to show the contrast between rural areas in the developed and developing world. Your headings should be:
Transport   Land uses   Building materials
Layout of buildings   Layout of farmland
Communications.

  b Use your table to give a brief written summary contrasting the regions.

2 Factors affecting farming

Farming is a very important primary industry especially in developing countries like Haiti where agricultural products contribute 38% to the Gross Domestic Product (GDP). (In the UK it is 2%.) Farming is a business and farmers have to ensure they make enough profit to pay for all the machinery, fertilizers, seeds and other products they may need. The term used to describe this type of farming is **commercial**. It is very different from **subsistence** farming found in many developing countries where little or no food is grown to sell. Rather, it is used to feed and support the farmer's family. The main differences between commercial and subsistence farming are shown in Figure 2.1.

A farmer has to decide what type of farming is going to give him the best returns on the land he farms. His decision can be influenced by *three* main groups of factors:

Physical

1 *Types of soil*: clay soils are often too wet for growing crops so are used as pasture, e.g. with dairy cattle. Loam, which is a mixture of clay and sand, is the best soil. It is suitable for arable or crop farming.
2 *Climate*: this includes daily and seasonal temperatures, the amount and distribution of rainfall in an area, and the hours of sunshine. All these climatic factors determine the length of the growing season.
3 *The relief* or height of the land: temperatures decrease with height and machines may be difficult to use.
4 *Aspect of slope*: south facing sides of valleys are warmer than those facing north (in the northern hemisphere). A farmer may grow crops on the south facing slope and use the north facing slope for pasture. This occurs in the Weald area of southern England.

Physical factors will be particularly important to subsistence farmers. They may not have the money or technology to overcome shortages of water or extreme changes in the weather.

Economic

1 *Price of land* or cost of renting buildings.
2 *Cost of machinery* needed; of new seed or stock; of fertilizers and pesticides.

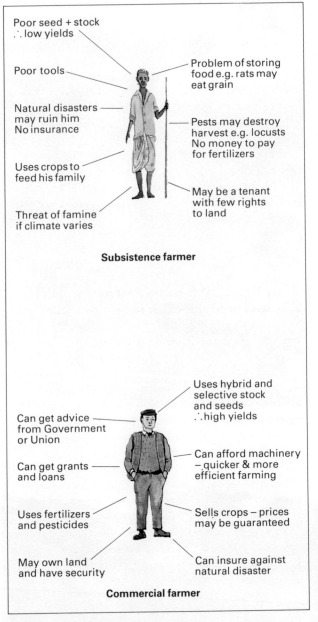

Figure 2.1 Differences between subsistence and commercial farmers

Subsistence farmer
- Poor seed + stock ∴ low yields
- Poor tools
- Natural disasters may ruin him No insurance
- Uses crops to feed his family
- Threat of famine if climate varies
- Problem of storing food e.g. rats may eat grain
- Pests may destroy harvest e.g. locusts No money to pay for fertilizers
- May be a tenant with few rights to land

Commercial farmer
- Uses hybrid and selective stock and seeds ∴ high yields
- Can get advice from Government or Union
- Can get grants and loans
- Uses fertilizers and pesticides
- May own land and have security
- Can afford machinery – quicker & more efficient farming
- Sells crops – prices may be guaranteed
- Can insure against natural disaster

3 *Government subsidies*: a subsidy is money given to a farmer by the government to ensure he gets a reasonable price for his product and makes a sufficient profit to stay in business. Farmers in the developing countries rarely get these as they are too costly for their governments.

Human

1 The amount and type of *labour* required.
2 *Personal choice*. This can be important as a farmer may decide to farm livestock, for example, when his land may be ideal for crop farming.

Some types of farming need extra labour at harvest time. Migrant workers may be hired, as occurs along the north eastern coast of Brazil on the sugar cane plantations. **Intensive** farming is where farms may be small but a large labour force is required to do all the different tasks, an example being market gardening (fruit and vegetables). Intensive farming produces a high yield from a given area of land. **Extensive** farming is the opposite where there is a large farm but few workers, for example cattle ranching in Argentina or hill farming in Britain. Here, a low yield per hectare results.

The farm system

Study Figure 2.2. It shows the **farming system**. Any farm needs **inputs** or resources to allow the farm to operate. The **processes** on a farm will include the day to day work and organisation. The **outputs** are the products of the farm but will also include waste. Some outputs like fodder crops may be used on the farm to reduce costs of inputs. With subsistence farming all outputs are used by the farmer. A farm system diagram is helpful as it shows very simply what happens on a particular farm. It is possible to draw a similar type of diagram for a farming type such as **pastoral**. Such a diagram would, of course, be more general as it does not refer to any single farm. Pastoral farming is where the outputs are mainly animal products. Crops grown on pastoral farms are usually fodder crops to reduce animal feed costs. **Arable** farming is where crops are main outputs and **market gardening** where fruit and vegetables are produced.

Coursework ideas

Find out what food products in your local supermarket come from the developing world. Decide if they are luxury goods. If so, why? Are they seasonal? Many countries grow export crops rather than food crops. Why do you think this is so? What problems might (and do) result?

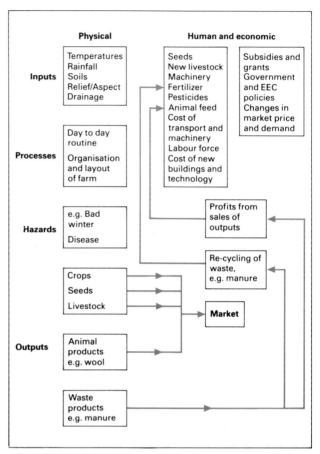

Figure 2.2 The farm system

ACTIVITIES

1 Most countries of the developing world have a high proportion of subsistence farmers. Use Figure 2.1 to suggest some of the problems facing these farmers.

2 Find out from your larder at home, or from your local supermarket, food products that come from the developing countries.

   a Make a list of products and name the country they come from.

   b Are these vital foods or luxury items?

   c If the prices of these products were raised, do you think people would still buy them?

   d If people were not prepared to pay higher prices, what effect would this have on the developing country?

3 How do the following factors affect farming?

   a labour;

   b relief of the land;

   c type of soil;

   d aspect of slope.

3 World pattern of farming

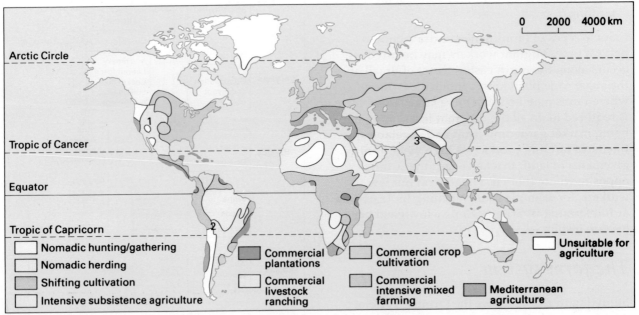

Figure 3.1 World map of farming

Figure 3.1 shows a map of the different world farming types. It is very difficult to classify every type of farming exactly but Figure 3.1 shows the general classification. Obviously farming will vary far more on a local scale.

Figure 3.2 shows the pattern of farming in the UK. Notice that the main arable farming areas are in the east, particularly in East Anglia. Pastoral and mixed farming occur in the west and highland areas. Market gardening is present near to urban centres where there will be a market for the perishable produce. An exception to this is the Fens region and Norfolk where large scale market gardens produce foods for the freezing and canning factories in the area.

Figure 3.2 Main types of farming in the UK

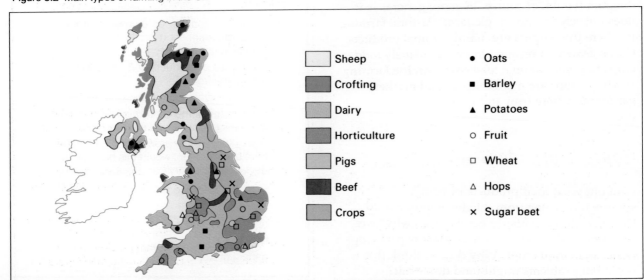

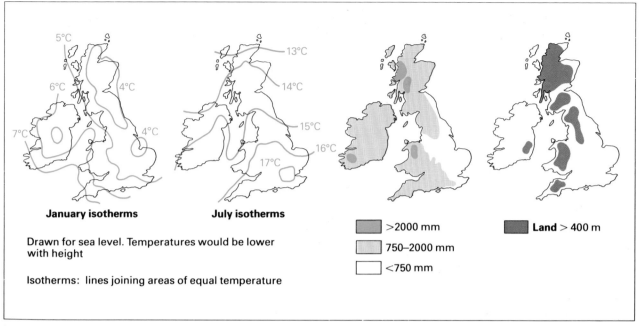

January isotherms   **July isotherms**

Drawn for sea level. Temperatures would be lower with height

Isotherms: lines joining areas of equal temperature

>2000 mm

750–2000 mm

<750 mm

Land > 400 m

Figure 3.3 January and July isotherms, rainfall and relief

Figure 3.4

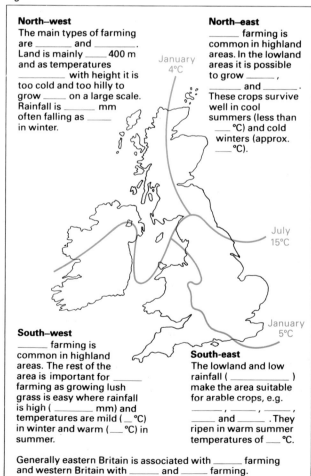

North–west
The main types of farming are _____ and _____.
Land is mainly _____ 400 m and as temperatures _____ with height it is too cold and too hilly to grow _____ on a large scale. Rainfall is _____ mm often falling as _____ in winter.

North–east
_____ farming is common in highland areas. In the lowland areas it is possible to grow _____, _____ and _____. These crops survive well in cool summers (less than ___ °C) and cold winters (approx. ___ °C).

January 4°C

July 15°C

January 5°C

South–west
_____ farming is common in highland areas. The rest of the area is important for _____ farming as growing lush grass is easy where rainfall is high (_____ mm) and temperatures are mild (__ °C) in winter and warm (__ °C) in summer.

South–east
The lowland and low rainfall (_____) make the area suitable for arable crops, e.g.
_____, _____, _____ and _____ . They ripen in warm summer temperatures of __ °C.

Generally eastern Britain is associated with _____ farming and western Britain with _____ and _____ farming.

1 Using Figure 3.1 and your atlas:
  a Study the types of farming found in tropical areas. List these under the following general headings, naming an area where each type of farming is found:
    i subsistence pastoral;
    ii subsistence crop;
    iii commercial pastoral;
    iv commercial crop;
    v mixed.
  b Using your atlas, suggest reasons why some areas in the tropics are unsuitable for farming.

2 a Which type of farming is carried out in each of the following places:
    i UK;
    ii Japan;
    iii Italy;
    iv Canada;
    v Greenland;
    vi Argentina;
    vii Eastern coast of Queensland (Australia).
  b Why do you think northern latitudes beyond the Arctic circle are unsuitable for commercial farming?
  c Suggest reasons why people living in areas 1, 2 and 3 might have difficulty earning a living from farming.

3 a Make a copy of Figure 3.4. Fill in the gaps using information from Figures 3.2, 3.3 and 3.4.
  b Suggest why your diagram provides a better understanding of British farming than Figure 3.1.

4 *Commercial ranching in Argentina*

Pastoralism is a term used to describe livestock rearing for products like wool, meat and milk. It can be intensive, like dairy farming in Western Europe which is very scientific giving a high yield per hectare. Or it can be extensive like the cattle and sheep **estancias** of South America, ranches in Australia and hill farming in Britain. In parts of Africa, for example the Sahel, pastoralism is still important for many nomadic people. They travel around with their animals looking for pasture and are often under threat from famine and drought.

CASE STUDY

Estancia San Ramon, Argentina: an extensive farm

Introduction

Estancia San Ramon is in an area called Patagonia within Argentina. With your atlas locate the estancia using Figure 4.1 to help.

The estancia is 22 000 ha in size, which is very big when compared with British farms (60 ha average), but is about average for this part of Argentina. Figure 4.2 helps explain why farms need to be so big. The grass vegetation is called **pampas** but it provides poor grazing so the farmer needs a large area to support his animals. The climate also makes it difficult for rich pastures to grow.

The estancia

The estancia is split into four mini companies which each contribute to the final profits made (see Figure 4.3). They are:

1. The ranch which includes livestock.
2. Forestry, including the woods and sawmill.
3. A nursery where tree seedlings are grown.
4. Hunting: the customers are visitors from cities who pay to shoot game.

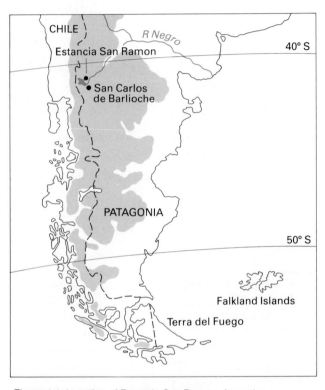

Figure 4.1 Location of Estancia San Ramon, Argentina

Figure 4.2 Pampas vegetation and Gaucho worker

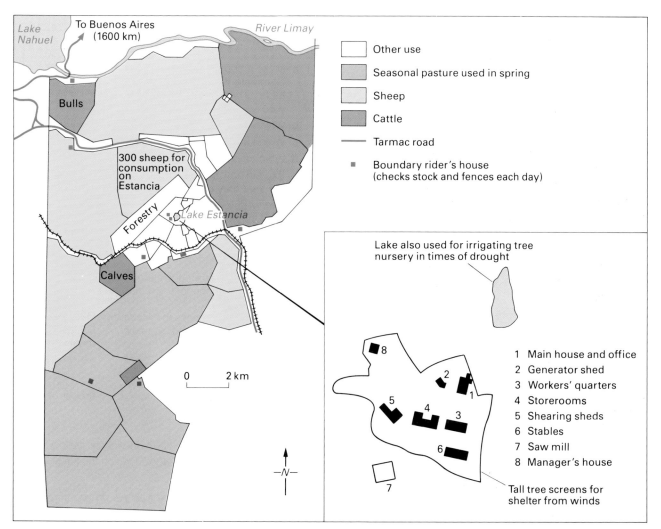

Figure 4.3 Plan of Estancia San Ramon

The labour force is made up of 25 farm workers, 2 managers, and 10 forestry workers. Many are migrant workers from Chile who stay 9 months a year and are provided with accommodation. They are paid higher wages than they would get in Chile and this encourages young men to leave their home areas.

The estancia has 10 000 sheep, 500 cows and 100 horses. The horses are important for travelling around the large fields as few roads exist.

Shearing takes place in October (see Figure 4.4) and takes 10 days with electric shears. Each sheep produces 4 kilos of high quality wool which is sold at $2 per kilo. It is an advantage that wool may be stored in the shearing shed until the market price is high. It is sold in Buenos Aires but meat products are sold in the nearby tourist town of Bariloche. Winter skiing boosts the population of this town to over 100 000.

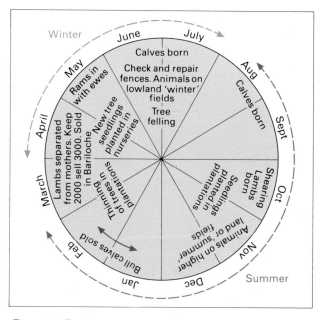

Figure 4.4 Farming year on estancia

Forestry has increased in importance recently. The Chilean workers are ideally suited as there are many timber companies in Chile. Seeds are taken from mature conifers and grown in a nursery for two years. They are transplanted to fields for a further two years and finally planted in 50 ha plots (approx. 50–70 000 trees) on poor rocky soils. After 10–12 years some thinnings are sold for firewood. After 30–40 years the trees are felled for boards and beams. None goes for paper and pulp – which would be more profitable – as there are no pulp mills for 1000 km.

The profits the farmer makes can be affected by international prices for meat and wool. It is difficult for the Argentinians to compete against subsidies given to European and American farmers.

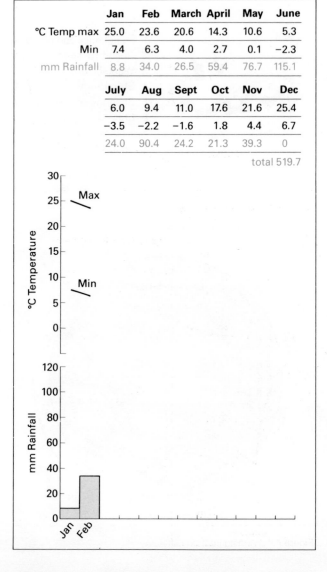

	Jan	Feb	March	April	May	June
°C Temp max	25.0	23.6	20.6	14.3	10.6	5.3
Min	7.4	6.3	4.0	2.7	0.1	−2.3
mm Rainfall	8.8	34.0	26.5	59.4	76.7	115.1

	July	Aug	Sept	Oct	Nov	Dec
	6.0	9.4	11.0	17.6	21.6	25.4
	−3.5	−2.2	−1.6	1.8	4.4	6.7
	24.0	90.4	24.2	21.3	39.3	0

total 519.7

ACTIVITIES

1 a Make a copy of the climate graph (Figure 4.5). Complete the graph using the figures in the table. (The first two months have been done for you.)

b Which months have the highest temperatures?

c Which months have the most rainfall?

d Which months have very little rain or none?

e Explain how this distribution of rainfall through the year may affect the pasture and therefore quality of stock.

f How might the climate of the San Ramon area be a problem for a farmer wishing to grow crops? Use your atlas to compare the climate with that of London.

2 Complete the following systems diagram using information from the text and diagrams.

Inputs	Processes	Hazards	Outputs
. . . .	shearing	snow	calves for meat
500 cows	. . . .	foxes	. . . .
. . . .	. . . .	birds of prey	. . . .
150 calves	planting trees	forest fires	timber
3000 lambs			

3 a How much money in US $ does the estancia earn from the sale of wool? Do you think this is an important output?

b With reference to Figure 4.3 how do you think the wool is transported? Give reasons to explain your answer.

4 a Using Figure 4.3 to help you, approximately how big are the largest two fields in the north of the estancia?

b How do they compare with the sizes of fields on a British hill farm (see Chapter 6)?

5 Why does the estancia need to cover such a large area?

6 With reference to Figure 4.3:

a Why do workers' houses have to be provided?

b What evidence is there to suggest that strong winds are common in this region?

Figure 4.5 Climate data for San Ramon

5 Dairy farming in Great Britain

Dairying is a form of pastoral farming where cattle are kept for milk production. In Britain dairying is **intensive** as a lot of technology is used to produce high yields. Study Figure 5.1. It shows the distribution of dairying in Britain. Notice that it is particularly concentrated in certain preferred areas. These are all areas with low lying plains or gentle hills which receive over 1000 mm of rainfall a year. Temperatures vary, but on average they are between 4°C and 16°C all year round. This means that grass growth is favoured and cattle can be kept outside for most of the year.

CASE STUDY

Old Hall Farm, Lancashire

Mr Bibby at Old Hall Farm in North Lancashire (Figure 5.2) owns a typical dairy farm. It is a small family run farm with a variety of animals (cows, sheep and poultry) to ensure maximum profit. Mr Bibby does not, therefore, rely entirely on milk production.

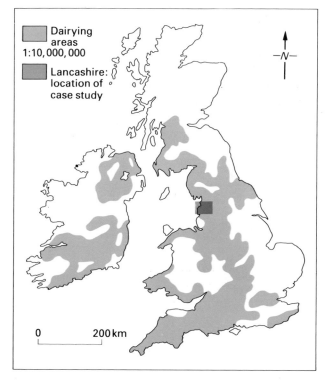

Dairying areas 1:10,000,000

Lancashire: location of case study

0   200 km

Figure 5.1 Dairying areas, UK

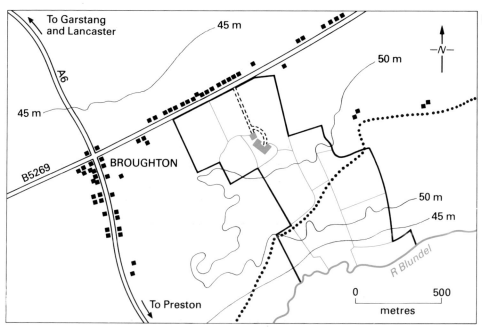

Figure 5.2 Old Hall Farm

To Garstang and Lancaster

45 m

50 m

A6

45 m

B5269   BROUGHTON

50 m

45 m

R Blundel

0   500

metres

To Preston

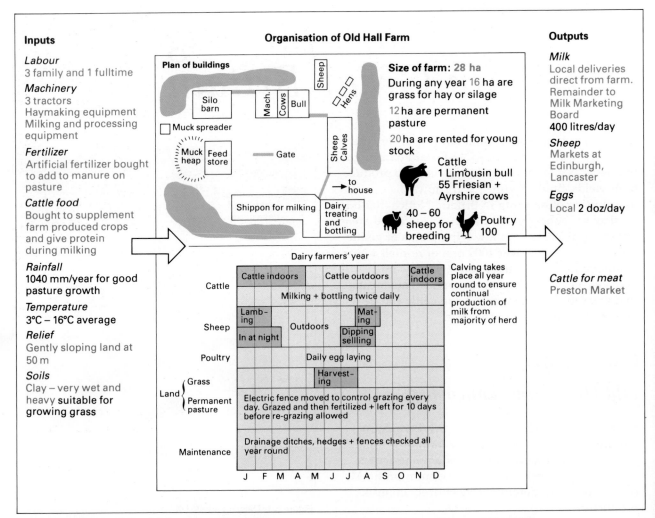

Inputs

Labour
3 family and 1 fulltime

Machinery
3 tractors
Haymaking equipment
Milking and processing equipment

Fertilizer
Artificial fertilizer bought to add to manure on pasture

Cattle food
Bought to supplement farm produced crops and give protein during milking

Rainfall
1040 mm/year for good pasture growth

Temperature
3°C – 16°C average

Relief
Gently sloping land at 50 m

Soils
Clay – very wet and heavy suitable for growing grass

Organisation of Old Hall Farm

Plan of buildings

Silo barn
Mach. Cows Bull
Sheep
Hens
Muck spreader
Muck heap
Feed store
Gate
Sheep Calves
→ to house
Shippon for milking
Dairy treating and bottling

Size of farm: 28 ha

During any year 16 ha are grass for hay or silage

12 ha are permanent pasture

20 ha are rented for young stock

Cattle
1 Limousin bull
55 Friesian +
Ayrshire cows

40 – 60 sheep for breeding

Poultry 100

Dairy farmers' year

	J F M A M J J A S O N D
Cattle	Cattle indoors / Cattle outdoors / Cattle indoors — Milking + bottling twice daily
Sheep	Lambing / In at night / Outdoors / Dipping selling / Mating
Poultry	Daily egg laying
Land { Grass / Permanent pasture	Harvesting — Electric fence moved to control grazing every day. Grazed and then fertilized + left for 10 days before re-grazing allowed
Maintenance	Drainage ditches, hedges + fences checked all year round

Calving takes place all year round to ensure continual production of milk from majority of herd

Outputs

Milk
Local deliveries direct from farm. Remainder to Milk Marketing Board
400 litres/day

Sheep
Markets at Edinburgh, Lancaster

Eggs
Local **2 doz/day**

Cattle for meat
Preston Market

Figure 5.3 Systems diagram of Old Hall Farm

Look at Figure 5.3 which summarises the organisation and layout of the farm. Notice that there is only a small labour force and machines are used wherever possible. For example milking, which is carried out twice a day, is done by machine (see Figure 5.4) with the 55 Friesian cows producing approximately 400 litres a day between them. This is possible because of the artificial feed and proteins which the farmer buys to supplement the cows' diet of grass. Some grass is also harvested and used as hay or turned into silage (grass which is cut and preserved in a silo). Both are fed to the cows especially in the winter months when they are indoors.

The Milk Marketing Board buys some of the milk from the farm at a guaranteed price. This milk is collected daily and taken to a local dairy where it is treated (pasteurised) and bottled along with milk from other farms in the area. Mr Bibby also sells a lot of his milk directly from the farm as he has his own treating and bottling machinery.

Since 1984 the EC has tried to reduce the amount of milk produced by farmers by setting **quotas** (limits) for each farm. The farmer will only receive payment for milk produced within these limits. In the past too much milk was being produced. Much of it, when turned into butter or skimmed milk powder, can be stored but this is very expensive and takes up a lot of space. This means that farmers have had to cut down their production of milk by selling some of the cows for meat. Mr Bibby has not suffered too much because he keeps other animals which provide him with extra income. He has also sold some of his land so the farm is now smaller than it was several years ago. He has bought a Limousin bull because a Limousin-Friesian cross produces a cow which is good for both meat and milk. He would eventually like to produce a pedigree Limousin herd for beef.

Figure 5.4 Milking parlour

ACTIVITIES

1 What is dairy farming?

2 Study Figures 5.2 and 5.3:

  a Describe the physical factors which are necessary for dairy farming.

  b List the human and economic inputs which are necessary for dairy farming.

  c Look at the diagram of the dairy farmer's year.
    i What activities does the farmer carry out every day?
    ii What extra activities does the farmer do in the summer (June, July and August)?
    iii What happens to the animals in the winter months?
    iv Why are electric fences used on the farm?

  d Why do you think fertilizers are necessary?

  e What are the outputs from Mr Bibby's farm?

3 a What is silage?

  b Why is this needed?

4 How do you think Mr Bibby could make extra money from his farm?

5 How has the EC affected dairy farming?

6 Why is Mr Bibby's farm **intensive**?

Coursework ideas

1 Find out if there is a dairy farm near you. You could base an inquiry around your local farm. Suitable topics would include:

  a How have human, economic and physical factors affected the local farm?

  b Have EC quotas had much effect?

2 Visit a dairy near you. What is the **sphere of influence** of this dairy in terms of:

  a the farms it collects the milk *from*;

  b the households it delivers milk *to*.

You could compare these spheres of influence with your local newspaper delivery or conduct a survey to find the spheres of influence of other products.

6 Hill farming in Great Britain

CASE STUDY

Ellerbeck Farm, Yorkshire

Figure 6.1 Ellerbeck Farm

Turn to the map extract, Figure 7.1, on page 18.
Locate Ellerbeck Farm at grid reference 732785.

At this height the weather (see Figure 6.2) has a
large impact on the success of the farm. The land is
classified as 'seriously disadvantaged'. Farming
here is therefore **marginal** as it is on the edge or
margin of cultivable land (see Figure 6.3).

The most profitable farming is to keep sheep and
beef cattle. Mr Brennand owns the 283 hectare
farm. He has three workers and keeps 1100 ewes
and 30 rams for breeding and producing 'fat' lamb,
i.e. lambs which can be bought by lowland farmers
for fattening. The Dalesbred, Swaledale (see Figure
6.4) and Blue Faced Leicesters not only provide
meat but wool, and on average produce 2002 kg for
the farmer who sells it to the North East Wool
Board.

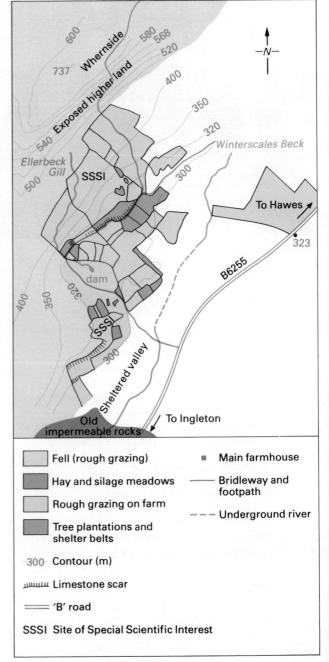

Figure 6.3 Plan of Ellerbeck Farm

Figure 6.2 Climate table for Ellerbeck Farm, 1987

	J	F	M	A	M	J	J	A	S	O	N	D
Av. Rainfall (mm)	88.9	20.8	131.8	152.4	65.0	133.8	239.8	335.8	261.1	113.0	124.9	243.6
Av. Temp °C	5	5	7	10	12	15	16	15	14	10	8	6

Figure 6.4 Mr Brennand worming the sheep in early summer

Hill farming in the future

As Ellerbeck is within the Yorkshire Dales National Park many people walk the footpaths around the farm. The footpath at 739773 (see map extract on p. 18) was used by 120 000 hill walkers between July and September in 1987. Footpath erosion (see Figure 21.6 and p. 19) is a big problem as it destroys pasture. Mr Brennand gets some help for repairs from the National Park Authority. Walkers also do damage to dry stone walls when they remove the top stones to sit on when having their picnics! Unless replaced the walls soon collapse and take many hours and considerable skill to mend. Time which farmers can ill afford!

The farm has two Sites of Special Scientific Interest (SSSI) which are areas to protect rare plants (see Figure 6.3). It is possible that the Government scheme of Environmentally Sensitive Areas may extend to this farm in the future. This may mean that farmers like Mr Brennand will be paid to farm traditionally, introducing few new farm methods so that the landscape is conserved for future generations to visit and study. Even if this were to happen the hill farms will still provide vital breeding stock of sheep.

The sheep population is 'rotated' every few years. For example, 100 new ewes are bred and 100 four year old animals sold. This way new healthy breeding stock is maintained (see Figure 6.6). As the weather conditions are so harsh, Mr Brennand receives a **subsidy** from the Government and the EC to help pay for winter feed (£6.45 for every sheep and £54.40 for every cow). Summers are often too wet (see Figure 6.2) to produce hay, so silage is made. Nevertheless he still needs to buy sugarbeet and sheep pellets for the winter. About 200 first year ewes are sent by Mr Brennand to lowland farmers in the east to winter on arable land. He pays £10 for every sheep and these return to Ellerbeck in the spring to lamb.

Sheep alone do not provide a good income so beef cattle are reared. 60 Friesian and Hereford cross suckler cows produce calves. Some are kept to replace older breeding cows but most are sold in the autumn for meat. The bulls are Marchigiana or Charolais which produce faster growing calves than Herefords — an important consideration if calves are to be sold after a few months.

Machinery is kept to the minimum as the relief is too steep. The two sheepdogs are vital but so too are the three wheeler motorbikes (see Figure 6.5) which can cope with rough terrain and can carry feed or livestock in small trailers.

Figure 6.5 The wide wheels of the motorbike do not damage the land

January	February	March	April	May	June
Medication given for parasites Medicine given to improve sheep condition to help with lambing		Inject against lamb pneumonia	Lambing Lambs marked 1st year ewes back from wintering in lowland Calves born		Walling Worm sheep Shear 1st year ewes Castrate male lambs Calves suckling →

July	August	September	October	November	December
Shear ewes Hay and silage made for cattle fodder	Dipping for sheep scab (Inspector visits)	Lambs weaned	Dipping New ewes to lowland for winter Sheep sales Sell and buy new rams → Calves weaned fed on silage and pea straw	Mating 50/60 ewes to 1 ram Calves sent to market at Hawes	Sheep on fell

Figure 6.6 Sheep farm year

ACTIVITIES

1 Study Figure 6.3 and the OS extract on page 18.
  a What height above sea level is Ellerbeck Farm?
  b Are the summer temperatures very warm? (Average summer temperatures in eastern Britain are 17°–18°C.)
  c Look at the rainfall figures in Figure 6.2. Is the rainfall in July high compared to April?
  d Study Figure 6.2. What main farm activity could be affected badly by a large amount of rain in July?

2 Using the text and Figure 6.2, what are the main outputs of the farm?

3 Why is it necessary to 'rotate' the breeding ewe population every few years?

4 What type of beef cattle is kept and why?

5 Study Figure 6.3. Why have shelter belts of trees (to protect livestock from cold winds in winter) been planted in a north/south orientation on the farm? (Think of the common prevailing wind direction in Britain.)

6 a Using the map extract on page 18:
    i Draw a cross-section from the summit of Whernside (738815) to the valley floor (725755) where the track to Dale House joins the main road. Use the vertical scale 2 cm equals 250 m, and a horizontal scale of 2 cm equals 1000 m. Use only the contours in **bold** colour, e.g. 300 m, 350 m, etc.
    ii Mark and label on your cross-section: Ellerbeck Farm, the River Twiss, the B6225, an area of limestone pavement, limestone scar, lowland pasture and upland pasture.
  b Using all the information provided, including the map extract, suggest why this area is classified as 'seriously disadvantaged' land. (Think in terms of steepness of land, drainage, the type of pasture etc.)

7 How could hill walkers cause problems for hill farmers? Suggest ways in which you think some of these problems could be reduced or avoided.

8 Why do you think it is necessary to conserve the landscape in this area for future generations?

7 Studying the countryside using maps and photographs

Uses of satellite images

The big advantage is that, unlike maps, satellite images can be taken often to show an area at different times of the year. For farmers, different crops being grown can be monitored and the Ministry of Agriculture, Fisheries and Food (MAFF) can estimate total hectarage. For shipping, sand banks which could cause a danger can be identified. For fishermen, plankton which fish feed on can be followed, so ensuring that the boats are in the right areas. The one area where satellites have helped the most is with weather forecasting, which in turn helps farmers. Satellite images will probably be used more and more in the future to monitor many human activities.

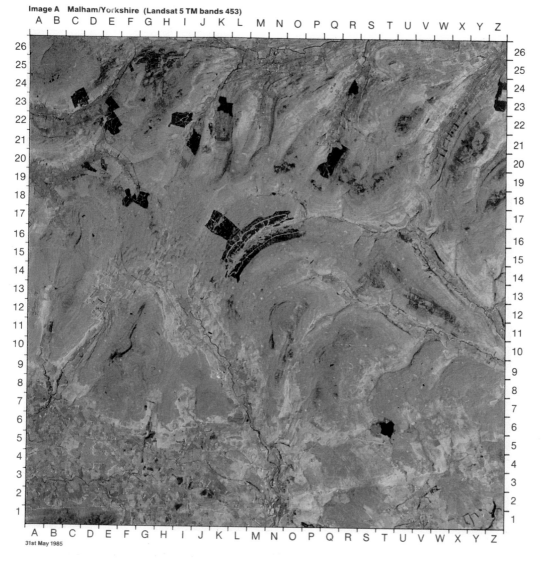

Image A Malham/Yorkshire (Landsat 5 TM bands 453)

31st May 1985

Key

Orange – hay meadows, e.g. E12

Greenish blue – rough hill grazing, e.g. C12

Dark brown blocks – conifer plantations, e.g. K17

Dark brown lines – rivers, e.g. B8

Turquoise blue – carboniferous limestone, e.g. H9

Royal blue – non-limestone quarries or urban areas, e.g. N2

White – limestone quarries, e.g. K9

Black – water

Figure 7.2

Figure 7.1

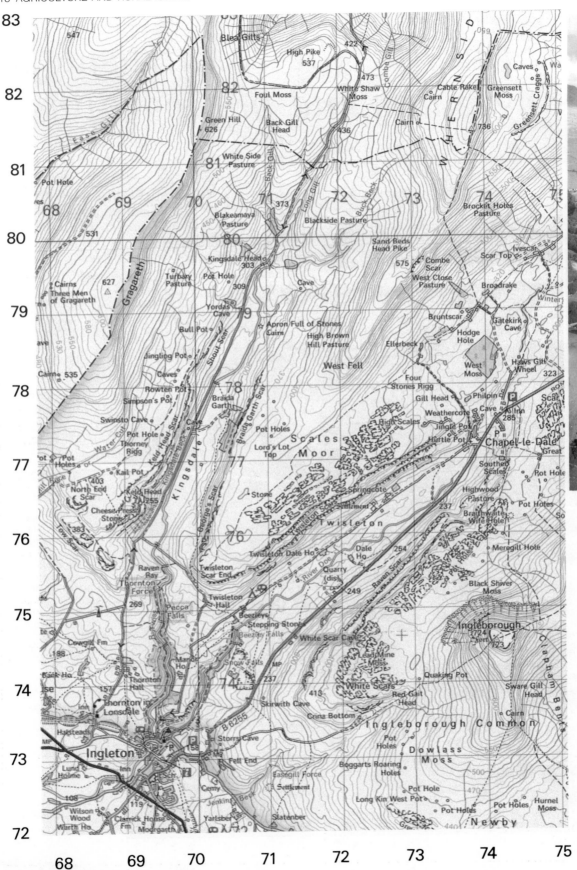

Figure 7.3

Using the aerial photograph in Figure 7.3, the satellite picture and the map extract (Figure 7.1), answer the following questions.

1 Which way was the camera pointing when the photo was taken?
2 a What is the feature to the east of the photograph (689728)?
 b What recreational use is being made of the land at this point?
3 What is found at D4 and E5?
4 Suggest why the housing in the foreground of the photograph is a more recent development.
5 What services does the village of Ingleton provide for the surrounding area?
6 Why might the area at C3 be interesting to tourists?
7 a Describe the shape of the valley in the background of the photograph.
 b Why is the settlement **dispersed** in this part of the valley?

ACTIVITIES

(Some of these activities involve reference to glaciation and limestone scenery — see *People and the Physical Environment*.)
You need to use the satellite image, the key provided and the OS map extract. Note that the map only covers a small part of the satellite image. To help you, the village of Ingleton (6973) is shown on the satellite image as a dark blue area at B7. The top of Whernside (738814) is at D15. The summit of Ingleborough (742746) is at E10. Now you have orientated the map and image, answer the following questions.

8 a Which river flows from E13 to B7?
 b What river joins it at B7?
9 What is the name of the limestone scar at D10?
10 What economic activity are people doing at B8?
11 What is the name of the conifer plantation at D16?
12 Find Ellerbeck Farm (732784). Where is it on the satellite image?
13 a What does the orange colour represent on the satellite image?
 b Are these areas on the highest land or on the lower slopes of the valleys? Why?
14 Find the railway in the north sector of the map and at F16. What happens to the railway at F17? Why might this railway be termed a scenic route?
15 a (Refer to *People and the Physical Environment*). The satellite image shows a lot of carboniferous limestone where glaciers in the past have stripped the soil away to expose the bare rock. Give map evidence of limestone features found in this area.
 b Does the satellite image show any rivers on the limestone? Why?
 c The dark area at T6 is Malham tarn. What happens to the river that flows south of the tarn at T5? Give reasons for your answer.
16 The footpath at D15 shows as a darker colour. This is where footpath erosion has damaged the dark peat. Why might this satellite image be of interest to the National Park Authorities?
17 The 'pock-marked' pattern at J15 is produced by drumlins. What are these glacial features? (Refer to *People and the Physical Environment*.)
18 Using the map extract only:
 a What does the broad yellow line represent?
 b What attractions for tourists are there?

19 Locate Ingleton in your atlas. Are there any motorways in the region? Where do you think most visitors will come from and why?

8 Nomadic pastoralism in the Sahel

This is a type of livestock ranching which involves the movement of tribes and their animals in search of suitable grazing areas. It occurs in the savanna areas of Africa in semi-arid (dry) conditions. Look at the position of the savanna in an atlas. Notice that it is close to the Sahara desert. Parts of this area are also known as the **Sahel** (see page 38).

Rainfall in this area varies. Study the climatic data of Sokoto and Timbuctu shown in Table 8.1. Notice that, although both places have different amounts of rainfall, they both have a dry season between November and March. This is when hot, dry winds blow from the desert to the savanna. Following this there is a wet season when winds blow in from the Atlantic Ocean bringing rain with them.

Rainfall in these areas is not always effective as much is lost by evaporation. Look at Figure 8.1 which shows there are only a few months of the year when rainfall is higher than the amount of moisture lost by evaporation! Where rainfall is

low, crop growing is impossible as the growing season is too short. Instead, people must move around to find grazing areas for their animals (see Figure 8.2). They tend to follow river valleys or move with the rains to newly watered pasture. However, the situation is made worse as rainfall is not reliable and may fail for one or several years. This has led to several serious droughts in the last 20 years in the Sahel region.

Figure 8.1 Soil Moisture Budget. Adapted from Hocking and Thompson

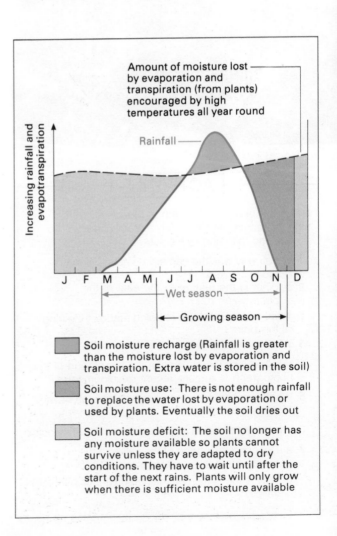

Soil moisture recharge (Rainfall is greater than the moisture lost by evaporation and transpiration. Extra water is stored in the soil)

Soil moisture use: There is not enough rainfall to replace the water lost by evaporation or used by plants. Eventually the soil dries out

Soil moisture deficit: The soil no longer has any moisture available so plants cannot survive unless they are adapted to dry conditions. They have to wait until after the start of the next rains. Plants will only grow when there is sufficient moisture available

Table 8.1 Figures from Thomson and Hocking: *Land and Water Use in W. Africa,*

	Sokoto		Timbuctu	
	Max Temp °C	Rainfall mms	Max Temp °C	Rainfall mms
J	33	0	31	0
F	36	0	34	0
M	40	0	38	2
A	41	10	42	0
M	40	51	44	5
J	36	89	44	23
J	33	147	40	79
A	31	236	36	81
S	33	145	40	38
O	37	13	40	3
N	36	0	37	0
D	33	0	32	0
	Av: 35	Total: 691	Av: 38	Total: 231

CASE STUDY

The Tuareg pastoralists, northern Mali

The Tuareg are a tribe of people who live in an area bordering the Sahara desert. Find the location of the desert and Mali in your atlas. Here annual rainfall totals 130 mm falling mainly in July,

Figure 8.3 The effect of rainfall on vegetation

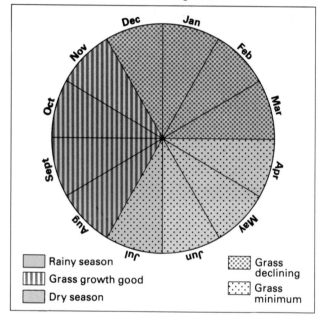

Figure 8.2 Tussock grass

August and September. Extra water comes from wells dug in wadi beds. (These are valleys formed by rivers which are dry for most of the year). Lack of rain and the unreliable rainfall pattern affect:

1 vegetation (see Figure 8.3);
2 the **carrying capacity** of the land, i.e. the numbers of animals which can be supported;
3 milk yields;
4 human populations.

Most breeding of cattle and camels takes place in the early part of the rainy season. Extra animals mean that the milk yields are high during and just after the rains. The Tuareg are, therefore, able to sell or trade some of their animals and milk for millet and other seeds. This will feed the people during the dry season. The tribe also keep goats which are hardier animals and breed after the rainy season. Some of the male kids are slaughtered for meat.

If the rains are high for several years in succession, the human and animal populations increase, creating pressure on the available resources. Eventually the carrying capacity is reached and the land can support no more. However, if there is a drought, the carrying capacity declines rapidly and the animals no longer survive in large numbers.

Low rainfall
↓
Reduced pasture
↙   ↘
Reduced fertility   Higher animal mortality
↘   ↙
Fewer animals
↓
More rapid recovery of pasture

The Tuareg keep as many animals as possible as they are dependent on the animals for survival. If half of a herd of 20 cattle dies in one year the remaining 10 will not feed a family ... but half a herd of 40 cattle, however thin, might. In the past Tuareg families that have lost all their cattle in a bad year have been able to rent part of a herd from more successful families. This is becoming less common as some of the Tuareg are migrating to towns. In order to stop this and the problems of overgrazing, some projects have been set up by the Mali Government and Oxfam to provide the families with land to farm. This will encourage permanent settlement of the Tuareg where tube wells provide necessary water. The families are also taught ways of breeding better quality cattle and how to grow crops both for the animals and for themselves. The cattle help by providing manure for fertilizer!

Problems of pests and diseases

There are two major pests in savanna environments.

Tsetse fly: These are flies which bite both humans and cattle, injecting a parasite which can cause *trypanosomiasis* or sleeping sickness. It kills cattle and makes humans weak, thus further burdening their nomadic lifestyle. Tsetse areas are widespread throughout Africa as can be seen in Figure 8.4. Tsetse flies, like cattle, prefer damp, sheltered areas. Unfortunately, some of the best grazing areas for cattle are in the tsetse zone. Therefore, they have to be avoided wherever possible.

Several attempts have been made to control the numbers of flies:

1 remove the livestock from the area and chop down the vegetation where the flies shelter. This is a large scale operation and environmentalists would disagree with the methods;
2 insecticides could be sprayed to kill the flies. This involves the use of planes so is expensive and spraying has to be timed correctly so that rains do not wash large quantities away;
3 using fly traps which smell like cows but which have been sprayed with insecticide. This attracts the flies and kills them. These traps can be set up on a small scale and are inexpensive but they only clear small areas;
4 sterile male flies are released to ensure that future generations are smaller in number.

Locusts: These are insects which fly in large swarms (Figure 8.5) and live on vegetation which appears after the rains. They leave the land totally bare, destroying any crops that may have been planted and any potential grazing areas.

Numbers are kept down by spraying insecticides from a plane. In 1986 Mali was given a crop spraying plane from some of the proceeds of 'Live Aid'. Locusts are very difficult to eliminate altogether and they thrive after the rains, covering thousands of kilometres.

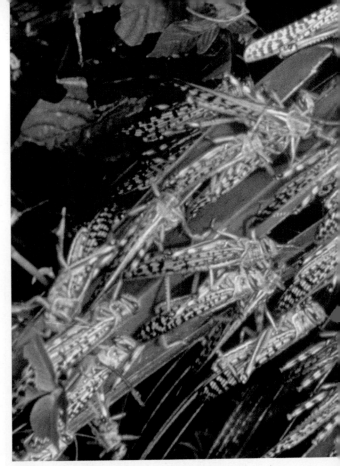

Figure 8.5 Locust swarm

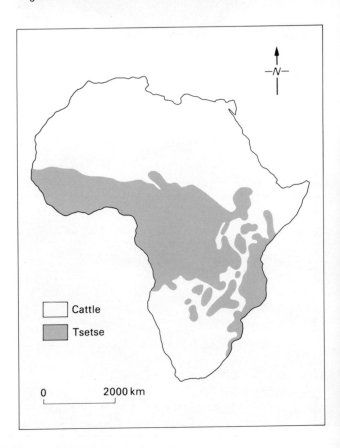

Figure 8.4 Tsetse fly areas in Africa

ACTIVITIES

1 a Using the figures in Table 8.1, draw a climate graph for either Sokoto or Timbuctu. (Page 10 shows you how to set out a climate graph.)

 b Describe the distribution of rainfall during the year. Try to explain this distribution.

 c Describe the temperature distribution.

 d i What is the range of temperature (difference between maximum and minimum)?

  ii Use your atlas to compare this with the temperature range in London, often given as KEW.

2 Study Figure 8.1.

 a i When is rainfall greater than evaporation and transpiration?

  ii What will happen to the soil at this time?

 b i When does the soil become dry again?

  ii How long does the soil remain too dry for plant growth? Give a reason for your answer.

3 a Using Figure 8.3, describe how the rainfall affects the vegetation during the year.

 b Look at Figure 8.2. Describe the typical savanna vegetation.

4 Why is it impossible to grow rain-fed crops in northern Mali?

5 Why are the Tuareg nomadic?

6 a When are the Tuareg's animal numbers greatest?

 b Why do the animal numbers vary during the year?

 c Why do the numbers vary so much from year to year?

7 What sources of food do the Tuareg rely on?

8 What problems do the Tuareg face?

9 On an outline map of Africa and with the aid of your Atlas and Figure 8.4:

 a Use colours to show the main cattle and tsetse areas;

 b Name the countries where cattle farming occurs

 c Label the Sahara and Kalahari deserts and the savanna;

 d Annotate the map to say why large numbers of cattle are absent from certain areas of Africa.

10 Imagine you are a nomadic pastoralist. What difficulties do you think you would have in trying to adapt to a settled, crop growing lifestyle?

11 Make a comparison between cattle farming in Mali and dairy farming in Britain. Use the following headings:

 a Physical inputs:
   i climate;
   ii soils;

 b Economic inputs;

 c Organisation of the farming:
   i size of farm;
   ii maintenance of grazing areas;
   iii farmer's year;

 d Outputs and their purpose.

12 Why is nomadic pastoralism **extensive**?

CROP FARMING

9 Arable farming in Great Britain

Figure 9.1 — Systems diagram of farm characteristics

Inputs

Labour
2 full time   1 part-time

Machinery
7 tractors   1 harrow
3 ploughs   4 seed drills
3 crop sprayers

Fertilizer
Artificial N, P + Na
organic – chicken manure

Seed
Herbicides (weeds)
Pesticides (aphids)
Fungicides (diseases)
Slug killer

Rainfall
635 mm per year, drier in summer for harvesting, irrigation adds water

Temperature
16°C July   3°C Jan – Average
High temperatures help ripening. Frosts break up the soil to allow aeration

Relief
Flat 15 m

Soils
Clay with sand – well drained
Heavy clay

Processes

Size of farm: 224 ha divided into 20 fields
Stock: 10 horses for riding stables run by farmer's wife

Farmer's Year

	J	F	M	A	M	J	J	A	S	O	N	D	Outputs
Grass			F			H hay	H hay						For horses kept on farm
Sugar Beet			S	SP	SP	SP				H	H	F, P	Contract to company at Wissington
Linseed			SP F, S	SP F				H	H	P			For oil and fodder locally
Winter barley		F	F	SP				H	H	P, S SP			For fodder locally
Winter wheat		F	F					H	H	P, SP F, S			For fodder and milling locally
Spring wheat		S	SP F					H	H			P	For fodder and milling locally
Peas		S, F	SP		H	H			L		P		For processing locally and export to Japan if high quality also for protein locally animal feed
Oilseed Rape		F	F					H	H	L, P F, S			For oil and fodder locally
Machinery	M	A	I	N	T	E	N	A	N	C	E	–	

F = Fertilizer   P = Plough   S = Plant seeds
H = Harvest   L = Lime   SP = Spray pesticides

Arable farming is the growing of either **grain** crops like wheat and barley or **root** crops like sugar beet and potatoes. These crops are grown to sell to feed us or they are used as fodder for animals. Arable farming may be **extensive** where vast fields of wheat are grown as in the Canadian prairies. The yield per hectare is low compared with **intensive** farming which is carried out in East Anglia in Britain.

Figure 9.3 Combine harvester at work

CASE STUDY

An arable farm at Ely in Cambridgeshire

Figure 9.1 summarises the characteristics of the farm as a systems diagram. Notice that the farm grows a mixture of grain and root crops. The farming is carried out with the aim of achieving high yields. To do this the farmer has to buy fertilizers (to encourage growth) and pesticides and herbicides (to kill pests and weeds which would harm the crop). If the farmer uses large amounts of these chemicals they may build up in the soil. Rainwater then washes these away into rivers and lakes causing pollution.

Figure 9.4 Stubble burning

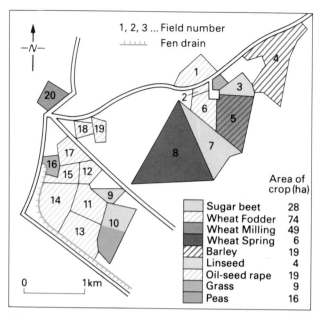

Figure 9.2 Plan of Fen Farm

Crop	Area of crop (ha)
Sugar beet	28
Wheat Fodder	74
Wheat Milling	49
Wheat Spring	6
Barley	19
Linseed	4
Oil-seed rape	19
Grass	9
Peas	16

Figure 9.5 Ploughing straw back in

The farmer also uses a system of crop rotation. He will not grow the same type of crop for more than two years running in the same field. This is because each type of crop takes different nutrients out of the soil and these would quickly be removed if the same crop was grown in one field year after year. Sometimes the farmer leaves the land in a field **fallow** and grows no crop at all for one year. Look at Figure 9.2 which is a plan of an arable farm based on crops grown in 1987. The plan for 1988 would look very different. Table 9.1 shows some typical rotations.

This type of farming is also highly mechanised (Figure 9.3). Machines can work faster and more efficiently than people and animals, so few labourers are needed. Fields are large so machines are able to manoeuvre easily. This means that farmers often remove hedgerows to make the fields as large as possible. The largest field on this farm is number 8 which is 100 hectares (a football pitch is one hectare.) This is much larger than the whole of the dairy farm mentioned on pages 11–13.

Farmers are sometimes criticised for removing hedgerows as this disturbs wildlife and can encourage soil erosion. They are also criticised for stubble (straw) burning (Figure 9.4). Straw is a waste product from grain production. It is not worth much but some is used for animal bedding and food. The rest is either ploughed into the soil directly (see Figure 9.5) or it is burnt in the fields along with weeds. Then it is ploughed into the soil as ash. This saves the addition of nitrogen fertilizer.

ACTIVITIES

1 Study Figure 9.1.

a Why is flat land suitable for arable farming?

b Explain why the climate is good for arable farming at Fen Farm.

c The soils are a mixture of sand and clay. Why is this good?

d How does the farmer make sure he gets the highest possible yields from his land?

e During which season(s) are the following activities most common:
   i ploughing;
   ii fertilizing and spraying;
   iii harvesting;
   iv maintenance.

2 Look at Figure 9.2.

a Give an example of two grain crops grown on this farm.

b Give an example of a root crop grown on the farm.

c Write down a list of the crops grown in order of importance.

d What percentage of the farm land is taken up by wheat?

3 a What local industries will be based on the outputs of the farm?

b Why does sugar beet have to be refined locally?

4 Look at Table 9.1.

a How many years does the typical rotation on this farm last?

b Which two crops are grown for two or more years running in the same field?

c Why is crop rotation necessary?

d Why do you think the rotation in each field is different?

5 a Why are most of the fields on the farm large?

b What problems are caused by the removal of hedgerows?

c Why might a conservationist object?

6 Look at Figure 9.4.

a Why do some farmers burn straw?

b Put yourself in the position of the following people and describe your feelings about straw burning:
   i a neighbour;
   ii a driver on a road by the farm;
   iii a person concerned about wildlife.

c Do you think straw burning should be allowed? Why?

7 a Why are there 'Grain Mountains' in Europe?

b How do you feel about the presence of grain mountains in Europe when many are starving in Africa?

Table 9.1 Crop rotations on Fen Farm

Field number	Year 1 (1987)	Year 2	Year 3	Year 4	Year 5
6	Fodder wheat	Spring barley	Sugar beet	Fodder wheat	same
12	Fodder wheat	Spring barley	Sugar beet	Spring barley	
13	Fodder wheat	Milling wheat	Peas	Sugar beet	as
14	Fodder wheat	Milling wheat	Oilseed rape	Winter barley	
15	Oilseed rape	Winter barley	Fodder wheat	Peas	year
16	Grass	Grass	Grass	Grass	
20	Milling wheat	Fallow	Grass	Grass	one

Grain mountains

Arable farmers in Britain are very successful and produce more grain than can be used or sold at home or abroad. The excess grain is bought by the EC at a guaranteed price. It is stored for the future as a 'Grain Mountain'. Looking after these grain mountains is very expensive and they take up a lot of space (see Chapter 17 for further information). Some people say that these enormous stockpiles of food should be used to feed the hungry in parts of the developing world.

10 Mixed farming in Great Britain

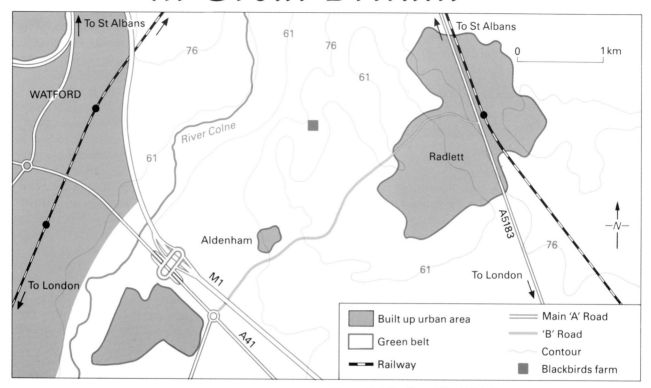

Figure 10.1 Position of Blackbirds Farm

CASE STUDY

Blackbirds Farm, Hertfordshire

Mr Bloomfield is part owner, part tenant farmer of the 545 ha which make up Blackbirds Farm (see Figure 10.1). The farm is a **mixed** farm because he not only has dairy cattle and pigs, but grows cereal, fruit and vegetable crops.

The farm

The soils on the farm are gravels and clay which are not very fertile. If high yields are to be gained large inputs of fertilizers and lime are needed. Irrigation is needed in the summer as the gravel soils dry quickly. Crop rotation is used to avoid pests and diseases building up in the soil. The main 5 year rotation is:
wheat→wheat→barley→barley or rape→forage maize.

Figure 10.2 Mixed farming

Farm data

Size   545 hectares

Arable land use   Temporary grazing 35 ha
Forage maize
Cereals 295 ha
Fruit and vegetables 30 ha
Orchard 5 ha
Potatoes 4 ha

Other land use   Roads and buildings 20 ha

Livestock   120 pigs
180 dairy cattle
350 sheep

Machinery   9 tractors
2 combine harvesters
several harrows and ploughs

Labour   17 full time
1 Youth Training Scheme trainee
25 seasonal workers

Other   3000 tonnes of lime every 3−4
years (to reduce the acidity of
the soil)
Fertilizers (nitrates) and
pesticides
Fungicides

Markets   Milk to Milk Marketing Board or
ice cream wholesaler
Meat to Meat Marketing Board
Cereals to the Farmers'
Co-Operative
Fruit and vegetables for 'pick
your own' or farm shop

Farm products

There are four main groups of farm products.

1   The dairy herd of 180 Friesian cows produces 1.2 million litres of milk per year. In the last 2−3 years Mr Bloomfield has made a 20% cut in milk production. He had hoped to expand his herd to 240 cows, but milk quotas imposed by the EC have meant this is unprofitable. To avoid selling surplus cows when prices for dairy cattle are low, he has reduced the output per cow from 900 litres to 600 litres per year. Some milk he can sell wholesale for Indian ice cream, the demand coming from the many restaurants in nearby north London.

2   The arable crops are mainly oilseed rape and wheat, but maize and grass silage are grown for cattle feed. A new grain store has been built and the farm has its own grain drier. Fields are big enough for the large machinery so few hedges have been removed.

3   The 120 Landrace and Large White cross pigs may not be profitable to keep in the future. Dutch pig breeders get cheap feed cereal and subsidies which means British farmers like Mr Bloomfield find it difficult to compete.

4   Fruit and vegetable growing and 'pick your own' (see Figure 10.3) are profitable for Mr Bloomfield as the farm is within easy access of a very large populated area (see Figure 10.1). Cabbages, onions, broccoli, courgettes, broad beans, raspberries, strawberries and blackcurrants are grown and sold in this way. Three hectares of orchard produce apples and pears. Up to 25 extra staff are employed at harvest time.

Figure 10.3 A 'pick your own' farm shop

Future changes

Surpluses of milk and grain in the EC, and a reduction in the price of these products, have led Mr Bloomfield to **diversify** his production to remain profitable. He is attempting to do so in three main ways.

1 He is selling more fruit and vegetables through a farm shop between Easter and October and has set up a small garden centre which visitors to the shop use.
2 Sheep are bought from hill farmers in the autumn (see Chapter 6) and Mr Bloomfield rents out some of his land to other farmers so sheep can be fattened to be sold at a profit.
3 The demand for grazing for privately owned horses is increasing as Blackbirds Farm is in the rural-urban fringe. This is termed 'Horsiculture' and can be profitable with relatively low inputs.

Figure 10.4 Farm diversification

ACTIVITIES

1 Study Figure 10.1 and describe the location of Blackbirds Farm.
2 Refer to Figure 10.2 and the text:
 a Why is Blackbirds Farm described as a mixed farm?
 b How many workers are there in the course of a year?
 c Why are so many workers seasonal?
3 Look at Figures 10.2 and 10.3. What crops are grown to sell through a farm shop?
4 a Study Figure 10.1. Where do you think Mr Bloomfield's customers for the 'pick your own' will come from?
 b Explain why being near to a large urban area is important in helping Mr Bloomfield to decide what to produce.
5 Why do you think a grain drier is used before the grain can go to market?
6 Explain how EC policies have affected Mr Bloomfield's decisions in farming over the last few years.
7 a How has Mr Bloomfield tried to diversify his farm production in order to stay profitable in the future?
 b Can you suggest any other things he could do in the future?
8 Using Figure 10.3 and the text, design a poster or newspaper advert for a local paper to attract more customers to Blackbirds Farm 'pick your own'.

11 Subsistence farming

Subsistence farming is very important in Nigeria. Look at Figure 11.1. Both **shifting cultivation** and **rotational bush fallowing** are types of subsistence farming. This means that crops are grown mainly to feed the families themselves. Notice that shifting cultivation only covers small parts of the country. This is because tribes are tending to remain more permanently in one area where they carry out rotational bush fallowing. Their villages (Figure 11.2) are surrounded by the farm land and by natural vegetation (**bush**).

Rotational bush fallowing

Bush fallowing can be carried out by farmers in the tropical rainforest or savanna areas. It involves growing subsistence crops (e.g. maize, cassava, yams, millet) in rotation for five years and then leaving the land fallow for three years before starting the rotation again. (Look at Figure 11.3.)

Figure 11.1 Nigeria: dominant farming systems

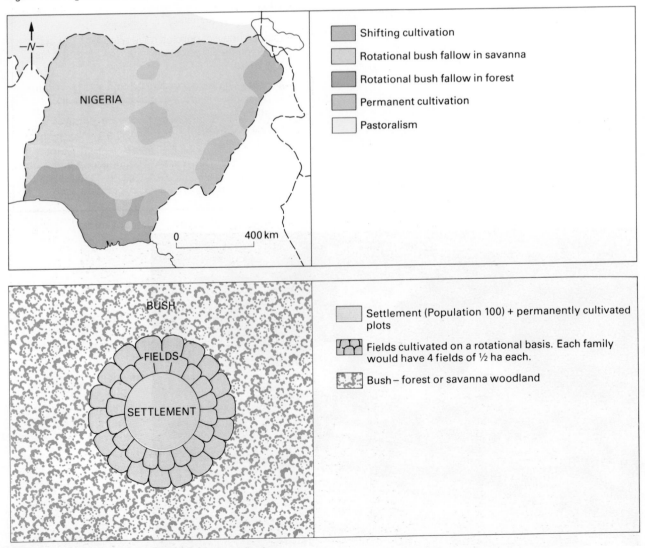

Shifting cultivation

Rotational bush fallow in savanna

Rotational bush fallow in forest

Permanent cultivation

Pastoralism

NIGERIA

0   400 km

BUSH

FIELDS

SETTLEMENT

Settlement (Population 100) + permanently cultivated plots

Fields cultivated on a rotational basis. Each family would have 4 fields of ½ ha each.

Bush – forest or savanna woodland

Figure 11.2 Subsistence farming: rotational bush fallowing

Descriptive model for
rotational bush fallowing
near **Nnewi**, eastern
Nigeria

– Use of one field over time

Year 1
Year 2
Year 3
Year 4
Year 5
Year 6
Year 7
Year 8

- Yam
- Beans
- Okra
- Maize
- Pepper
- Cocoyam
- Cassava
- Other vegetables
- Bush fallow
- Crops in diminishing order of importance

Shifting cultivation

Shifting cultivation involves clearing some of the trees and other vegetation by cutting and burning. The cleared land is planted with subsistence crops. The ash left over from burning is useful as a simple fertilizer. This is also known as **slash and burn farming** and it is summarised in Figure 11.4. The crops are grown on small plots with each family group farming about two hectares. Holdings are small because hand labour and increasing population pressure limits the area which can be farmed. After a few years the land gradually loses fertility and the yields from the crops decrease. The families abandon the land and move elsewhere in the forest, clearing a new site. Tribes generally have an understanding as to the areas they can farm. In some countries, like Brazil, tribes are often in conflict with authorities who are trying to develop the rainforests for timber, plantation agriculture or industrial and mining development.

Figure 11.3 Rotational bush fallowing

Figure 11.4 Shifting cultivation. With increasing population pressure the sequence would go 1,2,3,4,5,2,3, etc., not allowing regeneration of true tropical rainforest

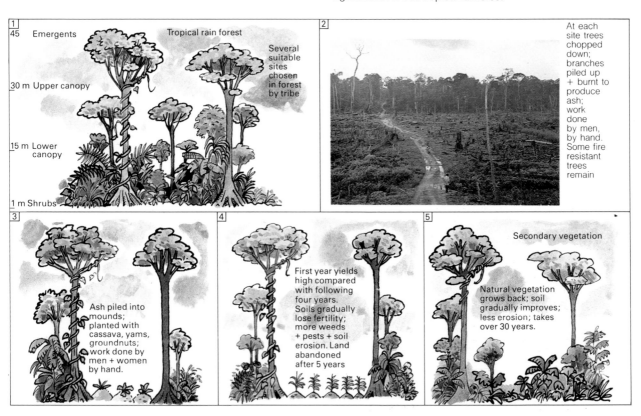

1
45 Emergents   Tropical rain forest
30 m Upper canopy
15 m Lower canopy
1 m Shrubs

Several suitable sites chosen in forest by tribe

2
At each site trees chopped down; branches piled up + burnt to produce ash; work done by men, by hand. Some fire resistant trees remain

3
Ash piled into mounds; planted with cassava, yams, groundnuts; work done by men + women by hand.

4
First year yields high compared with following four years. Soils gradually lose fertility; more weeds + pests + soil erosion. Land abandoned after 5 years

5
Secondary vegetation

Natural vegetation grows back; soil gradually improves; less erosion; takes over 30 years.

The abandoned land is left fallow and the natural vegetation is allowed to grow back. This new vegetation is called **secondary growth** as it is less dense and less complex than the original vegetation. In the tropical rainforest, for example, dense vegetation thrives in humid conditions where annual rainfall exceeds 2000 mm. High temperatures averaging 27°C encourage plant growth and, therefore, the rainforest is made up of a variety of trees, ferns and shrubs. It is a complex **ecosystem** which would take hundreds of years to develop again. After 20–30 years, however, the land will have recovered enough for the tribe to return, clear, and farm the land again (see *People and the Physical Environment* for more on the tropical rainforest ecosystem).

In areas where the population is rising rapidly and where the land is in short supply, the tribes may be forced to return to fallow plots too early. Yields are low and the future becomes one of ever decreasing yields. Clearing the rainforest on a large scale can also lead to many problems:

Deforestation

```
              ↙        ↘
Land exposed to sun's      Vegetation no longer
rays and is baked hard     binds soil or provides
          ↓                    humus and litter
Rainfall quickly runs              ↓
off taking loose top       Soil loses valulable
    soil with it           minerals and organic
          ↓                        matter
Flooding and soil erosion
              ↘              ↙
            Land can no longer
            support vegetation
                or wildlife
```

1 What is the difference between rotational bush fallowing and shifting cultivation?

2 a Why is 'slash and burn' a suitable alternative name for shifting cultivation?

   b Why do farmers burn some of the vegetation?

   c Explain the effect on shifting cultivation of a rapid rise in population.

   d What are the disadvantages of deforestation?

3 Draw a series of simple diagrams based on those in Figure 11.4 to show shifting cultivation. Below each diagram describe in your own words what it shows. Alternatively, you could add labels.

4 Turn back to Figure 3.1, the map of world farming types.

   a On a world outline, draw the areas of shifting cultivation.

   b Use an atlas to name the major regions or countries where this type of farming is carried out. Add the tropics of Cancer and Capricorn.

   c In your atlas, turn to a map of world natural vegetation types. What do you notice about the type of natural vegetation found in the areas where shifting cultivation takes place?

12 Plantation farming in Kenya

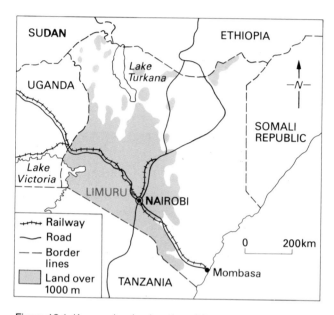

Figure 12.1 Kenya, showing location of the tea estate at Limuru

Figure 12.2 Layout of tea estate

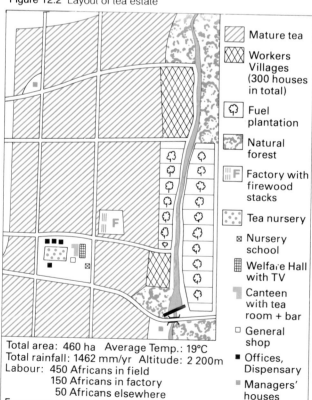

Mature tea

Workers Villages (300 houses in total)

Fuel plantation

Natural forest

Factory with firewood stacks

Tea nursery

Nursery school

Welfare Hall with TV

Canteen with tea room + bar

General shop

Offices, Dispensary

Managers' houses

Total area: 460 ha   Average Temp.: 19°C
Total rainfall: 1462 mm/yr   Altitude: 2 200m
Labour: 450 Africans in field
        150 Africans in factory
        50 Africans elsewhere
European management

Plantations are large scale farms or **estates**. They concentrate on the production of a single crop, e.g. sugar cane, coffee, tea or rubber. This is called **monoculture**. Plantations require huge quantities of land and a large labour force. They are commercial and have a large input of money and technology.

CASE STUDY

The Karirana tea plantation, Kenya

The Karirana Estate at Limuru is 30 km north west of Nairobi, the capital of Kenya (Figure 12.1). Figure 12.2 shows the layout of the estate. Notice that it is a self-contained unit with a variety of amenities for the labour force as well as a factory for processing the tea. The estate is a member of the Kenya Tea Growers Association (KTGA) which helps with training farmers, buying fertilizers and introducing new methods.

Figure 12.3 Picking tea; new leaves are harvested every 6–8 days all year round

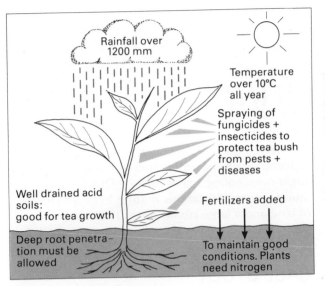

Figure 12.4 Tea bush requirements

- Rainfall over 1200 mm
- Temperature over 10°C all year
- Spraying of fungicides + insecticides to protect tea bush from pests + diseases
- Well drained acid soils: good for tea growth
- Fertilizers added
- Deep root penetration must be allowed
- To maintain good conditions. Plants need nitrogen

Figure 12.5 Spraying tea

Figure 12.7 Packing tea

Figure 12.8 Diffusion of methods

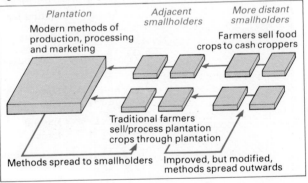

Plantation	Adjacent smallholders	More distant smallholders
Modern methods of production, processing and marketing		Farmers sell food crops to cash croppers

Traditional farmers sell/process plantation crops through plantation

Methods spread to smallholders

Improved, but modified, methods spread outwards

Tea is a crop ideally suited to the conditions in this part of Kenya. (Look at Figures 12.3, 12.4 and 12.5.) The leaves from the bushes can be plucked all year round. The soils on the estate are sometimes affected by soil erosion so planting of the tea bushes is done along the contour. (See Figure 12.6.)

As a result of the excellent inputs, tea production in Kenya has rapidly increased:

    1963   18 000 000 kg tea produced
    1986   147 000 000 kg tea produced

The processing of tea involves cutting up the tea leaves, allowing them to ferment, then drying and packing them in tea chests. (See Figure 12.7.) Tea leaves actually start off green in colour and end up black. The chests are then transported by rail to Mombasa and exported. 85% of the tea Kenya grows is exported to countries like the UK, the USA and Pakistan where it is blended and packaged for marketing. Therefore not *all* the processing is done in the countries which grow the tea.

Figure 12.6 Contour ploughing

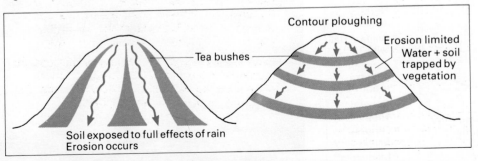

Contour ploughing

- Tea bushes
- Erosion limited Water + soil trapped by vegetation
- Soil exposed to full effects of rain Erosion occurs

Often the estates are owned by large **multinational** companies like *Brooke Bond*. This is a British company which owns several tea plantations in the tropical world and also blending and packing factories in the importing countries. In owning plantations, multinationals can be sure of obtaining tea at a low cost. Some people think such companies exploit countries like Kenya and that they should pay more. However, the local community may also benefit. In Kenya tea is also grown on small scale peasant farms:

	Area used for tea growing (ha)	% of all tea grown
Peasant farms (average size 5 ha)	56505	30
Plantations (average size 530 ha)	27319	70
Kenya total	83824	100

The peasant farmers cannot process the tea themselves so they sell it to local plantations. The Karirana Estate produces over 1 000 000 kg of tea a year with 19% of this coming from nearby small farms. Study Figure 12.8. It describes how small scale peasant farmers can sometimes benefit from being close to plantations.

ACTIVITIES

1 Using the information provided, describe the physical factors which favour tea growth in Kenya. (Consider rainfall, temperature and soils.)

2 a Make a list of all the things you would have to buy if you were setting up a plantation.

b Which of these things would you have to buy more of each year?

c Where do you think you would get the money from?

3 a What is the size of the Karirana tea estate?

b Study Figure 12.2.
  i How many workers are there on the estate?
  ii What types of jobs *could* they do?
  iii Where do they live?
  iv What amenities are there on the estate for the workers?
  v What do you think the firewood could be used for?
  vi Where might the firewood have come from?
  vii Why is there a factory on the plantation?
  viii New workers are needed by the company. Design a poster to advertise jobs on the tea plantation.

4 What happens to the packed tea chests when they leave the plantation?

5 What are the advantages of large scale production of crops on estates?

6 Although peasant farmers have more tea growing land compared with plantation farmers, why do they find it difficult to produce large amounts of tea?

7 Monoculture can sometimes be a risky business. Why might changes in world prices and the spread of disease be possible problems on plantations?

8 50% of Kenya's exports come from tea and coffee. Why is dependence on one or two products dangerous for the economy of a country?

9 a What is a multinational company?

b From the multinational companies' point of view:
  i What are the advantages of owning a plantation in Kenya?
  ii What are the disadvantages?

c From the Kenyan Government's point of view:
  i What are the advantages to the country of plantations run by foreigners?
  ii What are the disadvantages?

10 The table below gives details of major tea producing countries (% share of world tea production):

India	29.7%	Indonesia	4.7%
China	20.2%	Turkey	3.5%
Sri Lanka	9.1%	Bangladesh	2.2%
Kenya	6.0%	Argentina	1.9%
Japan	5.2%	Malawi	1.6%

On an outline map of the world name each country mentioned in the table above. Decide how you could show the information above on the map. You could use proportional circles or bars.

13 Soil erosion

Figure 13.1

Soil erosion is the removal of exposed soil by wind or water. It occurs in many areas of the world on a large scale as a result of natural and human causes. It is not a new problem. The USA suffered severe soil erosion in the 1930s when the 'Dust Bowl' developed in the central plains. This resulted from mismanagement of the land by farmers.

Where the land is left unprotected by vegetation, soil is easily washed or blown away leaving gullies (Figure 13.1). The problem is made much worse in hilly areas and in areas which experience drought conditions.

Soil takes hundreds of years to form. However, it is being removed at an alarming rate — much faster than the rate of formation! In Africa the rate of erosion has increased 20 times in the last 25 years. Farmers all over the world are concerned about the loss of **top soil**. (This is the portion of the soil that contains the mineral and organic matter that is necessary for the growth of crops.) For example, the state of Iowa in the USA has lost half of its top soil in the last 200 years. Once it has been removed soils are infertile and incapable of supporting crops. Soil erosion occurring on the fringes of deserts helps to encourage **desertification** (see Chapter 14).

Where has all the soil gone?

Fine particles moved by the wind (Figure 13.2) may travel hundreds of kilometres and are deposited as dust. Saharan dust sometimes appears in the UK giving the rain a reddish tint. Larger particles cannot be blown far, but they may travel long distances if they are washed away into rivers. Rivers normally carry some sediment but erosion of soil particles adds extra. This may give the river a muddy look. The sediment will eventually be deposited in the sea where it is of no use to the farmer unless it builds up into a delta. The Yellow River in China, for example, carries away 1.6 billion tonnes of soil per year to the sea.

If a river has a dam across it, silt may build up behind the dam. The Aswan High Dam in Egypt, constructed in the 1960s, created a large man-made lake (Lake Nasser) which is gradually being filled in with sediment that can no longer get past the dam. Silting up reduces the life of a reservoir.

Figure 13.2

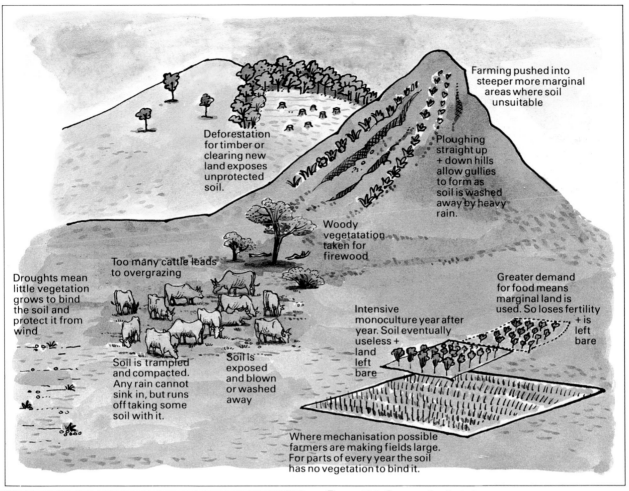

Figure 13.3 Causes of soil erosion

Within the diagram:

Deforestation for timber or clearing new land exposes unprotected soil.

Farming pushed into steeper more marginal areas where soil unsuitable

Ploughing straight up + down hills allow gullies to form as soil is washed away by heavy rain.

Woody vegetatation taken for firewood

Too many cattle leads to overgrazing

Droughts mean little vegetation grows to bind the soil and protect it from wind

Greater demand for food means marginal land is used. So loses fertility + is left bare

Intensive monoculture year after year. Soil eventually useless + land left bare

Soil is trampled and compacted. Any rain cannot sink in, but runs off taking some soil with it.

Soil is exposed and blown or washed away

Where mechanisation possible farmers are making fields large. For parts of every year the soil has no vegetation to bind it.

ACTIVITIES

1 a What is soil erosion?

  b Study Figure 13.3. List the causes of soil erosion under the following headings:
    i natural factors;
    ii human factors.

2 The following factors encourage soil erosion. Describe how this happens in each case:
  a deforestation;
  b steep slopes;
  c ploughing up and down the hills;
  d monoculture;
  e overgrazing;
  f drought.

3 Draw a diagram similar to Figure 13.3 to show the possible solutions to soil erosion.

14 Desertification

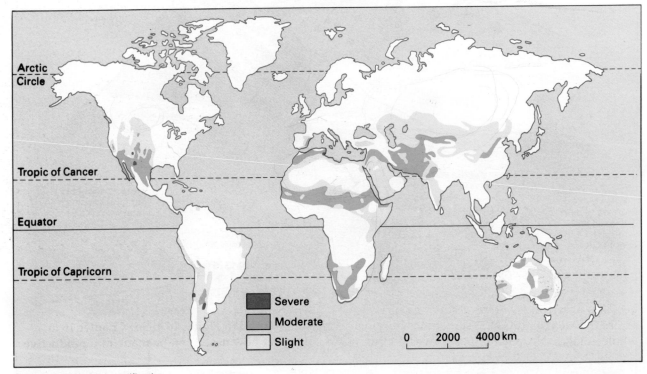

Figure 14.1 World desertification

Desertification is the extension of desert areas as a result of both natural and man-made activities. Every year an area twice the size of Belgium becomes a desert. The Sahara desert in Africa is advancing southwards at a rate of 10 km per year. Desertification is a worldwide problem. (See Figure 14.1.)

CASE STUDY

The Sahel, Africa

The Sahel borders the south side of the Sahara desert in Africa. (The word Sahel means 'fringe' in Arabic.) The Sahel stretches from Gambia in the west to the Somali Republic in the east and is the home for many nomadic people, like the Tuareg in Chapter 8. (Use Figure 14.1 and an atlas to locate the area.)

Several factors lead to desertification:

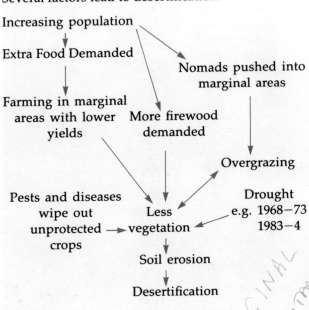

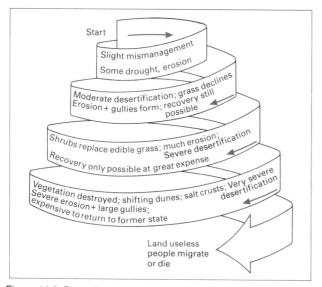

Figure 14.2 Desertification spiral

Figure 14.3 Dune encroachment

Desertification occurs gradually, but is accelerated by people's activities as you can see in Figure 14.2. Once it has started it is very costly and difficult to return the land to its earlier state. In very severe cases of desertification the land becomes desert-like and sand dunes actually move across the area (Figure 14.3) sometimes burying whole villages. No vegetation survives. In this situation people will not be able to grow food so they are forced to move. In Ethiopia large numbers of people have migrated to refugee camps or urban areas where they hope to find food. Food in refugee camps is usually provided free by donations from foreign countries. However, such solutions can only be short term.

Several long term solutions to the problem of desertification have been suggested. These are summarised in Figure 14.4. See Chapter 16 to discover how desert can be made into productive farmland. Some of these solutions involve the use of techniques or equipment. If the equipment is cheap, easy to make from local materials, and simple to use or understand, it is referred to as **appropriate technology** (see Figure 14.5 and *Energy and Industry*).

Figure 14.4 Solutions to desertification

Figure 14.5 In Burkina Faso a new clay stove saves up to 70% of the fuel wood consumed by the old stove: it can be made in a day with free materials

1 On a world outline map and using Figure 14.1:
   a Copy the areas suffering from severe desertification;
   b Use an atlas to name the world's desert areas;
   c Label the Sahel;
   d Name the continents.

2 a Describe the location of the main areas suffering from severe desertification.
   b How wide is the Sahel north to south, and east to west?
      i How large an area is the Sahel?
      ii Name the Sahel countries.
      iii How many countries are included?
      iv How might the scale of the problem and the number of countries involved make it difficult to find a solution?

3 a How does drought cause desertification?
   b Explain how people help to cause desertification?
   c What are the results of desertification?

4 Look at Figure 14.4. What are the solutions to:
   a drought;
   b sand dune encroachment;
   c overgrazing;
   d removing vegetation for firewood;
   e lack of food?

5 Why are the tube well, the slow burning stove, and the grain store seen as appropriate technology?

6 Why would a tractor be seen as **inappropriate technology** for helping to solve the problem of food supply?

7 Define short term and long term aid. Give an example of each.

15 The Green Revolution

In the 1960s new techniques were developed with the aim of increasing the yields of staple crops to help solve some of the problems of famine. This led to what has become known as the **Green Revolution**. One of the centres involved was the International Rice Research Institute in the Philippines which developed the famous IR8 or 'miracle' rice seed. (See Figure 15.1.) High-yielding wheat and other grains were also developed.

All these high-yielding varieties (HYV) of seeds require the use of fertilizers, pesticides and often irrigation to achieve the high yields. Therefore these techniques do not often help the very poor farmers who cannot afford the inputs necessary. Without these techniques, though, hunger and famine would be more widespread than they are at present. In India millions of lives have been saved and the country is now able to export some food, e.g. to Ethiopia in 1985. However, malnutrition amongst certain groups is still a problem in India.

CASE STUDY

The Green Revolution in India: the experiences of three farmers

Mr Shah

(See Figure 15.2) I own 1.5 hectares of irrigated land. I had to borrow some money from my local bank to buy a pump to help distribute the water for my rice crops. I am able to produce two rice crops a year because of the favourable climate, the extra water I have through irrigation and the HYV seeds I use. I also buy fertilizers and pesticides which I use on the crops. I learnt all about these new methods from a demonstration farm which has been developed near my farm using money from the Overseas Development Association in Britain. There were expert farmers who taught us all about these new methods. I sell most of my crop in my nearest town after keeping enough to feed my wife and seven children. With the money I earn I buy other types of food for my family but sometimes I have sufficient to be able to buy extra land. In 1982 I bought 0.5 hectares of land nearby. Most of the work on my farm is done by hand with help from my family, but we do have two water buffalo and a plough.

Figure 15.1 Traditional: tall crop, grown far apart so as not to shade each other, extensive root system takes many minerals out of the soil. IR8 rice: short, strong crop, able to support large head of rice, high yielding, grow close together, short root system to make maximum use of fertilizers and pesticides

Figure 15.2 Indian rice farmer

Figure 15.3 Rice farmer working by hand

Mr Patel

(See Figure 15.3.) I own 0.1 hectare of land. I bought it with money I borrowed from a money lender as I could not get credit through the bank. Although the temperatures in India are high all year round, I can only produce one rice crop a year because my planting is dependent on the monsoon. This is the time of year when the rains come and it is the only time that my crops are watered. I cannot afford irrigation equipment. All the work on my farm is done by hand. I have a simple plough and my family also helps. Most of my rice crop is kept by my family so I also work for another farmer to earn some extra money. I use this to help repay the money lender. I also buy some seeds. These seeds are not the best ones I could buy, but they don't need large amounts of fertilizers which I can't afford anyway!

Mr Bagchi

I do not own any land at all. To earn a living I work for a farmer who has 5 hectares of land. I live on the farm with four other families and we work the land as though it was our own. I do not get paid in wages but I do get a share of the crop that is produced. It is enough to feed my family but there is very little left over to sell. The owner is thinking of buying more land to farm. This will not make me any better off because we will need more people to help with growing the crops. They will have to have a share of the produce too.

In Asia, where the Green Revolution originated, there have been both good and bad effects. Generally the richer farmers have been the ones to benefit as they have been the ones who could afford the necessary inputs. The gap between the richer farmers and the poorer farmers has therefore increased.

In Africa, Green Revolution techniques have not yet been so successful. Here there are desert, mountain and humid environments, all of which require different techniques. Increased yields have not been widespread and problems of drought and famine have been experienced in many countries. Migration is common amongst the poorer farmers who cannot obtain a livelihood from the land even at subsistence levels. The techniques have, however, been successful in countries like Egypt.

Figure 15.4 Farmer's year

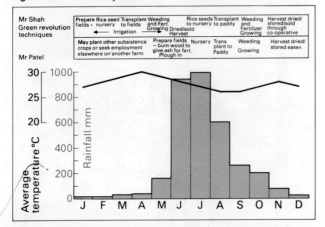

Mr Shah Green revolution techniques	Prepare Rice seed Transplant Weeding fields · nursery to fields and Fert Growing Irrigation Dried/sold Harvest	Rice seeds Transplant to nursery to paddy	Weeding and Fertilizer Growing	Harvest dried/ stored/sold through co-operative
Mr Patel	May plant other subsistence crops or seek employment elsewhere on another farm	Prepare fields Nursery – burn wood to give ash for fert. Plough in	Trans plant to Paddy	Weeding Growing	Harvest dried/ stored eaten

ACTIVITIES

1 a What do you understand by the phrase 'Green Revolution'?

  b Why have the richer farmers benefited more from the Green Revolution than poorer farmers?

2 Study Figure 15.1. Draw a sketch of the traditional rice and the miracle rice. Annotate these to show the advantages and disadvantages of both types of rice.

3 a Which of the three farmers in the case study has gained the most from the Green Revolution? Why?

  b Which farmer has gained the least from the Green Revolution? Why?

  c How could Mr Patel and Mr Bagchi improve their outputs and become wealthier?

  d Imagine you were in a similar position to Mr Bagchi. Describe your feelings about the Green Revolution and the way it has helped the wealthy become even richer. Suggest ways in which peasant farmers like yourself could have been better served by the scientists.

16 Egypt: greening the desert

Background

With the population of Egypt being 47 million in 1985 and the growth rate 3%, there was concern over the need to increase food production. Traditionally, crop production was based on the flood waters of the river Nile. Rainfall, at less than 125 mm per year, was insufficient. Most of the agricultural land was along the banks of the river and the fertile delta. (See Figure 16.1.)

In 1964 the Aswan High Dam was completed. This allowed the water level of the Nile to be controlled so water was available all year round. As Egypt has a hot desert climate, average temperatures of 25°C were suitable for the growth of maize, wheat, cotton, etc. with two harvests per year. The demand for food has been increasing and, as the more fertile areas are densely populated, the only alternative was to expand into the desert.

Sand is capable of supporting crops if the right methods are used, e.g.:

1 Windbreaks have to be provided to prevent burial of crops by sand.
2 Water has to be provided in just the right amounts − enough to allow the crop to grow, but not too much as large quantities are lost in evaporation. If the soils are kept too moist the high rate of evaporation gradually draws salts to the surface. These form a whitish crust and create conditions unsuitable for crop growth. This is called **salinisation** (Figure 16.2).

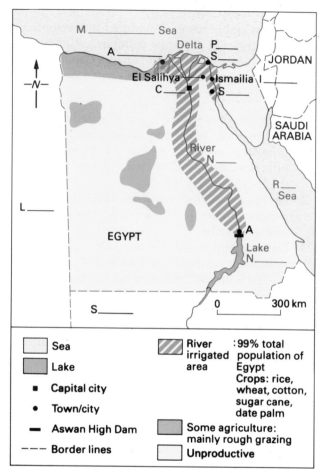

Figure 16.1 A map of Egypt

Legend	
Sea	
Lake	
■ Capital city	
● Town/city	
▬ Aswan High Dam	
--- Border lines	

River irrigated area : 99% total population of Egypt
Crops: rice, wheat, cotton, sugar cane, date palm
Some agriculture: mainly rough grazing
Unproductive

Figure 16.2 Salinisation

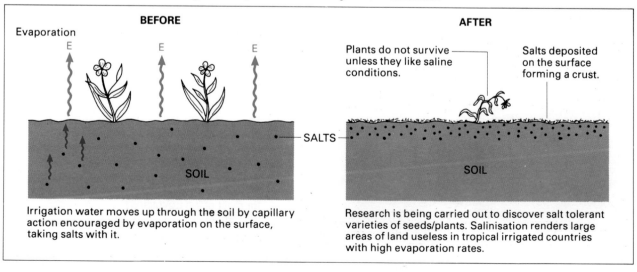

BEFORE

Evaporation

Irrigation water moves up through the soil by capillary action encouraged by evaporation on the surface, taking salts with it.

AFTER

Plants do not survive unless they like saline conditions.

Salts deposited on the surface forming a crust.

Research is being carried out to discover salt tolerant varieties of seeds/plants. Salinisation renders large areas of land useless in tropical irrigated countries with high evaporation rates.

CASE STUDY

The El Salhiya Project (100 km north east of Cairo)

This is an ambitious project aimed at developing 30 000 hectares of desert by using irrigation water from the newly constructed Ismailia Canal. Experimental **mixed farming** is carried out involving:

1 Orchards of citrus fruits are irrigated using the drip method. Water continually drips from a pipe onto the soil at the base of the trees. Very little is wasted as all the water goes directly to the roots.
2 Greenhouses grow tomatoes, cucumbers and peppers mainly for export to Europe in the winter months.
3 Dairy farming is carried out. Large quantities of milk are produced to be sold locally. 1000 cows are fed with fodder grown in circular fields of 60 hectares each (see Figure 16.3). Alfalfa and sorghum are grown using sprinkler irrigation (see Figure 16.4). This controls the amount of water sprayed on the crops. Also, irrigation water can be mixed with fertilizers and pesticides. This project shows that given enough money and the right equipment, even deserts can become green!

Figure 16.3 Circular fields with sprinkler irrigation

Figure 16.4 Sprinkler irrigation: Centre-pivot sprinkler farms

ACTIVITIES

1 Why is the demand for food increasing in Egypt?
2 Why is most of Egypt's population concentrated near the Nile?
3 Make a copy of Figure 16.1. Using your atlas, complete the names of the seas, river, cities and neighbouring countries. The first letter of each name has been given to help you.
4 a Use your atlas and information in the text to suggest reasons why the yellow areas in Figure 16.1 are unproductive.
   b What could be done to make these areas more productive?
5 Why is the El Salhiya project an example of mixed farming?
6 a Describe the two modern methods of irrigation used at El Salhiya.
   b What are the advantages of these systems?
7 Several problems are associated with large scale irrigation projects of this kind.
   a Make a list of all the equipment and materials you would need to set up a similar project. Where do you think you would get the money and equipment from?
   b Why do you think that large scale 'agribusiness' of this kind is criticised as being inappropriate for most of the Egyptians?
   c Why might the soil eventually become useless?

17 Britain and the European Community

The Common Agricultural Policy

Figure 17.1 a

In 1984 British farmers produced £12.4 billion of farm produce (£57 000 per farm). This was 6% more than in 1983. However, it cost the Government approximately £3 billion in subsidies to the farmers to produce this food. The result is an increasing supply of surplus farm products which builds up into butter mountains or milk lakes (see Figure 17.1a). How has this situation come about? The answer lies in an agreement between the member countries of the EC called the **Common Agricultural Policy** (or **CAP** for short).

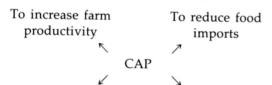

In 1973 Britain entered the EC. Until then farmers had set a minimum price for a crop they produced. Once in the Common Market, when farm produce fell below a certain price the EC bought what was left over at a price set in advance. This is called the **intervention** price. This led to farmers growing as much as they could, knowing they could get a guaranteed price.

Study Figure 17.1b. This shows the effect of CAP in the EC.

Figure 17.1b Effect of CAP in Europe

- Cereals encouraged rather than livestock because grain prices were set high.
- Investment in technology such as new high yield cereals, using fertilizers and machinery.
- A decrease in the agricultural workforce, e.g. France 28% in 1950, 8% in 1982.
- Increase in size of farms and amalgamation of fields.
- Some self-sufficiency achieved to overcome unreliable imports.
- Rural depopulation has been slowed down due to subsidies to farmers.

- Land ploughed up and drained with the aid of grants; wildlife habitats destroyed (see p. 52).
- Pollution of water by slurry and fertilizers (see p. 54).
- Creation of food surpluses and high cost of storage.
- Hedgerows removed which may increase soil erosion risk.
- Self-sufficiency in Europe may affect economies of developing countries whose exports are reduced.
- Inefficient farmers encouraged to remain in farming. High cost of subsidies.

The future of CAP is not certain. Spain and Portugal have 18% and 27% of their workforce in agriculture but with a very low standard of living. It will take at least ten years for these countries to be fully integrated with the rest of the EC. Grants and aid will need to be given. At present 70% of the EC budget is spent on agriculture and governments are trying to find ways of reducing this figure to allow money to be spent on other projects like industrial development.

Food surpluses

Figure 17.2 British farm products 1986

Product	Production a	Consumption b	% Self-sufficiency* $\left(\frac{a}{b}\right) \times 100$
Wheat	14957	10553	141.7
Other cereals	26590	19171	
Potatoes	7310	7636	
White sugar	9314	2062	
Beef and veal	1090	1251	
Pigmeat	978	1381	
Sheepmeat	292	402	
Poultry meat	876	917	

* Answer over 100 = surplus

Figure 17.3 Crop statistics (GB)

	1973	1974	1975	1976	1977	1978	1979	1980	1981	1982	1983
Oilseed rape	31	55	61	111	142	155	198	300	325	581	563
Milk (skimmed)	156	105	105	170	246	272	233	237	251	296	302
Butter	97	54	48	89	134	164	161	170	172	216	241

(Figures are 000s of tonnes harvested.)

These surpluses show that British farmers have been too successful in the production of some crops. It must be remembered, however, that some of these surpluses result from the EC's subsidised prices.

With recent health concerns about fats in our diet, the consumption of butter and other dairy products has declined. This means spare butter and milk is produced that no one wants. Giving it away to the developing world can cause problems.

(See Figures 17.4 and 17.8.) The information in Figure 17.3 shows the dramatic increase of oilseed rape production, compared to butter and skimmed milk. Rape oil is used in making margarine, which means there is increasingly more spare butter and skimmed milk. Other European countries like Italy and France have surplus wine or olive oil. In 1985 the EC spent £23 million on storage or disposal of surpluses. Figure 17.5 shows us the scale of surpluses in Europe.

Figure 17.4 Possible solutions to surpluses

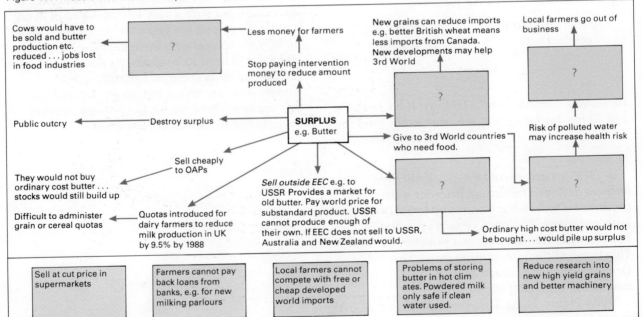

Why is there a surplus?

Over-production is a way of ensuring against risks like pests or diseases, or even bad weather which could destroy crops. It is very difficult to predict what our needs will be. Figure 17.4 helps show some questions and answers raised about food surpluses. People have been concerned about the effect that the entry of Spain and Portugal will have on the EC. These two countries have large farm populations. Their climates encourage production of early fruit and vegetables. It may well be that they will provide these products at a time when north European farmers will not be competing in the market. Will more wine and olive oil be added to the lakes, however?

Figure 17.5 The scale of surpluses

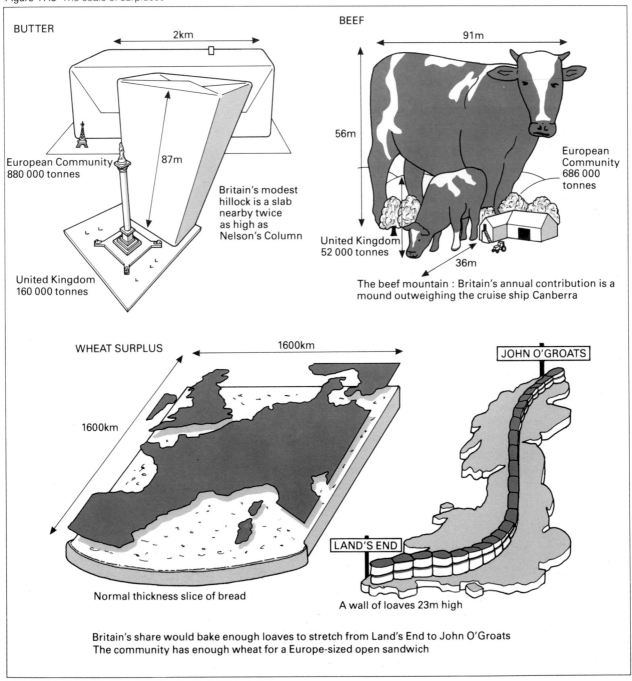

BUTTER

2km

European Community
880 000 tonnes

87m

Britain's modest hillock is a slab nearby twice as high as Nelson's Column

United Kingdom
160 000 tonnes

BEEF

91m

56m

European Community
686 000 tonnes

United Kingdom
52 000 tonnes

36m

The beef mountain : Britain's annual contribution is a mound outweighing the cruise ship Canberra

WHEAT SURPLUS

1600km

1600km

Normal thickness slice of bread

JOHN O'GROATS

LAND'S END

A wall of loaves 23m high

Britain's share would bake enough loaves to stretch from Land's End to John O'Groats
The community has enough wheat for a Europe-sized open sandwich

Solutions?

Although Figure 17.4 raises some possible solutions, what we have discovered in studying the farm systems in this book is that **diversification** seems to be the most popular way of avoiding excess production. This is discussed further in Chapter 19.

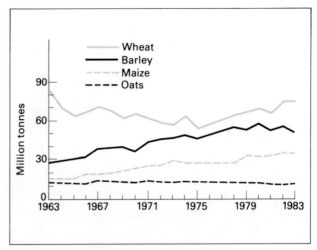

Figure 17.6 Cereal production in west Europe

Figure 17.7 Average farm size in Europe

ACTIVITIES

2 a What is meant by CAP?

  b What are the main aims of CAP?

  c How has CAP influenced:
    i farm size and layout?
    ii methods of farm production? Use Figure 17.1b to help you.

3 a Using Figure 17.6 describe the pattern of cereal production in Europe between 1963 and 1983.

  b Why do you think oat production has hardly increased during this time?

4 a Study Figure 17.7. Which areas of Europe have the largest percentage of farms over 30 ha?

  b Which country has most of its farms sized 7.4 ha or below?

  c Where will the most efficient farming be? Give reasons to support your answer.

5 a Make a copy of Figure 17.4, and fill in the spaces with the appropriate caption provided.

  b What is Figure 17.4 suggesting about the effect of short term food aid on traditional peasant farmers in the developing world?

6 Using the information provided in Figure 17.3 draw a line graph using a different colour for each crop.

  a When did the production of oilseed rape suddenly increase?

  b Why did farmers suddenly switch to producing more oilseed rape than other farm products like milk?

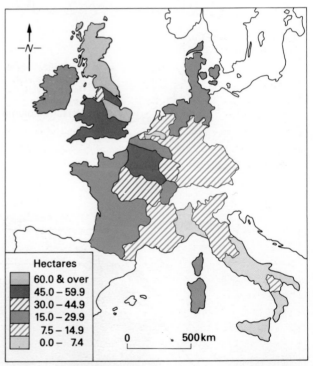

Figure 17.8

18 Co-operative farming in Denmark

In a **co-operative** a group of farmers join together for their mutual benefit. There are several co-operative societies in Denmark. They are of different sizes, but all are owned and run by the people themselves. Their aim is to help the individual family farmers run their farms as efficiently as possible by making available modern methods. Look at Figure 18.1. The success of Danish farming is based on these societies.

A study of farming concentration in Denmark

Farming is becoming increasingly specialised so there are certain areas in Denmark where one type of farming is more common. For example, grain farming is more common in the eastern areas of Jylland and the islands. This can be illustrated using a **location quotient**. This is a measure of the concentration of one activity in an area and it is calculated using the following formula:

$$\frac{\text{Local figure (e.g.\%) for a particular activity}}{\text{National figure (e.g.\%) for the activity}}$$

If the answer is 1 the local percentage is the same as the national percentage. However, if the answer is greater than 1 the local area has a higher percentage than the national average. If the figure is high, that activity can be said to be **highly concentrated** in that particular area of the country.

Look at the figures for dairy cows in Table 18.1 Location quotients using these figures have been worked out for you below:

Copenhagen: $\dfrac{1 \text{ (local figure)}}{32 \text{ (national figure)}} = 0.03$

Ribe: $\dfrac{52}{32} = 1.62$

This shows that there is a concentration of dairy farming in Ribe, but it is not very concentrated and therefore not very important in Copenhagen.

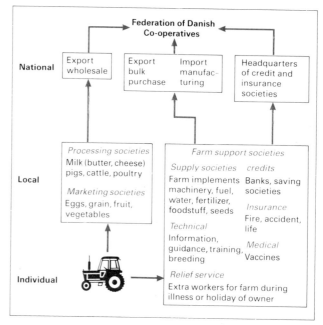

Figure 18.1 The co-operative organisation in Denmark

Table 18.1

County	Dairy cows per 100 ha of agricultural land	Percentage distribution of farms with:			
		a c&p	b c	c p	d no c&p
1 Copenhagen	1				
2 Frederiksborg	13				
3 Roskilde	8				
4 Vestsjaelland	14	18	15	27	40
5 Storstrøm	10				
6 Bornholm	18				
7 Fyn	28	20	20	23	37
8 Vejle	33				
9 Århus	25	26	22	23	29
10 Sønderjylland	40				
11 Ribe	52	26	40	14	19
12 Ringkøbing	40				
13 Viborg	41	33	25	21	22
14 Nordjylland	38				
Denmark average	32	26	24	22	28

c = cattle, p = pigs

(Figures are for 1985)

Animal farming

Cattle

Most cattle farming is carried out on family-run farms of between 10 and 100 ha. There are 28 cows in the average dairy herd, the favoured breeds being Black and White Danish Dairy and Red Danish. Milk quotas have reduced the size of several of the dairy herds but high milk quality is still very important. Cattle can graze outside for five to six months of the year. Large amounts of money are spent on improving the farm buildings to keep animals in the best conditions and there is constant research in breeding techniques.

Pigs

There are 9 000 000 pigs in Denmark. They are kept indoors all year round with farms having an average herd size of 220 Landrace pigs. A National Breeding and Monitoring Programme maintains the high quality of the animals and products which Denmark is famous for.

Poultry

Production is concentrated on a few farms with over 10 000 hens each. These farms have modern automated methods. This is an example of **factory farming**, an intensive type of farming with high outputs. Poultry are kept in large buildings where they have individual compartments. Food appears automatically at regular intervals and heating and lighting are carefully controlled. The hens are bred either to produce eggs or for selling directly.

Mink

Denmark is one of the world's leading producers and the rearing of these animals for fur is increasing.

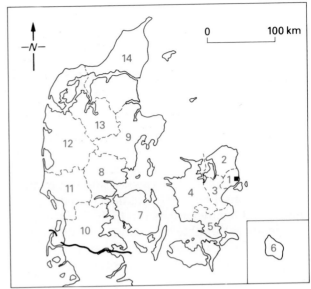

Figure 18.2 Counties of Denmark

Look at the pie chart (Figure 18.3) which shows that approximately 75% of farm income comes from animal produce. Most of these products are exported, mainly to EC countries, Japan and the USA. For example, Britain and West Germany are large importers of pig meat, while Italy imports mostly veal. Denmark also exports large amounts of cheese to Iran and Saudi Arabia.

Crop farming

Crop farming requires maximum inputs in the form of fertilizers, pesticides, irrigation and machinery. Look at Figure 18.4 which shows the percentage of different crops produced. Notice the importance of grain crops. They are mainly grown in the east of the country because of the favourable conditions (Figure 18.5). Denmark is self-sufficient in grain, sugar, potatoes and seeds. It also exports crops and seeds.

Figure 18.3 Danish animal farm income by value

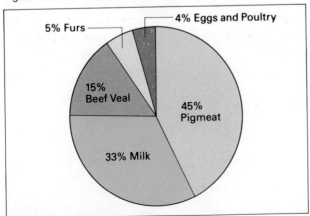

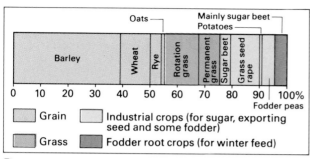

Figure 18.4 Crop production (%)

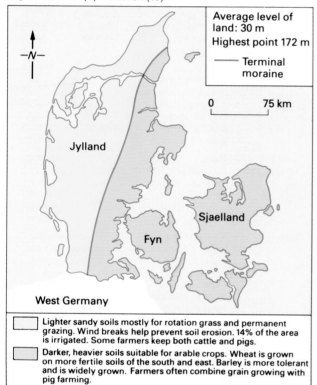

Lighter sandy soils mostly for rotation grass and permanent grazing. Wind breaks help prevent soil erosion. **14%** of the area is irrigated. Some farmers keep both cattle and pigs.

Darker, heavier soils suitable for arable crops. Wheat is grown on more fertile soils of the south and east. Barley is more tolerant and is widely grown. Farmers often combine grain growing with pig farming.

Figure 18.5 Danish soils

Table 18.2 Farm information

	1960	1970	1985
No. farm units	196 076	140 197	100 000
Ave. farm size (ha)	18	21	31
No. farms <10 ha	44 000	–	16 000
10–30 ha	68 600	–	40 800
30–50 ha	18 900	–	18 500
50–100 ha	8 700	–	14 100
Full time workers	300 000	161 200	114 000
Hired workers	128 300	34 100	22 000
% population in agriculture	16	9	6

ACTIVITIES

1 a Look at Table 18.1. Work out the location quotients for farms with only cattle (Column b) and farms without cattle or pigs (Column d).
  b On a copy of the map of Denmark (Figure 18.2) use one colour to shade in all counties with a location quotient greater than 1 for cattle. Use another colour to shade in all the counties with a location quotient of 1 or more for farms without cattle and pigs.
  c Describe the distribution of cattle farming in Denmark.
  d Describe the distribution of farms without cattle and pigs. What sort of farms do you think these will be?
  e With the help of Figure 18.5 try to explain why dairying is concentrated in the west and arable farming in the east.

2 Look at Figure 18.3. What % of the animal farm income comes from:
  a pigs?
  b cattle?

3 Look at Figure 18.4. What % of the crop production is taken up by·
  a grain crops?
  b fodder crops?
  c grass?

4 Using the figures from Table 18.2 comment on the changes in:
  a farm size;
  b number of farms;
  c number of farmers from 1960.

5 What is factory farming?

6 Why is Danish farming so successful?

Coursework ideas

You could do a survey to assess the importance of Danish products in your local supermarket or at home.

19 *The future use of agricultural land*

Can wheat fields be turned into woods?

Figure 19.1

The butter, grain, beef and other food surpluses are continuing to grow. The Government and the EC are trying to find other uses for the land but want to protect farmers' incomes so they don't go out of business. Some estimates suggest 5½ million hectares of agricultural land could be taken out of production in Britain by 2015. Study Figure 19.2a which shows some of the advantages and disadvantages of the alternative uses of agricultural land.

In the next twenty years the changes outlined may well alter the look of the countryside. The amount of cereals grown on poor quality land will decrease. However, eastern Britain is likely to remain mostly arable as conditions here are very suitable. There will be less specialisation of farming and areas of mixed woodland will increase. Some of these changes may improve the landscape.

Land use	Advantages	Disadvantages
Farm less intensively	Yields lowered by reducing inputs like fertilizers. Less land taken out of production	Less profit for farmer. Still high cost of rent, labour, machinery and bank loans
Grow other crops	Oats, beans and oilseed rape are possible but market is limited. Pasture only increases surplus animal products	May be surplus crops now but what of future? Once agricultural land is taken or abandoned it is expensive and difficult to re-establish
Take land out of production	Paying farmers for not cropping part of their land reduces surpluses	High cost. Probably poorest land set aside and may crop other land more intensively to make up for loss in overall production
Planting trees	Reduces timber imports. Climate of GB suitable for conifers. 1987 Government scheme to plant 36 000 ha of farmland in 3 years	May plant on good lowland areas. Long time before there is a profit (40 years). Grants not enough to encourage farmers to try
Recreation	Demand for rural recreation increasing as people become more mobile	Often poorest farmland taken for golf courses etc. Not much effect on surpluses. Problem of access to this land too
Housing and industry	Shortage of land is said to be causing high house prices. Demands from new residents may create new opportunities for extra income for farmers	Much of suitable land for building is again not very productive so little effect on surpluses. Most demand is in SE of UK where there is already much pressure on rural areas.

Figure 19.2 a Alternative land uses

Figure 19.2b Table to show lost or damaged wildlife habitats in postwar Britain

Lowland meadow	81%
Chalk downland	78%
Lowland bog	59%
Lowland marsh	51%
Limestone pavement	43%
Ancient woodland	30%
Lowland heath	39%
Upland woodland	24%

Source: Adapted from bar graph *Geog Review* p36 Vol. 1 No. 1

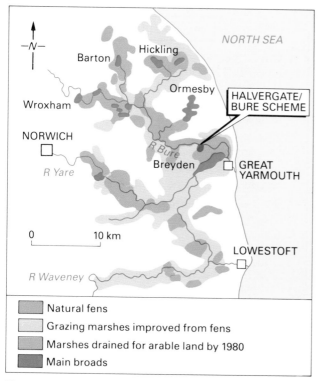

Figure 19.3 Location of Halvergate Marshes

Legend:
- Natural fens
- Grazing marshes improved from fens
- Marshes drained for arable land by 1980
- Main broads

Agriculture and conservation

Figure 19.2b shows the massive loss of wildlife habitats that has resulted from farming becoming more intensive. People are becoming more aware of their environment and some farming methods have caused controversy, e.g. removing hedgerows, burning stubble or adding fertilizers. In order to protect some of the unique habitats left, **Environmentally Sensitive Areas** or ESAs were set up in 1986, based on a scheme started in the Norfolk Broads at Halvergate Marshes.

CASE STUDY

Halvergate Marshes — an example of an Environmentally Sensitive Area

Halvergate Marshes cover an area of 3000 ha between the Rivers Yare and Bure in eastern Norfolk (see Figure 19.3). Although marshland, it is not very important for wildlife but does form an attractive and traditional grazing landscape which is used for fattening dairy and beef cattle. Figure 19.4 shows this unique landscape.

In 1980 the marshes could have been drained to become another arable landscape. Instead, local

Figure 19.4 Grazing on Halvergate Marshes

Figure 19.5 Environmentally Sensitive Areas in England and Wales

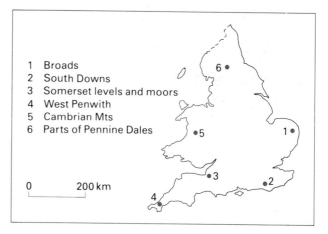

1 Broads
2 South Downs
3 Somerset levels and moors
4 West Penwith
5 Cambrian Mts
6 Parts of Pennine Dales

conservationists and the farming unions agreed on a scheme. Any farmer who volunteered not to change the land use of the marshes could receive a subsidy of £123 per ha. Farmers had to promise not to graze animals on the marshes between November and March to avoid compaction of soils by the animals. They had to limit the use of fertilizers and not use modern herbicides so that water pollution would be avoided (see Figure 19.6). Silage was only to be cut once a year to allow the wild grasses a chance to grow.

This project was very successful (108 out of 110 farmers joined) and meant the farmers could still gain a reasonable income. In 1986 five other ESAs were established as shown in Figure 19.5. ESAs are likely to be more attractive to farmers today as farm incomes fall and milk quotas have their effect. ESAs show agriculture and conservation are not incompatible.

Figure 19.6 River pollution

Coursework ideas

Find out if there is a farm near you which you can visit. (The local National Farmers Union may help.) Carry out a survey or questionnaire to see if the farmer has changed his land uses over the last few years. Try and find out reasons why. What does he propose to do in the future, and why?

ACTIVITIES

1 Why is there a need to reduce the amount of agricultural land in production?

2 Out of the possible future alternative land uses shown in Figure 19.2a which do you think will best reduce surpluses? Give reasons for your answer.

3 How would you try to avoid the disadvantages of this alternative?

4 Which of the proposals would have the most harmful impact on the countryside? Why?

5 a Explain what the term ESA means, using the text to help you.

  b In an atlas find a physical and political map of the British Isles and, with the help of Figure 19.5, identify:
    i The areas in which the six ESA sites are situated.
    ii Two other ESA sites, apart from Halvergate, shown on Figure 19.5.

6 a Using Figure 19.3 draw a sketch map to show the location of Halvergate Marshes.

  b Explain how the ESA scheme at Halvergate not only protects the environment but also ensures a reasonable income for the farmer.

  c Do you think the ESAs are a good idea? Write a short paragraph giving your opinions and ideas.
  OR

  d You have been given the job of producing a single page document promoting ESAs. The document is to be distributed to the members of the National Farmers Union. You should describe what ESAs are, and make specific reference to Halvergate Marshes using a sketch map and brief written summary. Give your document a heading and make it attractive as well as informative.

20 The Norfolk Broads — to conserve or develop?

The Norfolk Broads consist of 290 km² of marshland and reclaimed fen and woodland (see Figure 19.3 on page 53). The landscape is partly the result of human activity in that the Broads are flooded peat pits. The peat was dug as a valuable fuel supply in the Middle Ages. If nature was left alone the Broads would soon disappear. They would become silted up and vegetation would take over. It is only through human activities such as farming, reed cutting and fishing that they have survived. Today, they provide a unique landscape which is enjoyed by the many visitors who come each year for boating and other leisure activities (see Figure 20.1).

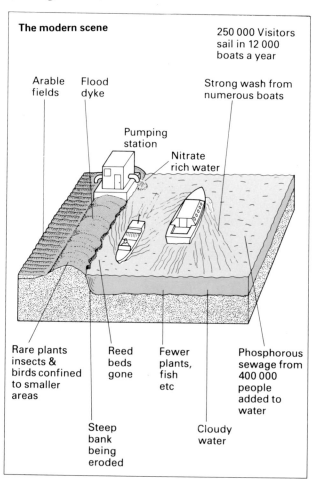

The modern scene

250 000 Visitors sail in 12 000 boats a year

Arable fields

Flood dyke

Strong wash from numerous boats

Pumping station

Nitrate rich water

Rare plants insects & birds confined to smaller areas

Reed beds gone

Fewer plants, fish etc

Phosphorous sewage from 400 000 people added to water

Steep bank being eroded

Cloudy water

Figure 20.2 Summer scene on the Broads

Figure 20.1

If the Broads are to survive in the future there needs to be careful planning and conservation. Increasing demands are being made by farmers, tourists and a growing resident population. Look at the block diagram (Figure 20.2) and the flow chart (Figure 20.3) which show how modern activities are affecting the Broads. The Broads cannot be fenced off from the polluted rivers but the solutions shown in Figure 20.4 can help prevent erosion and reduce the scale of pollution. The setting up of a Broads Authority to oversee the conservation is a step forward and it is possible that in the future the Broads may become a National Park (see Chapter 21).

ACTIVITIES

1 Using your atlas and Figure 19.3 on page 53:
  a Describe the location of the Broads;
  b Name at least two main Broads.

2 How does Figure 20.2 explain how modern day use of the Broadland area can harm the environment?

3 There are several suggested solutions to the problems in the area. Using the information given, make up a table with two columns — one headed **Problems** and the other **Solutions**. Match each of the problems outlined to the solutions in Figure 20.4. Can you think of other solutions you could add?

Figure 20.3 Environmental impact on the Broads

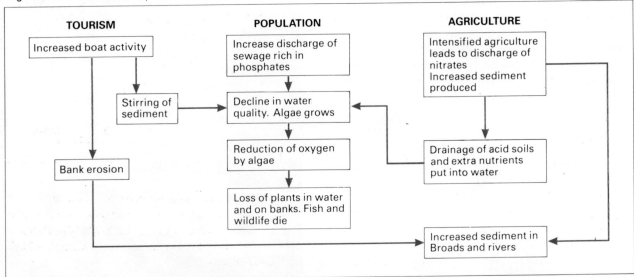

Figure 20.4 Solutions to problems

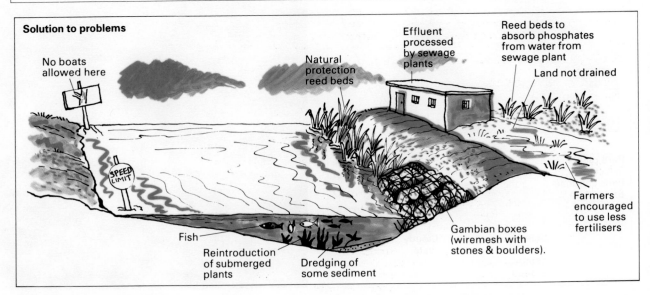

21 National parks

The ten protected rural areas in England and Wales shown in Figure 21.1 are national parks. They were set up in 1949 following a Countryside Act in Parliament. Their main aims are as follows:

1 to preserve the beauty of the landscape;
2 to allow access to the public for open-air activities such as walking or climbing;
3 to protect wildlife;
4 to protect places of historic interest;
5 to allow farming to continue in order to maintain the landscape.

Most land is privately owned in national parks but planning is strictly controlled by a National Park Authority, not local councils as elsewhere.

National parks can be found all over the world, at Yellowstone in the USA, Tsavo in Kenya and Serengeti in Tanzania.

Country parks and Areas of Outstanding Natural Beauty

Country parks and Areas of Outstanding Natural Beauty (AONBs) are smaller in size than national parks. They too are protected from development. Many, like Queen Elizabeth Country Park near Portsmouth in Hampshire, were set up to attract

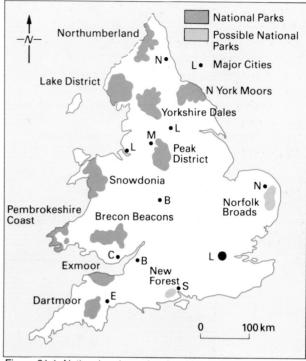

Figure 21.1 National parks and major cities

visitors from surrounding urban areas and to reduce pressure on other countryside areas. Study Figure 21.2 which compares the Yorkshire Dales National Park with Queen Elizabeth Country Park in Hampshire.

Figure 21.2 Comparison between country park and national park

	Country park	National park
Size	566 hectares	1761 km²
Average length of visit	2–3 hours	1–6 days
Popular visiting times	Weekends, public holidays	Summer and weekends
Ownership	County Council and Forestry Commission	Most land privately owned
Cost of entry	50p–£1 for cars	No fee
Catchment area	Local region	National and international
Amenities	Iron Age farm with rare animal breeds	Nature reserves
	Hang gliding	Museums
	Nature trails	Wide range of active and passive pursuits.
	Grass skiing	e.g. climbing and picnicking
	Pony trekking	
	Guided walks	Camping and caravan sites
	Picnic sites	Historic settlements

ACTIVITIES

1 a On an outline map of Britain mark on and label the ten national parks using Figure 21.1 to help you.

   b Using an atlas name the major cities shown.

   c Which area of England and Wales is furthest from a national park?

   d Do you think there is a need for a national park here? Give reasons for your answer.

   e Which two areas are being proposed as future national parks?

   f Which area would you propose for a national park? (Use Figures 21.1 and 21.3 to help you.) Give reasons for your answer.

   g Look at Figure 21.3. What impact do you think motorways will have on national parks?

   h Which national park will be the one most likely to be under greatest pressure from visitors coming by car? Explain your reasoning.

2 Look up the words **conservation** and **preservation** in a dictionary. How do they differ? Why would conservation be a better term to protect the countryside than preservation?

3 Study the text and Figure 21.2.

   a Where do visitors to a country park mainly come from?

   b What attractions are there for visitors to Queen Elizabeth Country Park?

   c Do you think these attractions apply to all age groups?

   d Why might the activity areas in the country park have to be separated from each other?

4 Study a local OS map near your home or school. Decide on a possible location for a country park which will serve your local town. Give reasons for your choice. What will need to be provided for people wishing to make use of your country park?

Coursework ideas

Visit a local park near you. Find out what facilities it has to offer and if it caters for all age groups. Does its use vary through the week? What is its catchment area? You could compare two very different parks.

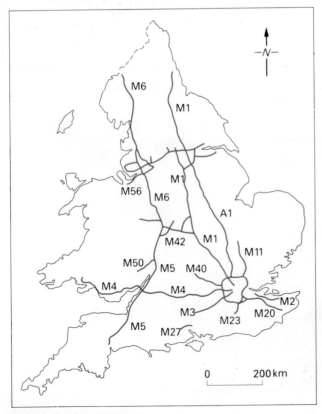

Figure 21.3 UK motorways

CASE STUDY

Yorkshire Dales National Park

The Yorkshire Dales National Park covers a large area of the Pennine Hills. It includes some spectacular limestone scenery with caves and potholes which attract many visitors. Figures 21.4a–21.4d show information about the park and its visitors in 1986.

Figure 21.4a Reasons for visit to Yorkshire Dales National Park

Visit friends/relatives	5.9%
Educational	0.8%
General tour and sight seeing	66%
Walking less than three miles	17.6%
Climbing and caving	0.9%
Fishing	0.8%
Passive pursuits e.g. picnicking	2.9%
Active pursuits	1.2%
Other	3.9%

Figure 21.4c Traffic survey along B6255 Easter Sunday 1988

Time	Number of vehicles
8 am	25
9 am	51
10 am	116
11 am	235
11.30 am	351
12 noon	260
1 pm	248
2 pm	249
3 pm	264
4 pm	264
5 pm	200
6 pm	223
7 pm	70
8 pm	53

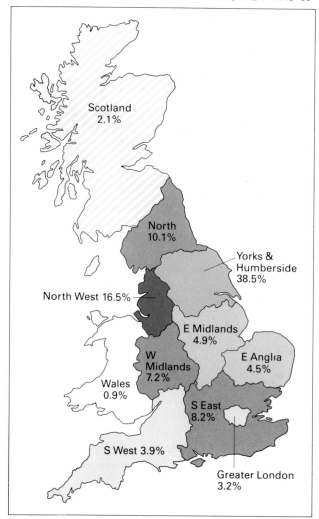

Figure 21.4b Origin of visitors

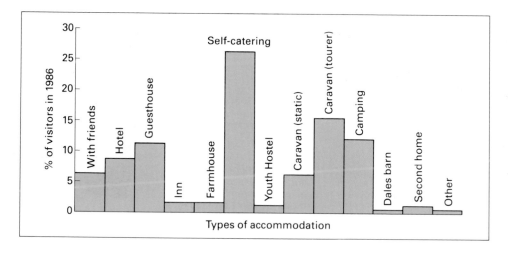

Figure 21.4d Visitor accommodation

ACTIVITIES

5 Look at Figure 21.4b. This shows the idea of **distance decay**. What do you think this means?

6 Using the information given in Figure 21.4c draw a line graph to show the traffic flow along the B6255 Hawes-Ingleton road. (705735, page 18.)

  a Label the three peak times of traffic flow.

  b Explain why these peaks occur at these times.

  c Why is there less traffic between midday and 3 pm?

  d What problems may be caused by this uneven flow of traffic?

7 a Using Figure 21.4d assess what type of holiday accommodation visitors prefer.

  b What impact will this have on the local economy?

  c Why might provision for self-catering holidays cause an environmental conflict?

8 a Draw a pie graph to illustrate visitor transport to the park using the figures below:
Cars 86% Coach 9% Train 2% Other 3%

   i Why are cars so popular and trains so little used?

   ii What effect will this have on small narrow road often found in the park?

  b Look at Figure 21.4a. Which is the most popular reason why people visit? How does this relate to your answer in 8a?

Figure 21.5 Overflowing car park at Ingleton

Figure 21.6 Footpath erosion on Whernside. Note the gravel and mesh which is laid to form a new hard wearing surface

The data from the visitors' survey show that the environmental, physical and social capacity of the park can be quickly reached. The **environmental capacity** is the ability of an area to withstand public use. Figure 21.5 shows the **physical capacity** of a car park at Ingleton has been reached. Parking of cars on grass verges may cause erosion. Figure 21.6 shows the problem of foothpath erosion along the Pennine Way which crosses the park. People may end up destroying the very environment they are coming to see, and cause loss of pasture land for farmers as moorland turns to mud. Solutions could include netting with gravel laid on top to form a hard wearing surface, as shown in Figure 21.6. However this is quite expensive and difficult to do in remote areas.

Social capacity depends on the individual: for some visitors a motorbike scramble on the moorland is an ideal way of using the area, for others it shatters the peace and quiet they are seeking.

Maintain buildings that may otherwise be left empty or remain in disrepair

Travelling shops may set up e.g. Butcher van to visit all villages in order to gain enough trade

Deserted village midweek and out of season. Feeling of loss of community for those residents remaining.

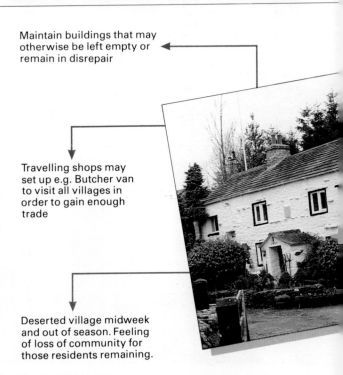

Figure 21.7

Popular sites are termed **honeypots**, e.g. Malham village. Too many visitors cause congestion, noise, and litter. Solutions to the many demands made by visitors are not easy. The use of clear signs and footpath maintenance help. New roads and car parks are not only ugly but often only encourage more visitors. The introduction for a trial period of a Dales bus service at weekends failed as people still preferred to use their own cars.

Main land use demands in the park

1 Tourism

Tourism is important as it brings in revenue for the local economy. Hill farmers whose farms are not very profitable may earn money from caravan and camp sites or from farm 'Bed and Breakfast'. Jobs for local people have also been created in the service sector, e.g. as assistants in cafes and shops. The growth of second homes in the park has increased in recent years. Second homes are houses which are bought by urban dwellers for use at weekends or over holidays. In the parish of Aysgarth 11% of houses are second homes and at the village of Starbotton in Wharfedale 15% are second homes. Figure 21.7 shows the effect of second homes on a Dales village.

2 Quarrying

The eleven quarries in the park extract either limestone or high quality gritstone which is used in the road and construction industry. The quarry at Ingleton, shown in Figure 21.8, quarries 400,000 tonnes of gritstone per year out of a reserve of 8 million tonnes. It employs 27 local people and has to seek permission from the park authority before extending its boundary. Trees to screen quarries in national parks have been tried but problems such as visual impact, dust pollution, heavy lorry traffic and possible damage to unique cave systems remain. It is unfortunate that national park areas such as the Yorkshire Dales, Brecon Beacons and the Peak District are the main sites where such high quality rock can be found.

3 Farming

The major land use of the park is for agriculture. Hill farming is common (see Chapter 6) and, although visitors can cause erosion and trespass, they can provide extra income with the demand for accommodation. It is important to maintain farming as the landscape is the result of human activities and would soon change if left unattended.

Figure 21.8 Quarry at Ingleton

Figure 21.7 Impact of second homes in a national park

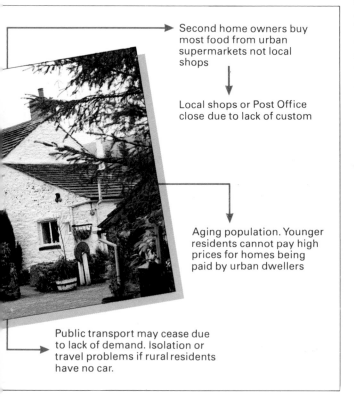

Second home owners buy most food from urban supermarkets not local shops

↓

Local shops or Post Office close due to lack of custom

Aging population. Younger residents cannot pay high prices for homes being paid by urban dwellers

Public transport may cease due to lack of demand. Isolation or travel problems if rural residents have no car.

4 Forestry

Large areas of moorland have been planted with conifers to supply the increasing demand for softwood. Forests cover only 2% of the Yorkshire Dales National Park but in Northumberland it is 19%. Forests may provide an amenity for visitors such as forest trails but conservationists are opposed to further large scale afforestation.

5 Other uses

Small areas may be used by the Ministry of Defence for training. Water authorities often use national parks for reservoirs due to high rainfall and suitable valleys for damming. In the Brecon Beacons 4% of the park is under Water Authority ownership. Reservoirs may be used for fishing or other such activities but farmers object to the loss of better farmland in the valley bottoms.

CASE STUDY

Game reserves and national parks in Kenya

In the past, wildlife has been exploited by people for skins, ivory and other organic substances. Animals were hunted for pleasure or killed when they were a threat to human dwellings or cultivated plots.

Now many governments are concerned to conserve wildlife in their natural environments for the following reasons:

1 the potential for tourism;
2 the high numbers of animals that are naturally supported by the environment. Each type of animal has adapted to the environment. For example, the zebra grazes on the grassland, the black rhino grazes in the denser bush vegetation and the giraffe is able to browse the tops of trees leaving lower vegetation to other animals.

ACTIVITIES

9 a Using the text explain briefly what is meant by the physical, social and environmental capacity of a rural area.

   b How does Figure 21.5 illustrate that the physical capacity of the area has been reached?

   c How has the problem of footpath erosion been tackled in Figure 21.6?

   d Do you think this will be successful?

   e What is meant by the term **honeypot**? What services would the park authority have to provide in such popular areas?

10 The term 'national park' is said to be misleading. Land is not owned nationally but privately and they are not like urban park areas. What alternative name would you suggest for the national parks and why?

11 a Complete the matrix (Figure 21.9) showing the possible land use conflicts in national parks.

   b Should quarrying be allowed in national parks?

   c Write a short paragraph explaining how other land uses can cause conflict in national parks.

   d What social and economic effects do second homes have in national parks?

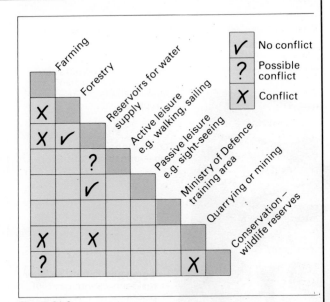

Figure 21.9

Kenya Major Game Reserves ☐ **& National Parks** ☐

1 Marsabit Game Reserve
2 Mt Elgon National Park
3 Samburu National Park
4 Meru National Park
5 Lake Nakuru National Park
6 Aberdare National Park
7 Mt Kenya National Park
8 Nairobi National Park
9 Ol Doinyo Sapuk National Park

10 Lambwe Valley Game Reserve
11 Masai Mara Game Reserve
12 Amboseli Game Reserve
13 Tsavo National Park East
14 Tsavo National Park West
*1 Malindi Watamu Marine
   National Park
* Marine National Parks

Northern Tanzania – Some Game Reserves ☐ **& National Parks** ☐

1 Serengeti National Park
2 Ngorongoro Game Controlled Area
3 L Manyara Game Reserve
4 Arusha National Park

5 Ngurdoto National Park
6 Kilimanjaro Game Reserve
7 Mkmomazi Game Reserve
○ Ngorongoro Crater

Figure 21.10 Major national parks and game reserves in Kenya and northern Tanzania

Domesticated animals like cattle do not survive in such large numbers. In the future the natural animals may be a useful source of food. Sun-dried meat or **biltong** is already produced.

At present there are 15 national parks in Kenya (see Figure 21.10). They have been designated to ensure the survival of animals natural to the savanna environment (see Figure 21.11). Some are very small and are dedicated to the survival of one species, e.g. Lake Nakuru National Park is 46 km² and is famous for its flamingoes. Tsavo National Park spreads over 20 000 km² and contains a variety of animals. They are important tourist areas and contain small lodges, hotels and camps where people can stay whilst on **safari**.

Game reserves, of which there are 22 at present, are areas where animal life is dominant and human activity is much more limited.

Hunting animals for skins and trophies is now illegal but controlled culling (hunting) of elephants, for example, still takes place. Elephants are destructive animals and trample ground and pull up trees leaving land prone to soil erosion. Therefore, their numbers need to be controlled.

ACTIVITIES

12 a What is the purpose of national parks and game reserves in Kenya?

  b How do these differ from national parks in the UK?

13 a Describe the main features of savanna vegetation using Figure 21.4 to help you.

  b How have animals adapted to this environment?

  c Why do you think this environment cannot support such large numbers of domesticated animals?

14 What sorts of accommodation are provided for tourists 'on safari'? (You could find out extra information from a travel agent's brochure.)

15 Explain carefully why the following people might welcome or oppose tourists to Kenya:

  a tour operators;

  b game wardens;

  c nomadic herders living in the park;

  d souvenir makers;

  e road builders;

  f conservationists;

  g hotel owners;

  h the Government;

  i farmers.

16 Imagine you are the Minister for Tourism in Kenya. You have been asked to plan a new tourist complex in a game park. The following are a list of possible amenities and services you could include:

  i hotel;
  ii camp site;
  iii water supply;
  iv electricity;
  v shops;
  vi surfaced roads;
  vii unsurfaced roads;
  viii airstrip;
  ix swimming pool;
  x water hole for the animals;
  xi sewerage system.

  a For a luxury complex:
    i List the amenities you would like to develop in priority order;
    ii Where do you think you would get the money from?
    iii Do you think a large scale luxury complex is suitable for the environment?
    iv What are the advantages of such developments?

  b For a basic complex list the amenities you think are essential.

  c Which would you prefer to visit? Give reasons for your answer.

  d Design a poster to advertise the attraction of spending part of a safari holiday in one of these complexes.

Figure 21.11   Savanna vegetation and typical wildlife

Index

LONGMAN
CO-ORDINATED
GEOGRAPHY

Series editor - Simon Ross

Energy and Industry

Paul Warburton

The Times 5.84, & headline 'Why not open arms to cheap steel imports' in *The Times* 16.12.82. All (c) Times Newspapers Ltd 1982, 1986, 1987, 1984 & 1982.

We have unfortunately been unable to trace the copyright holder of the article 'Tyneside loses 1000 more shipyard jobs' by David Simpson in *The Guardian* 3.6.86. and would appreciate any information which would enable us to do so.

Alta Holidays Ltd, fig 24.3; British Home Tourism Survey International Passenger Survey reproduced with the permission of the controller of Her Majesty's Stationery Office; Courtesy of BP Exploration, fig 4.1; Commission for the New Towns, Glen House, Stag Place, Victoria, London, fig 26.7; Corby Industrial Development Centre, fig 27.7; National Grid Division of the Central Electricity Generating Board, fig 8.8; *The Financial Times*, figs 7.9, 19.2; 'Geographical Education in Secondary Schools' by N J Graves. Published by The Geographical Association, 1980, fig 8.7; c Mary Glasgow Publications, London Geofile, January 1987, no 84, fig 9.3 and 1986, fig 11.3; International Broadcasting Trust, figs 13.1, 13.3; extract from *British Isles* by Brian Nixon, reproduced by kind permission of Unwin Hyman Ltd, fig 17.11; Ordnance Survey 1:50 000 Middlesbrough & Darlington area, Landranger 93 and Newquay & Bodmin area, Landranger 200, with the permission of Her Majesty's Stationery Office. Crown Copyright, page 46 and page 33. We are unable to trace the copyright holder of fig 22.3 and would be grateful for any information that would enable us to do so.

We are grateful to the following for permission to reproduce photographs: Aerofilms Limited, pages 42, 44, 51, 69 *below*; Austin Rover, page 50; BP, page 7; Barnaby's Picture Library, page 11 *above*; British Coal, pages 2 *below right*, 36 *above*, 37 *below*; Camborne School of Mines, page 25 *above right*; J. Allan Cash Ltd, pages 3 *below right*, 5, 25 *above left*, 36 *below*; Central Electricity Generating Board, pages 9, 15, 22, 37 *above*; Bruce Coleman Ltd, pages 11 *below* (photo: Nicholas Devore), 17 (photo: Norman Myers), 26 *above*, 35; Daily Telegraph Colour Library, pages 2 *below left* (photo: Francis Corbineau), 23 *above right*; English China Clays Group, pages 32, 34 *above and below*; Sally & Richard Greenhill, pages 71, 78 *centre below*, 78 *below right*; Bernard Hales Photography, page 69 *left and right*; Robert Harding Picture Library, pages 29 (5), 30 *above* (photo: Sarah King), 30 *centre* (photo: David Lomax); B Hofmeester, page 48; Hutchison Library, page 78 *below right* (photo: R Ian Lloyd); ICI, page 47 *centre*; Intermediate Technology, pages 76, 77 *above and below*; Tony Morrison, pages 28, 79; National Remote Sensing Centre, page 20; North of Scotland Hydro-Electric Board, page 16; Panos Pictures, page 78 *above* (photo: Sean Sprague), 78 *centre above* (photo: Alain le Garsmeux); David Pratt, page 36; Reflex, pages 29 *below right* (photo: Philip Gordon), page 62 *right* (photo: Piers Cavendish); Sealand Aerial Photography, pages 19 (photo: Institut Geographique Nationale), 47 *above*, 55; Shell, page 45; Skyscan, page 62 *left*; Frank Spooner Pictures, page 12 (3); Times Newspapers, page 14; Tropix Photographic Library, pages 2 *above right*, 26 *below* (photo: D Charlwood); Wind Energy Group Ltd, page 2 *centre*.

Acknowledgements

Alton Towers Limited; Birds Eye/Walls; British Coal; British Petroleum Limited; British Steel Corporation; British Tourist Authority; Camborne School of Mines; CEGB; Department of Energy; Department of Trade and Industry; East Anglian Examinations Board; European Schoolbooks Limited; Ford Motors Limited; Intermediate Technology Manpower Services Commission; *New Scientist*; Ross Frozen Foods Limited; Shell International Petroleum Company Limited; *Sunday Telegraph*; University of Cambridge Examinations Syndicate; University of Oxford Delegacy of Local Examinations; Vauxhall Motors Limited.

We are grateful to the following for permission to reproduce copyright material:

Financial Times Syndication London for article 'Sombre mood as Ghana celebrates 30 years' in *Financial Times* 26.11.82; Guardian News Service Ltd for three headlines in *The Guardian* 27.11.82 & 3.12.82; Monographs in International Studies for an extract from *Island of the Blest* by J.P. Mason in 'Monographs in International Studies, Africa Series No 31' (c) The Center for International Studies, Ohio University; Times Newspapers Ltd for articles 'Oil men fight in forest' by Elizabeth Grice in *Sunday Times* 10.1.82, 'Ukraine struggles to cope with Chernobyl radiation risk' by Chris Walker in *The Times* 16.12.86, 'Severn Bridge barrage scheme raises fears for wildlife' by David Cross in *The Times* 2.87, 'Technology in Wales' by Tim Jones in

Energy and Industry

Paul Warburton

Contents

My thanks to Simon Ross for all his advice and assistance. This book is dedicated to my wife, Eileen. Eileen exercised great patience in typing the early drafts of this book and also provided support and encouragement during its compilation.

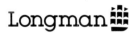
Longman

1 Renewable and non-renewable resources

There are many different ways of producing energy but they all need some form of input or fuel. Unfortunately for us many fuels, such as coal and gas, are irreplaceable except on a time-scale of millions of years. Coal, gas and oil are collectively known as **fossil fuels**. They are called **non-renewable** resources as they are not replaced once they have been used.

Many natural forms of energy that we use, mainly to produce electricity, are **renewable**. Wind power is a good example. A windmill can be used to produce energy, but the use of the windmill does not stop wind from blowing.

ACTIVITIES

Study the photographs in Figure 1.1 which represent many different forms of energy that we now use to produce power.

1 a Draw up a table with one column headed **Renewable energy sources** and the other **Non-renewable energy resources**.

  b Now identify the energy sources in the photographs and place them in the correct column. Add any other sources of energy to your lists that you can think of.

Figure 1.1 Sources of energy

2 Trends in energy use

Global

Studying changing patterns of energy consumption at the world scale can be misleading. There are considerable differences between individual countries, and between those that are developed and less developed (Table 2.1).

There is a very uneven distribution of energy resources and an uneven pattern of energy consumption. The availability of energy has been one of the major factors affecting economic development in different parts of the world. It might be expected that the largest producing countries would also be the largest consumers. However, this is not always the case, as shown in Tables 2.2 and 2.3. There are significant differences in patterns of energy use between developed and less developed countries.

Table 2.1 Energy consumption per head of population – 1984 (kilograms of coal equivalent)

USA	9577	Saudi Arabia	3640
USSR	5977	Brazil	656
W Germany	5564	India	237
UK	4760	Sudan	77
Sweden	4703	Chad	21

Table 2.2 Major coal, oil, and gas producing countries – 1984 (rank order)

Coal	Oil	Gas
USA	USA	USSR
China	USSR	USA
USSR	Japan	Canada
Poland	W Germany	UK
W Germany	UK	W Germany
India	France	Romania
Japan	China	Japan
E Germany	Italy	Italy
S Africa	Canada	Netherlands
UK	Mexico	E Germany
Czechoslovakia	Brazil	Mexico
Korea	India	Venezuela
Australia	Iran	Argentina

Table 2.3 Major coal, oil, and gas consuming countries – 1984 (rank order)

Coal	Oil	Gas
USA	USSR	USSR
China	USA	USA
USSR	S Arabia	Canada
Poland	Mexico	Netherlands
India	UK	Romania
W Germany	China	UK
S Africa	Iran	Norway
Australia	Venezuela	Mexico
E Germany	Canada	Algeria
Czechoslovakia	Nigeria	Indonesia
Canada	Kuwait	Venezuela
Korea	United Arab Emirates	W Germany
UK	Iraq	Italy

ACTIVITIES

Study Tables 2.2 and 2.3.

1 It is easier to interpret the tables if the information is put onto maps. Using an atlas and three world outlines, draw maps to show the main consuming and producing countries for coal, oil and gas. Use one colour for consumers, like red and another for producers, like green. Brown could be used where a country is both a consumer and producer. There has been little significant change in the patterns since 1984.

2 In a group, or in pairs, discuss the patterns shown by your maps. Then write a descriptive and explanatory account of what these maps show you. The following questions will help you:

a What types of countries – developed or less developed – have most of the world's coal, oil and gas?

b What type of countries are the world's major energy consumers?

c Several less developed countries are in the list of consumers. What do countries like Kuwait, Venezuela, Iran, United Arab Emirates, Iraq and Algeria have in common that might help to explain this?

3 Study Table 13.1 on page 26.

a What remains one of the major energy sources in developing countries?

b Why is this a problem for developing countries?

British energy trends

Changes in energy consumption and production in Britain are reasonably typical of many developed countries. Only a century ago, energy demand was almost entirely met by coal. Oil has been the single most important energy source in Britain since the early 1970s. However, since various political events in the Middle East and the OPEC (Organisation of Petroleum Exporting Countries) oil price increases of the 1970s, oil has become a more expensive source of power. The discovery of oil and natural gas in the North Sea in the 1950s has been of enormous benefit to the country (see Chapter 3).

In the twentieth century many new energy sources have been developed to diversify Britain's energy base. Table 2.4 shows these trends. Governments have been keen to encourage alternative energy sources like nuclear power, and the possibility of energy from wind and tides (see Chapters 7, 9 and 10). This will be increasingly important as coal and oil supplies slowly run out and become more expensive.

Energy in the Third World

A few developing countries have benefited from large discoveries of oil and have based industrial and economic development on these resources. Examples of such countries include, the Gulf States, Libya, Venezuela (see Tables 2.2 and 2.3).

Others have been able to develop hydro-electric power (see Chapter 8) but few countries have suitable physical environments. Some developing countries have been able to take advantage of modern technology to develop nuclear power stations and alternative energy sources, like solar power. However, this technology is expensive and

ACTIVITIES

4 Draw a graph to show the information in Table 2.4. Label the vertical axis **Energy consumption** (million tonnes of coal equivalent) and the horizontal axis **Year**. Use different coloured lines for each energy type. Make up a title for your graph.

5 Using the graph, answer the following:

a What was the most important energy source in Britain in the 1950s/1960s?

b What became the single most important source of energy in 1971?

c Why did Britain find oil an increasingly attractive energy source during the late 1950s/1960s?

d Explain why oil has lost some of its importance since the early 1970s.

e Why was there a sudden increase in the use of natural gas after 1959?

f What other energy source has shown a rapid increase in importance since the late 1950s?

g Give some reasons why **HEP** does not feature very stongly in the graph. (See Chapter 8 for some help.)

has to be imported from developed countries.

For most developing countries, the lack of energy resources is one of the major limiting factors on economic development. The majority, such as Kenya, Brazil, India and Sri Lanka, have to use valuable income from exports to pay for imported energy, usually oil. Another serious problem is the pace at which trees are being cut down in many parts of the world as wood remains one of the most important sources of fuel (see Chapter 13). This has many serious side effects, encouraging desertification, soil erosion and local climatic change. Of course, the energy source itself is slowly disappearing.

Table 2.4 UK energy consumption (million tonnes of coal equivalent)

Year	Coal Actual	%	Petroleum Actual	%	Natural gas Actual	%	Nuclear electricity Actual	%	Hydro electricity Actual	%	Total gross energy Consumption
1950	203	90	22	10					0.9	0.4	226
1953	208	88	28	12					1.0	0.4	237
1956	218	85	38	15					1.3	0.5	256
1959	189	77	56	23	0.1	0.1	0.5	0.2	1.5	0.6	248
1962	191	70	79	29	0.1	0.1	1.5	0.5	2.1	0.8	274
1965	185	62	103	35	1.2	0.4	6.0	2.0	2.3	0.8	297
1968	165	54	127	41	4.7	1.5	9.1	3.0	2.2	0.7	308
1971	137	42	149	46	28.4	8.7	10.0	2.8	2.0	0.5	326
1975	118	36	134	42	54.5	17.3	11.5	3.5	2.0	0.4	320
1978	123	36	137	40	63.0	18.5	14.3	4.2	2.0	0.6	339
1981	118	35	110	35	72.0	23.0	13.7	4.3	2.3	0.7	317
1985	105	32	115	35	82.3	25.2	22.1	6.8	2.1	0.6	327

Source – Department of Energy

3 North Sea oil and natural gas

We have seen in Chapter 2 how important oil has been as a source of energy in the twentieth century. In the late 1950s natural gas was discovered onshore at Groningen in the Netherlands. This was followed in the early 1960s by important discoveries of natural gas in the North Sea. This led to exploration in the northern part of the North Sea for oil. Here, geological conditions seemed to indicate possible oil reserves. Many discoveries of oil were eventually made as Figure 3.1 shows. The first oil was produced from the Ekofisk oilfield in 1976.

Conditions for the exploration and extraction of oil and natural gas are very hostile. The North Sea often experiences violent storms and drilling involves establishing rigs and platforms in very deep water (Figure 3.2). Long pipelines also have to be laid on an uneven sea bed to bring the oil and gas onshore.

Discoveries in the North Sea were made just at the right time as world oil prices were rising (see Chapter 2 page 4). Britain now had its own supply of oil and gas and was no longer so dependent on foreign supplies. Many homes and factories were converted to using oil and natural gas.

Figure 3.1 shows how North Sea oil and natural gas also led to the development of new oil and gas terminals on the east coast of Britain and new refinery capacity at Teesside. (The growth of Aberdeen is discussed in *Settlement*.)

At present, North Sea oil and gas serves almost all our needs. A little Middle Eastern oil is, however, still imported. North Sea oil contributes about five per cent of Britain's GNP; some is exported. The industry has had a multiplier effect (see Figure 24.3 on page 59) on others such as construction and service industries. In all, oil has been a great boost to the economy of Britain.

Figure 3.2 Oil drilling platform

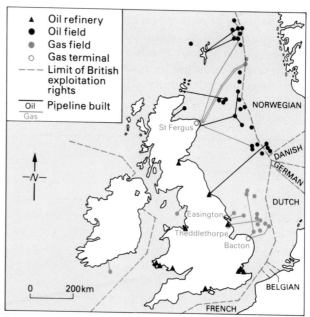

Figure 3.1 The North Sea — oil and gas fields and terminals and refinery location

ACTIVITIES

1 Study the text and Figure 3.1

  a When were discoveries of natural gas made in the North Sea?

  b When was the first oil produced from a North Sea oilfield? Name the oilfield.

  c In which part of the North Sea are most of the natural gas fields located?

  d Are the oilfields located in the northern or southern end of the North Sea?

  e How many oil terminals are there on the east coast of Britain?

  f How many gas terminals are there on the east coast of Britain?

2 a How many oil refineries are located on the
    i east coast?
    ii south coast?
    iii west coast?

  b Why are most of the modern refineries on the east coast?

  c Why were most of the earlier refineries located on the west and south coasts?

3 How have North Sea oil and natural gas affected the location of industry in Britain? Chapter 13 will help you.

4 Onshore oil in Britain

The existence of onshore oil has been known about for centuries. Oil in certain parts of Britain often seeped to the surface as a black, sticky substance. Oil was also extracted from shales in Dorset from the late eighteenth century. This died out when oil imports rose from the Middle East. Figure 4.1 shows the location of known onshore oilfields.

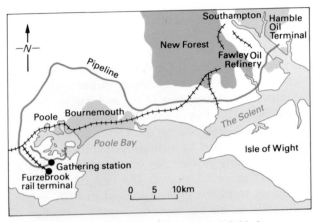

Figure 4.2 The location of the Wytch Farm oil field site

The most successful area for onshore oil extraction in Britain is in Dorset. Geological surveys in the 1950s of the Weymouth Anticline led to the discovery of oil in 1959 at Kimmeridge. A major oil discovery was made in 1974 to the south of Poole harbour. The Wytch Farm site (Figure 4.2) is now the largest onshore oilfield in Western Europe with recoverable reserves of more than 200 million barrels. Surprisingly perhaps, this is greater than some of the fields in the North Sea.

Figure 4.3 shows the main features of the Wytch Farm oilfield development. The oilfield is located within an **AONB** (Area of Outstanding Natural Beauty). It includes several **SSSIs** (Sites of Special Scientific Interest) and various nature reserves. BP, the operator at the site, has taken great care to minimise the environmental impact of the oilfield.

Figure 4.3 Wytch Farm land use

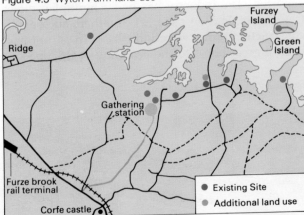

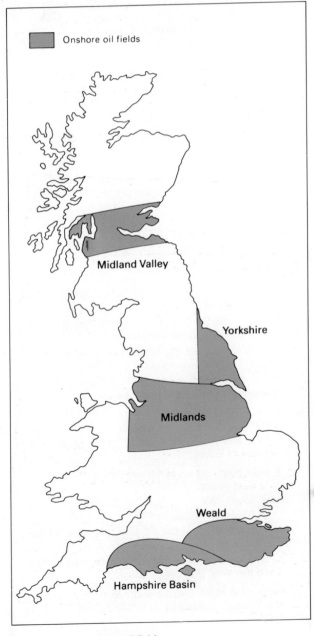

Figure 4.1 UK onshore oil fields

Figure 4.4 Tree screening at Wytch Farm

Figure 4.4 shows how they have screened the permanent drilling site on Furzey Island. Sites are restored when BP has finished with them.

Similar developments have been planned in other parts of the country. In the early 1980s Shell UK planned to extract oil in the New Forest about 32 km away from Wytch Farm.

ACTIVITIES

1 Why are onshore fields more cost effective for drilling than those in the North Sea?

2 Study Figure 4.1. List the main areas of the country where onshore oilfields are found.

3 a Describe BP's operations at Wytch Farm.

   b How do you think the oil is transported from Wytch Farm to the refinery at Fawley near Southampton?

4 Carefully read through the newspaper article in Figure 4.5.

   a Draw a table with two columns, one headed **Arguments for** and one **Arguments against**. Now list the main arguments for and against oil exploitation in the New Forest. Give the table a title.

   b Imagine that Shell UK's application had been put to you. Discuss as a group the various arguments for and against the application and reach a joint decision.

Figure 4.5 Drilling for oil in the New Forest (*Sunday Times* 10 January 1982)

'Medieval' England at risk, say naturalists
Oil men face fight in forest

by Elizabeth Grice

OPPONENTS of Shell UK's plan to sink an oil well in the New Forest compare it to drilling in the middle of the nave of Westminster Abbey.

But at the public inquiry that opens on Tuesday at Lyndhurst in Hampshire, "capital" of the New Forest, they will deliberately not say so. Their case against oil exploration will turn—powerfully and unemotionally, they believe—on what has happened less than 20 miles away, at Wytch Farm in Dorset.

Wytch Farm, near Corfe Castle, began as a single exploratory well like the one Shell wants to bore at Denny Lodge Inclosure, a five-acre site in the heart of the New Forest.

It proved a spectacular find. In a series of rapid advances, it began to eat into Dorset's "protected" countryside. Eight years later, it has spawned 15 drilling sites, a gathering station and a railhead.

Last year it produced 165,000 tonnes of the total 240,000 tonnes of oil extracted from the British mainland—and, according to the Nature Conservancy Council's chief advisory officer, Professor Norman Moore, is responsible for "the semi-industrial atmosphere pervading the Isle of Purbeck" today.

The Nature Conservancy Council, Shell's most formidable opponent in the coming six-week inquiry, fears similar encroachment in the New Forest—a complex and beautiful mosaic of Hampshire woodland, heath and bog, described as a living piece of medieval England.

Local amenity groups will be represented by a Plymouth solicitor, John Saulet, who will make the point that under the New Forest Act 1877, the forest's enclosures are protected from any activity other than that of growing trees. Without an enabling Act of Parliament, he will argue, drilling cannot take place.

Tubbs will argue that Shell, despite its impressive record for reinstating land, could never repair the damage to the forest's 278 species of tree lichens that test-flaring of gases would cause. He is supported by Dr Keith Barber, of Southampton University's geography department. "One man with a machine can destroy 10,000 years of history in a day," says Barber.

The New Forest is an extraordinary survival, unique in Western Europe. Because of its vastness, it is the only home left in lowland Britain where certain vulnerable flora and fauna can survive.

Almost the whole of southern England from Somerset to Kent is now covered by exploration or production licences. It has been described by an American geologist, Dan Williams, as "one of the most exciting oil plays anywhere in the world."

5 Electricity – an introduction

Electricity is one of the most important ways in which we use energy in Britain. Its main advantage is its wide range of applications. Electricity can be used to provide heat in an electric oven or in a blast furnace in the making of steel. It can provide light for your home or floodlights for a football match. It can power a model railway or the real thing. Electricity is also clean and easily transported by cable.

Most electricity is generated by power stations burning fossil fuels (see Chapter 6). The amount of electricity produced in Britain by different types of power station has changed greatly in the last few decades as Figure 5.1 shows.

Each type of power station has different locational needs. This will be seen in the chapters (6 to 13) dealing with the various ways in which electricity can be generated. Sometimes large cities and markets are not close to the power stations. London's needs are so great that power has to be supplied from more distant stations, particularly now that old ones like Battersea in London, have been closed down. The fact that electricity is available almost everywhere in the country is due to the **National Grid**. The lines which carry the electricity can be seen near Eggborough power station in Figure 6.2 on page 9.

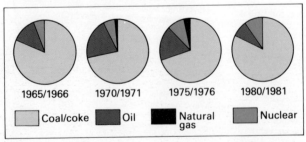

Figure 5.1 Fuel consumed in power stations in England and Wales 1965/1966

Coal/Coke	Oil	Natural	Nuclear
*79.0	5.1	0.3	16.4
(282°)	(18°)	(1°)	(59°)

*Figures in millions of tonnes of coal or equivalent

There is also a very small proportion from others including diesel, gas, hydro-electric, and pumped storage.

Table 5.1 Fuel consumed in power stations in England and Wales 1985/1986

ACTIVITIES

1 a Draw a pie diagram, like those in Figure 5.1 for 1985/1986, based on the statistics in Table 5.1. The angles for each segment have been provided for you. Colour the segments to make the diagram clearer.

 b Which was the most important type of power station for the production of electricity in Britain in 1965/1966?

 c Which was the most important type of power station for the production of electricity in Britain in 1985/1986?

 d Which type of power station has grown most rapidly in importance between 1965/1966 and 1985/1986?

 e Which new energy source was used for the generation of electricity starting in the late 1960s? Can you suggest why?

2 Study Figure 5.2 and use a map of the British Isles in an atlas to help you to answer the following questions about the National Grid.

 a Why are there so many lines and sub-stations at the locations lettered A to E?

 b Suggest why there are very few lines in the areas lettered F and G.

 c Some lines seem to radiate from isolated locations like Sizewell and Wylfa. Try and explain these two lines in particular.

 d Why are there several lines radiating from Fawley on the South Coast?

3 Why is a National Grid necessary?

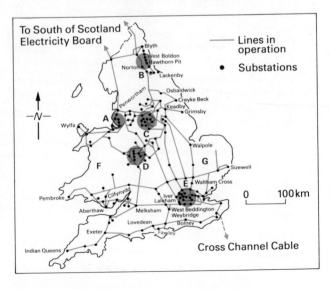

Figure 5.2 The national grid in England and Wales

6 Thermal power stations

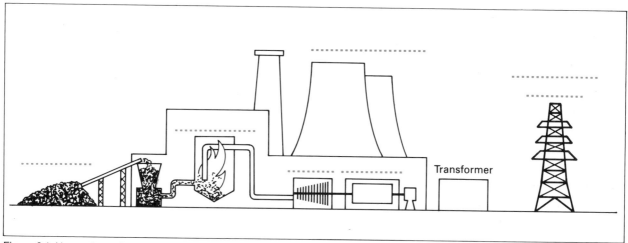

Figure 6.1 How a thermal power station works

Most of Britain's electricity is generated by burning coal or oil in huge furnaces. These are called **thermal power stations**. Coal or oil (some power stations can use either) is used to change water in pipes around the furnace into steam. Pipes carry the steam at very high pressure to the turbine. The curved blades of the turbine rotate and turn a shaft connected to the generator which produces the electricity. Once the steam has turned the turbine blades it is cooled and used for cooling the machinery in the power station (Figure 6.1). The cooling is done in the huge **cooling towers** — most of us recognise a power station from these. Often a white plume of water vapour can be seen as some of the water vapour escapes from the towers and condenses (Figure 6.2). Power stations are linked to the National Grid and power is sent all over the country to wherever it is needed. The supply of electricity is very reliable; if one power station breaks down enough power can be supplied from other stations.

ACTIVITIES

1 a Make a copy of Figure 6.1 and write the correct labels into the missing spaces.
   coal   furnace   generator   turbine
   cooling towers   electricity pylon
  b Add arrows showing the paths followed by the coal, steam and electricity. Use different colours for each arrow.

2 Explain in your own words how a thermal power station works.

Figure 6.2 A thermal power station — Eggborough near Selby.

CASE STUDY

The Trent Valley

Figure 6.3 shows the distribution of thermal power stations in Britain. One of the main concentrations of thermal power stations is in the valley of the River Trent near Nottingham. Figure 6.4 shows the main power stations and some of the other features of this area.

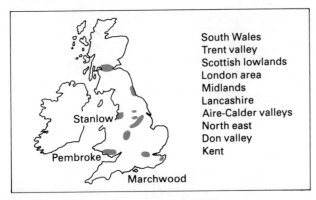

South Wales
Trent valley
Scottish lowlands
London area
Midlands
Lancashire
Aire-Calder valleys
North east
Don valley
Kent

Figure 6.3 Major areas of thermal power stations in Britain

Figure 6.4 Power stations in the Trent valley

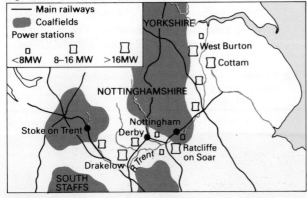

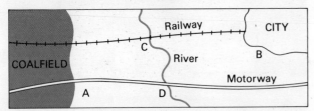

Location	Transport	Water	Market	Labour	Pollution	Total
A	*1*	*2*				
B						
C						
D						

ACTIVITIES

3 a Using an outline map of the British Isles make a copy of Figure 6.3

  b Add all of the areas listed beside Figure 6.3 to your map. An atlas will help you to complete this correctly.

4 Compare the map you have drawn with Figure 17.11 on page 37. What are the differences/similarities?

5 a Make a copy of Figure 6.4 and then answer the following questions.

  b There are so many power stations in this area that some of the local people call it 'Kilowatt Alley'. Do you think this is a good nickname?

  c What is the fuel for the power stations in the Trent Valley and where does it come from?

  d How is the fuel for the power stations transported?

  e Explain the importance of the River Trent to the power stations.

6 In this activity you will be asked to decide on a location for a new power station.

  a Copy Figure 6.5 and the table underneath.

  b All the main locational influences are given below. You must now complete the table by scoring each location, A to D, on a scale of 1 to 5. 1 is lowest, 5 highest. A has been started for you. Add up the scores for each location and put your answers in the **Total** column. The location with the highest score is the best site.

    A Pollution could be a serious problem near to a town or city.

    B Quite a large labour force is employed so it should be close to a town or city.

    C Most of the electricity will be used in industries and houses in cities – the main markets.

    D Large quantities of coal are needed so it helps to be close to a coalfield.

    E Large amounts of water are needed for cooling.

    F It is most economical for coal to be transported and delivered by rail.

  c Which site appears to be the best location? Briefly state why this site is better than the others.

Figure 6.5 Planning the location of a thermal power station

7 Nuclear power

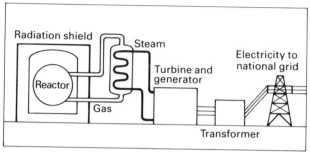

Figure 7.1 Simplified operation of a nuclear power station (AGR type — Advanced gas cooled reactor

Figure 7.2 Britain's nuclear power stations

Nuclear power is likely to be an increasingly important source of energy in the future. Already it makes a major contribution to the energy demands of many countries, including Britain (see Table 2.4 on page 4). The splitting of atoms, a process called **nuclear fission**, involves the release of huge amounts of energy. The fuel that is commonly used is an element called **uranium** obtained from certain rocks.

Figure 7.1 shows how one type of nuclear power station works. Nuclear heat energy is used to convert water to steam which then turns a turbine to generate electricity. Nuclear power stations are similar in this way, to thermal power stations.

The process of nuclear fission takes place in a **reactor** which is where the atoms are split. In the power station shown in Figure 7.1 a gas is used to convey the heat energy to water which is then converted to steam. The process of nuclear fission is dangerous. Harmful radiations are released which can affect people and animals, causing burns, cancer and even death. The reactor, therefore, is encased by a shield of steel in concrete, often several metres thick. Nuclear power stations are thus made safe places to visit and work in.

The future of nuclear power in Britain

Nuclear power technology is already highly developed. In 1988, Britain had 15 nuclear power stations. More are planned for the future (Figure 7.2). Most developed countries like Britain, France and the United States have nuclear power stations and there are many in the developing countries such as Brazil and India (Figures 7.3 and 7.4).

Figure 7.3 A British nuclear power station, Oldbury

Figure 7.4 A nuclear power station in a developing country, Brazil

Nuclear power is a very controversial subject. There are many arguments both for and against. You will probably be familiar with these from magazines and the television. The main arguments are listed below. It is for you to make up your own mind whether or not you think nuclear power is a good idea.

Advantages

1 Huge amounts of energy can be obtained from a small quantity of fuel. One kilogram of uranium contains nearly three million times the energy of one kilogram of coal. 'Breeder' reactors can produce their own fuel.
2 Nuclear power stations can provide electricity more cheaply than many alternatives.
3 The only raw material usually required, other than uranium, is large quantities of water for cooling.
4 Only very small amounts of waste are produced for the electricity generated.
5 With fossil fuels like coal, oil and gas running out, nuclear power could provide cheap electricity into the next century and perhaps beyond.
6 Unlike some alternative energy sources, the technology is already available.

There are also risks associated with nuclear power. Several countries like Denmark and Austria have decided not to build any nuclear power stations.

Disadvantages

1 Leakages do happen. Releases of mildly radioactive water have taken place from power stations in Britain. Higher than normal levels of radioactivity exist in the Irish Sea associated with the nuclear waste treatment plant at Sellafield, in Cumbria (Figure 7.5).
2 Major accidents or explosions may occur. It is true that these are rare, but when they do occur the results can be very serious. In 1979, for example, a technical failure at the Three Mile Island nuclear power station in Pennsylvania, USA led to the release of a small amount of radioactive material into the environment. Some operator error was involved. Fortunately, the problem was solved before a more serious leak occurred and no one was seriously affected by the accident. (There is a section following the list of disadvantages about the Chernobyl incident in 1986).
3 Many people have shown their concern about the transport of nuclear waste. Nuclear waste is, on occasion, transported from nuclear power stations to be disposed. This is done by rail; the trains pass through some of Britain's densely populated areas. The public want to know what might happen if there was a rail crash. The waste, however, is in sealed and extremely strong containers. Figure 7.6 shows a 160 km (100 miles) per hour impact between diesel locomotive and a nuclear fuel flask in a demonstration in 1964. The flask was dented, but remained sealed. The locomotive was destroyed.

Figure 7.6 Train crash (mock) — to demonstrate the strength of a waste container

Figure 7.5 Sellafield nuclear waste treatment plant

4 Disposing of radioactive waste permanently is a difficult problem. The danger from the waste can last for thousands of years. One method of disposal has been to dump the waste in sealed containers far out to sea. Early in 1986 NIREX (the Nuclear Industry Radioactive waste Executive) suggested several sites in Britain where waste could be buried in the ground. Members of the public and conservation organisations like Greenpeace, for example, have expressed concern about what might happen if waste ever leaked from those sites or into the sea.

5 Some fear that terrorists might obtain radioactive material.

The debate about nuclear power is largely to do with risk and the probability of an accident occurring. It is important to realise that we accept risks as part of our everyday lives, but have learnt to take many for granted, like crossing a busy road. Sometimes, death does occur as shown in Table 7.1. Unless there is a major nuclear accident, few people are likely to die in Britain or to have their health seriously affected by radiation. In fact the health of many people who live near thermal power stations is affected by gases and acid rain.

Table 7.1 Accidental death statistics for England and Wales 1983

Cause of Death	Total Deaths	Number per million of resident population*
All deaths	579 608	11 677.64
Road traffic accidents	5 059	101.93
Falls at home	2 909	58.61
Accidental poisoning	626	12.61
Railway accidents	105	2.12
Machinery	104	2.10
Electric current	82	1.65
Air transport	47	0.95
Firearms and explosives	33	0.66
Lightning	6	0.12
Radiation	0	0.00

*Population 49 634 000 approximately

CASE STUDY

Chernobyl 1986 — a nuclear accident

The world's worst nuclear accident occurred at Chernobyl near Kiev in the Soviet Union in April 1986 (Figures 7.7 and 7.8). Following an explosion and a fire in the reactor building, a cloud of radioactive dust and gas escaped into the atmosphere (Figure 7.7). The newspaper article in Figure 7.8 describes the impact of the disaster.

Figure 7.7 The Chernobyl disaster

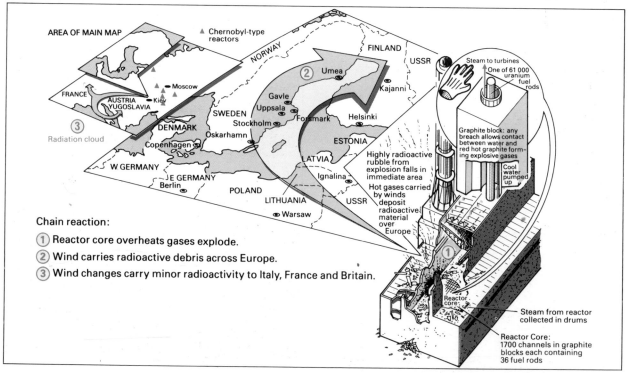

Although we cannot be certain about numbers, many people died in the Soviet Union. Many more will die in the future as a result of radiation related illnesses. People were also affected in Western Europe as the cloud spread westwards. Many animals like reindeer in Sweden and sheep in parts of Britain had to be slaughtered too, as they were contaminated by radiation which fell in rain and affected the grass.

Ukraine struggles to cope with Chernobyl radiation risk

From Christopher Walker
Zdvyzhevka, Ukraine

Nearly eight months after the Chernobyl nuclear disaster authorities in the Ukraine are still facing huge but little publicized medical, financial and logistical problems coping with the contamination risks which it posed to several million Soviet citizens.

Last week, I was one of the first Western correspondents permitted to tour the region and to inspect the costly efforts being undertaken to minimise the human cost of the April 26 explosion.

At the new village of Zdvyzhevka 60 kilometres (about 38 miles) from the crippled reactor, hundreds of families are beginning the difficult process of resettling into new brick-built homes (each costing £30,000 to construct).

Mr Anatoly Romanenko the Ukranian Health Minister told *The Times* that the continuing fear of "the invisible enemy, radiation" was one of the main difficulties with which his officials were now faced.

Mr Romanenko said that various techniques including massage and music were being used to calm Chernobyl evacuees with psychological problems.

Figure 7.8 The impact of the Chernobyl disaster on Kiev the *Times* 16.12.86

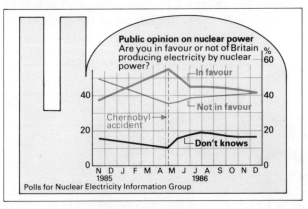

Figure 7.9 Nuclear Power — trends in public opinion before and after the Chernobyl disaster

Figure 7.10 A peaceful demonstration against NIREX engineers at Bradwell-on-Sea in 1986

It is important to realise though, and supporters of nuclear power are keen to point this out, that there are no reactors of the Chernobyl type in Britain. Human error was very much to blame for the accident, and the power station was of poor design.

ACTIVITIES

1 Explain briefly what you understand by the term **nuclear fission**.

2 Copy Figure 7.1 and describe how a nuclear power station works.

3 Study Figure 7.7. On an outline map of Europe, mark the location of Chernobyl. Add other details such as the major cities, the countries and the spread of the nuclear cloud. Your final map will be a simplified version of part of Figure 7.7.

4 a Describe in your own words what happened at Chernobyl in 1986.

  b Why could an accident like this be described as an 'international problem'?

5 The graph in Figure 7.9 appeared in the *Financial Times* in January 1987, after the Chernobyl accident. Describe and try to account for the trends in the graph before and after the accident.

6 Figure 7.10 shows the demonstration held at Bradwell-on-Sea, one of the sites NIREX proposed for the dumping of nuclear waste in 1986. The waste would have been buried in sealed containers which would have been surrounded by concrete. Explain briefly why there was so much opposition to the dumping of waste at Bradwell-on-Sea. How would you feel if this waste was to be disposed of in your local area?

7 The statistics in Table 7.1 formed part of a table produced by the Central Electricity Generating Board (CEGB).

  a Do the figures support the case for nuclear power?

  b Do you think they give a true impression of the likely deaths caused by a nuclear disaster?

8 Nuclear power was an important issue in the General Election of 1987. One of the parties was against and one planned to build more nuclear power stations in Britain. Hold a debate in your class with one or two representatives for each of the two political parties. At the end have a vote on nuclear power in Britain to see what your class feels.

9 Carefully read through the advantages and disadvantages of nuclear power. Make up *your mind* whether or not you support nuclear power. Produce a poster or advertisement aimed at people of your age giving your point of view, either for or against nuclear power. Try to use an illustration and a slogan.

The location of nuclear power stations

Figure 7.2 on page 11 and Figure 7.11 show the distribution of nuclear power stations in Britain and France. There are clearly differences and similarities in their distributions. It is also obvious that there are a greater number of nuclear power stations in France. A very high proportion of electricity in France (about 70%) comes from nuclear power stations. France is unusual in the world in this respect. This is largely because the country lacks oil reserves of its own and only has small reserves of natural gas and coal.

One of Britain's nuclear power stations is at Wylfa (Figures 7.12 and 7.13); it was commissioned in 1987. Built on the north coast of Anglesey, it is the largest of its kind. It was the last 'magnox' nuclear power station, so called because of the magnox metal casing containing the uranium fuel. It can supply enough electricity to the national grid to meet the power demands of two cities the size of Liverpool.

ACTIVITIES

10 Study Figure 7.11
   a How many nuclear power sites have been built in France? Briefly suggest why there are so many.
   b What locational factor is shared by France's nuclear power stations? Try to explain this.

11 Study Figure 7.2
   a All of Britain's nuclear power stations except one are in very similar locations. Describe and explain in a few paragraphs the distribution of nuclear power stations in Britain.
   b Which nuclear power station is the odd one out? Can you think why — an atlas will help you.

12 a Approximately how far is Wylfa from the large settlements of Caernarvon and Llandudno?
   b Why do you think Wylfa is so far away from any large cities?

13 Wylfa needs 4.5 million litres of water per hour for cooling. This is about twice the amount needed by a conventional power station. Where do you think Wylfa gets its water from?

14 Wylfa is built on very solid and stable rock. Why is this important?

Figure 7.12 Wylfa power station

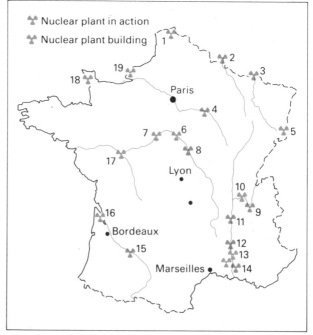

Figure 7.11 Distribution of nuclear power stations in France, 1986

Figure 7.13 Location of Wylfa power station

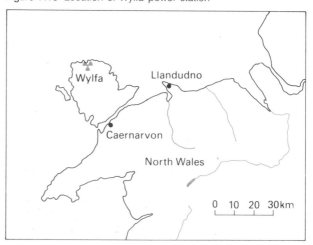

8 Hydro-electric power

HEP technology has existed and been developed since the nineteenth century. Today large HEP plants produce vast amounts of electrical energy using the power of fast flowing rivers or the speed of water flowing down steep slopes (Figure 8.1 and 8.2). The height that the water falls is called the **head** of water. In the power station the force of the water hitting blades on turbines makes them rotate very fast. The turbines work generators which produce electricity. HEP plants are expensive to build but they do produce power more cheaply than coal or oil fired power stations.

HEP plants have special locational needs and require careful planning so that they do not spoil the environment.

ACTIVITIES

1 Study Figures 8.2, 8.3, 8.4, 8.5 and an atlas. Explain why HEP is a renewable energy source.

2 a Describe the location of HEP stations in Britain.

  b Are the areas that are suitable for HEP production in Britain uplands or lowlands? Give a typical height. Explain your answer.

  c Do the areas have high or low rainfall? Give a typical annual total in millimetres. Explain your answer.

  d Why do you think there are no HEP stations in south east Britain?

3 After studying the photograph of the scheme shown in Figure 8.2 and using what you have learnt about HEP:

  a Copy the cross-section of a typical HEP scheme (Figure 8.2) and add the labels given below in the missing spaces
       Dam   Power station and turbines
       Reservoir

  b In a few sentences explain how such a scheme works.

  c What extra advantages might there be from a scheme like that at Cruachan.

4 When HEP schemes are developed and large valleys are flooded a landscape is completely changed. List some different groups of people who might object to such a project and explain why for each group.

Figure 8.1 An HEP scheme in Cruachan, North Scotland

Figure 8.2 Cross-section through a typical HEP scheme on a river

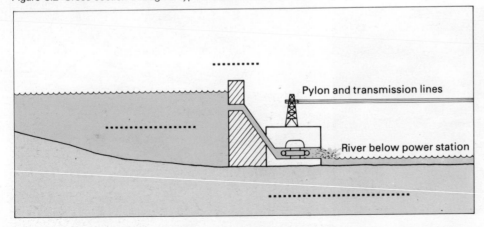

Pylon and transmission lines

River below power station

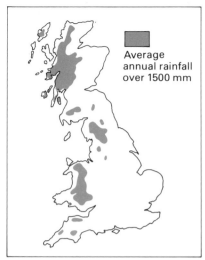

Figure 8.3 Rainfall – Britain

Figure 8.4 Main areas of HEP schemes – Britain

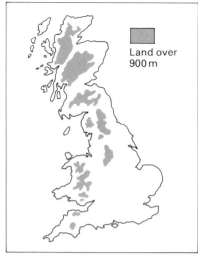

Figure 8.5 Relief – Britain

CASE STUDY

The Volta Dam project, Ghana

HEP is used in many developing countries where suitable physical conditions can be found. The Volta Dam project was completed in Ghana in the 1960s. Figure 8.6 shows the dam which was built across a narrow gorge at Akosombo. An area of 8200 square kilometres (nearly half the area of Wales) behind the dam was flooded to form Lake Volta (Figure 8.7). This transformed both the physical and human geography of this part of Ghana. For a third world country this was a very expensive and adventurous project. It was built with financial aid and expertise from the USA. An important part of the project was the completion of an aluminium smelter at Tema, which is a new port on the coast completed in 1962 (Figure 8.7). The smelter needed large amounts of cheap electricity.

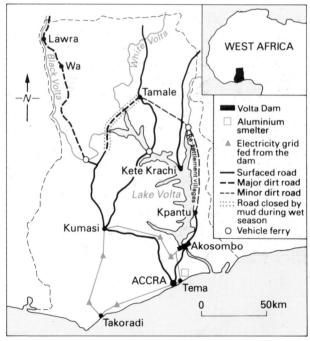

Figure 8.7 Map of the Volta HEP scheme

Figure 8.6 The Volta Dam

ACTIVITIES

5 Study Figures 8.6 and 8.7.

   a Draw a sketch map showing the location of the Volta Dam project. Clearly label the dam at Akosombo, the aluminium smelter at Tema, the River Volta and Lake Volta.

   b Explain why the site at Akosombo was chosen for the dam.

   c Suggest what most of the electricity from the dam was used for.

A project like this has many other valuable advantages.

   d How could the lake be used to increase food supply in this part of Ghana?

   e Try to list and explain as many other benefits as you can that could come from the project.

Pumped storage HEP schemes

One of the problems with electricity is that it cannot be stored in large amounts and sometimes there are sudden large increases in demand. Dinorwig in north Wales (Figure 8.8) was built to cope with such events and in case of failure of other power stations or transmission plants. Dinorwig can produce vast amounts of power very quickly and more reliably and cheaply than thermal power stations.

Electricity is generated when water falls from the upper lake to the turbines inside the mountain. The water eventually finds its way into Llyn Peris. During off-peak periods of the day electricity from the national grid can be used to pump water back to refill the upper lake. Water is always available in this way to generate electricity when it is needed.

ACTIVITIES

6 Study the maps and details about Dinorwig HEP scheme (Figure 8.8).

   a Make copies of the map showing the location of Dinorwig power station in North Wales, and the cross section of the scheme.

   b On the cross section name the lakes at A and B, and label the buildings at C.

   c Add arrows on the cross section to show the movement of water during night and day — use two different colours.

7 The generating station was built inside a mountain. How was this achieved? Why do you think it was a good idea for it to be hidden away out of sight?

8 Dinorwig is in an area that was once glaciated (see *People and the Physical Environment*). Write a few sentences explaining why Dinorwig is an ideal site for an HEP station, referring to some of the glacial features which have been used in the scheme.

9 Explain why Dinorwig was built and how a **pumped storage scheme** works.

Figure 8.8 Dinorwig HEP scheme

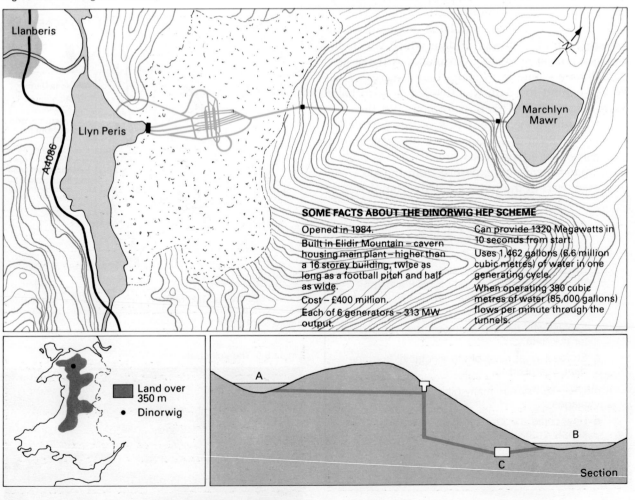

Llanberis

A4086

Llyn Peris

Marchlyn Mawr

SOME FACTS ABOUT THE DINORWIG HEP SCHEME

Opened in 1984.

Built in Elidir Mountain – cavern housing main plant – higher than a 16 storey building, twice as long as a football pitch and half as wide.

Cost – £400 million.

Each of 6 generators – 313 MW output.

Can provide 1320 Megawatts in 10 seconds from start.

Uses 1,462 gallons (6.6 million cubic metres) of water in one generating cycle.

When operating 390 cubic metres of water (85,000 gallons) flows per minute through the tunnels.

Land over 350 m

• Dinorwig

A

B

C

Section

9 Tidal power stations

Figure 9.1 Tidal power station, Rance estuary, Brittany, France

A different form of hydro-electric power uses the tides. At present only one tidal power station has been built in the world across the Rance Estuary in Brittany, northern France (Figure 9.1).

Figure 9.2 shows how the movement of the tides can be used to produce electricity. As the tide comes in (**flows**) the depth of water increases as shown in the first diagram. Gates in the dam are eventually opened to allow water to turn the turbines which generate electricity. The estuary then fills with sea water. As the tide goes out (**ebbs**) the level of the sea is allowed to drop, as shown in the second diagram. When the gates in the dam are opened water flows out driving the turbines and pours into the sea.

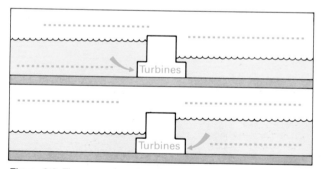

Figure 9.2 The operation of a tidal power station

Tidal power in the United Kingdom

In 1986 the Government decided to extend the UK's tidal power programme. Five million pounds were invested in the next phase of development. It should be clear from what you have already learnt about tidal power that a high **tidal range** is important in locating barrages. This is the difference between high and low water that occurs with high and low tides. Figure 9.3 indicates that there are a number of locations in the UK where there is a high tidal range.

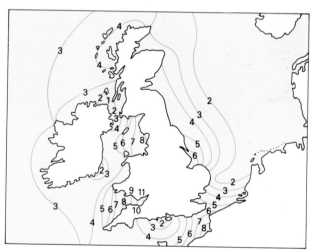

Figure 9.3 Tidal range at mean spring tides. British Isles (in metres)

ACTIVITIES

1 Make a copy of Figure 9.2. Add appropriate labels to the diagrams in the spaces provided using information from the text.

2 Where at present is the only large tidal power station in the world located? Find this in an atlas and draw a simple sketch map to show its location.

3 Explain in a few sentences how a tidal power station works.

ACTIVITIES

Assuming that a difference of *at least* five metres is needed:

4 Using an outline map of the UK shade in blue those estuaries and inlets where tidal power stations could be built.

5 Use an atlas to name the locations you have marked on your map.

6 Explain in a few sentences why you think the government decided to invest money in developing tidal power.

CASE STUDY

The Severn Estuary

One of the best sites in England is the estuary of the River Severn shown in the satellite photograph (Figure 9.4). Figure 9.5, a map based on the satellite photograph, shows some of the features of the estuary and the surrounding area. Carefully locate the features on the satellite photograph. A tidal barrage would probably be built at the position shown in the map. It has been estimated that a scheme at this site would produce about 7200 megawatts each year. This would meet approximately six per cent of the electricity needs of England and Wales.

Figure 9.4 A satellite image of the Severn estuary and parts of south Wales and south west England

Figure 9.5 The proposed location of the Severn tidal power station – a map based on the satellite photograph (Figure 9.4)

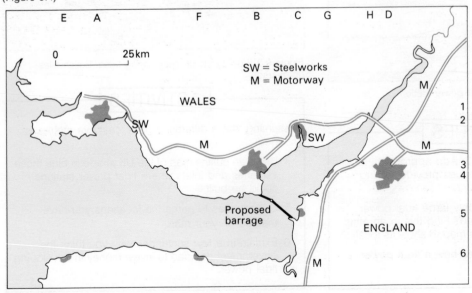

ACTIVITIES

7 a Make a copy of Figure 9.5 without the letters and numbers round the edge.

b Using an atlas identify and label the following on your map:
 i the settlements at A1, B3, C2 and D3;
 ii the holiday resorts at E6, F6, C5, and G4;
 iii the bridge at H2;
 iv the water area at F5;
 v the river flowing into the sea from the north east.

c Using your atlas find out which motorways are shown on Figure 9.5 and label them on your map.

8 Road transport in this area relies heavily on the Severn Bridge where the M4 crosses the River Severn, and the bridge is often congested. How might the completed barrage help to lessen this problem?

9 At present ships use the estuary to reach the ports of Avonmouth and Cardiff, and ore carriers need access to the steelworks at Llanwern. Clearly a barrage would be a major barrier to navigation. How could this problem be overcome?

10 Some industrial waste and cooling water from power stations are put into the Severn and carried out to sea. Pollutants could collect behind a barrage and would not be able to escape so easily. State briefly any ways you can think of in which this problem could be solved.

11 While the barrage was being built, how could the people who live in the area benefit from the project?

12 If a barrage were to be built across the Severn Estuary there would be a number of important environmental and economic issues to think about. The effects of a barrage on birds and fish are as yet not fully understood. Clearly though, the estuary environment would be changed ecologically and visually.
Study the extract in Figure 9.6.

How would birds and fish be affected by the project?

13 The barrage would create a huge lake on the upstream side. Describe some of the uses that such a lake could have.

14 You should by now be aware of many of the advantages and disadvantages of the proposed Severn Tidal Power Scheme. You may be able to think of some more with help from your teacher. Write an imaginary report to the Department of Energy outlining the arguments for and against the scheme. You must decide whether, in your opinion, the project should go ahead or not, and explain your decision.

Severn barrage scheme raises fears for wildlife

Conservationists yesterday expressed deep concern about the possible impact of the proposed Severn barrage tidal power scheme on birdlife and fish in the estuary.

Mr Stan Davies, south-western regional officer for the Royal Society for the protection of Birds (RSPB), said the area was one of the top six or seven in the country for wading birds and wildfowl.

"Any tidal barrier will keep water levels inside the barrage much higher than they are now and reduce the area of mudflats which can support birdlife". he said. "The whole concept of losing mudflats fills organizations like ours with a great deal of concern."

Supporters of the scheme were arguing that new mudflats which could be created behind the barrier would be more stable, because of less variable currents, and would be able to sustain larger populations of small shellfish and worms for birds to feed on.

"But this argument is not proven and the new mudflats would have to be more densely populated with birdlife than any other part of Britain", Mr Davies said.

"The problem for inter-tidal wading birds, in particular, is that they are highly specialized in their eating habits and have nowhere else to go."

According to the RSPB. 102,000 waders and wild-fowl use the estuary during January and February.

They include some 47,000 dunlins, small brown waders which migrate south from the Arctic during the winter. There are also important colonies of red-shank, curlew, ringed plovers and shelducks.

There would cleary be problems for migrating fish like salmon and some species of trout, and for fish which used sheltered sea-grass beds for spawning. Many of these beds would disappear. Miss Fiona Corby, assistant regional officer for the Nature Conservancy Council, said it was clear that the whole ecological balance would change once the barrage was built. "What we need now is a good idea of what will happen so that we can take compensatory measures in advance."

Figure 9.6 The environmental impact of a Severn Barrage *The Times*, February 1987

10 Wind power

Wind power may provide windfall

Wind power is one of the most promising sources of energy for development in Britain. In 1987 a report by the British Wind Energy Association said that wind turbines can produce electricity more cheaply than either nuclear or coal fired power stations. Electricity from wind could cost about 2p per kilowatt/hour compared with 3p for electricity from nuclear or coal power stations. However, many wind turbines would be needed. About 700 would supply ten per cent of Britain's electricity.

Figure 10.1 shows that Britain has already built several wind turbines, for example at Carmarthen Bay (Figure 10.2) and in the Orkneys. If wind power was developed on a larger scale **wind farms** would have to be built. One of the earliest examples in the world is at Altamont Pass in California (see Figure 1.1 on page 2). Here, 44 machines generate electricity which is then fed into California's grid meeting a small percentage of the state's energy needs.

The Central Electricity Generating Board has recently announced plans for three wind farms in Britain (see Figure 10.1). Each wind farm would have about 25 turbines covering an area of about 3 km². Each farm would produce enough electricity for about 5000 people.

Wind turbines only operate within certain wind speeds. As wind speed drops, less power is produced. At very low speeds, no electricity is generated. At very high speeds, the turbines must be stopped to prevent them being damaged.

Figure 10.2 A wind turbine at Carmarthen, South Wales

Figure 10.3 The scale of wind turbines

ACTIVITIES

1 a Use an outline map of Great Britain and make a copy of Figure 10.1. Label clearly the main areas with potential for wind power.

   b Name the areas of Britain with potential for wind power. You can write these on your map or put them in a key.

   c Write a brief description of the best types of location for wind power stations – remember that regular high wind speeds are needed, and that wind turbines can be built in shallow water off the coast.

2 The heading to this section 'Wind power may provide windfall' appeared in a newspaper article about wind power. Why might wind power provide a windfall and what are the advantages of this source of energy?

3 Look at the photographs of Carmarthen wind turbine (Figure 10.2) and the wind farm in California (Figure 1.1) and Figure 10.3. Think about your answers to activities **1b** and **1c** and list what you think are likely to be the main objections to wind turbines.

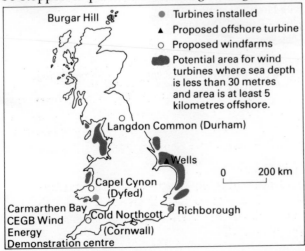

Burgar Hill

- ● Turbines installed
- ▲ Proposed offshore turbine
- ○ Proposed windfarms
- Potential area for wind turbines where sea depth is less than 30 metres and area is at least 5 kilometres offshore.

Langdon Common (Durham)

▲ Wells

0   200 km

Capel Cynon (Dyfed)

Carmarthen Bay CEGB Wind Energy Demonstration centre

Cold Northcott (Cornwall)

Richborough

Figure 10.1 Wind power stations in Britain, and potential offshore sites

11 Solar power

The sun represents a vast source of energy that will last for millions of years — it can be seen as another renewable energy resource. People have learnt to use the power of the sun in various ways. Not only can it provide heat for buildings, but it can also be used to generate electricity. One well publicised example is the space shuttle where solar collection panels convert the sun's energy into electricity to power equipment on the shuttle. Many satellites use solar power in the same way (Figure 11.1). Some of you may have calculators that contain small solar cells to power them (Figure 11.2).

Where you live is important as some parts of the world receive more of the sun's energy than others. We are well aware in Britain of how infrequently the sun seems to shine sometimes. The map of the world in Figure 11.3 shows how amounts of solar energy vary from place to place throughout the year.

Solar power can be used for both large and small scale projects.

Large scale uses of solar power

There are three large scale methods that have shown particular potential. Some of these are now used in power stations that feed electricity into national grids.

Figure 11.1 A satellite in space showing solar panels

Figure 11.2 Solar cell powered calculator

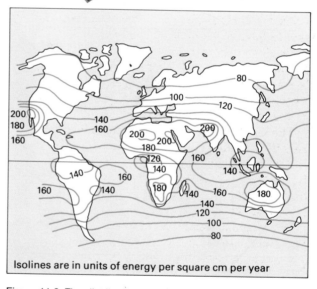

Isolines are in units of energy per square cm per year

Figure 11.3 The distribution of the sun's energy at the Earth's surface

Focusing mirrors and lenses
Mirrors or lenses concentrate the sun's rays: temperatures of several thousand degrees centigrade may be reached. The heat is used to convert water into steam, which in turn drives a turbine and a generator then produces electricity which is fed into a national grid.

Solar ponds
Israel has been successful in developing this method of solar power. Dense layers of saline water absorb the sun's energy; again electricity can be produced.

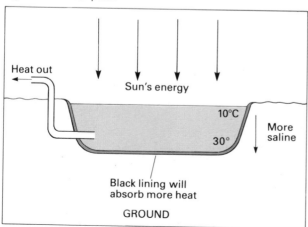

Figure 11.4 Solar pond

Heat out

Sun's energy

10°C

30°

More saline

Black lining will absorb more heat

GROUND

Photovoltaic cells
The technology involved in these devices is very complex; however, at its simplest photovoltaic cells convert light from the sun straight into electricity. One such power station using this method is in California, USA. It produces about 16 MW of power each year, which is fed into California's grid.

Small scale uses of solar power

The list of small scale applications of solar power is almost endless. Even in countries where the sun is often covered by cloud, solar power can be used instead of oil or other fossil fuels at times when the sun is shining. This would not only save fossil fuels but also large sums of money that would otherwise have to be spent, perhaps on imports of coal or oil.

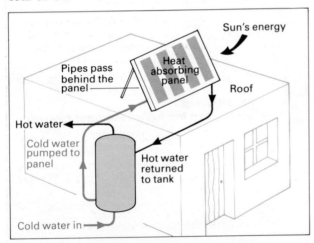

Figure 11.5 Use of a solar panel to power a heating system

One of the most common ways of using the sun's energy in many countries involves solar panels which are used to provide hot water in buildings (Figure 11.5).

Photovoltaic cells are also useful where fairly small amounts of electricity are needed. Ambitious projects in India and Pakistan, for example, involve the use of cells in villages to produce electricity for lighting, television and radio. Hundreds of villages have already been equipped in this way. A Japanese firm is manufacturing photovoltaic cells as roof tiles to produce a domestic electricity supply. These could be widely used in the third world as well as in other countries.

The future

Solar power technology is with us now and clearly has many valuable uses already. However, it is of limited use in some parts of the world where the sun is not always shining. One other disadvantage is that electricity cannot easily be stored in batteries. This is both very expensive and can take up a lot of space. Again, this is a problem in areas where the sun is not always shining. Some other system would be needed to fall back on.

More research is needed to extend the potential of solar power. Some dramatic possibilities for the future include solar powered cars, and a scheme involving satellites beyond the earth's atmosphere. The sun's energy would be collected in space and sent to earth to special receivers. Such a system would not be affected by cloud cover and vast amounts of energy could be produced for anywhere on earth.

ACTIVITIES

1 Study Figure 11.3. Which parts of the world are suitable for collecting solar power? Make a list of ten different areas or countries.

2 Draw a copy of the diagram in Figure 11.5 and explain in your own words how the system works. Do not forget to mention how the hot water in a building could be used.

3 List some other situations where water heated by solar panels could be useful and save other forms of energy.

4 Solar power – either collected by solar panels or photovoltaic cells – has a number of applications in the third world, for example for lighting, cooking and powering pumps. Such uses can be seen as **appropriate technology** (see Chapter 30). Briefly explain why.

5 Several examples have been given of how solar power can be used to produce electricity for quite small scale uses. Try and think of some more, particularly in hot countries. Write them down as a list.

12 Geothermal power

It is well known that huge amounts of heat are stored beneath the surface of the Earth. Anyone who has been into a coal mine will have experienced rising temperatures with increasing depth. With the development of electricity in the last one hundred years, the idea of using geothermal (meaning 'Earth heat') power as an energy source has grown.

The early development of geothermal power was limited to unstable parts of the world near crustal plate boundaries, like the North Island of New Zealand. In these areas, underground heat is already close to the surface. Water is naturally converted to steam which can be used to produce electricity. Geysers (Figure 12.1) are a visible result of these natural processes.

In stable areas like Britain where there is no naturally heated water, the aim is to use the heat associated with radioactive elements. These are found in rocks like granite deep beneath the Earth's surface. These are referred to as **hot dry rocks** (HDR). Only recently has the technology to tap this form of geotherml power been developed.

A disused granite quarry at Rosemanowes near Penryn in Cornwall (Figure 12.2) was selected in 1977 as a pioneer site for UK HDR research. Figure 12.3 shows how the system works. Two wells are drilled deep into the granite. The natural joints in the rock are expanded by controlled explosions to produce an interconnecting network of joints between the ends of two wells. Water is then circulated down the injection well. It absorbs heat in the granite and the hot water and steam is brought to the surface in the recovery well. Similar to a thermal power station, steam is then used to drive a turbine and a generator to produce electricity.

Similar projects are being carried out in Sweden, Japan, France and the United States. Although considerable success has been achieved, much research remains to be done. It is quite possible though, that a plant in Cornwall could in the future be the world's first HDR geothermal power station producing electricity for Britain's national grid.

Figure 12.1 A geyser erupting

Figure 12.2 Rosemanowes Quarry site

ACTIVITIES

1 a Make a copy of Figure 12.3 Add a blue arrow indicating water descending the injection well. Add a red arrow representing hot water and steam ascending the recovery well.

   b Add the following labels in the spaces provided:
      Power station   Injection well
      Recovery well
      Joint system forming underground reservoir

   c Explain briefly how the geothermal power project in Cornwall works.

   d How would a geothermal power station close to a plate boundary as in New Zealand, differ from the project in Cornwall?

2 Is geothermal power a renewable or non-renewable energy resource? Explain your answer.

3 It is estimated that geothermally produced electricity from Cornwall could be generated very economically. Consider other power sources such as coal or nuclear that are already important in Britain. Why might many people be attracted to using geothermal power rather than coal or nuclear power in the future?

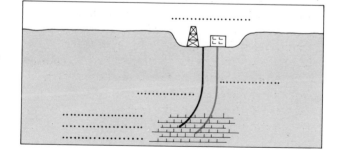

Figure 12.3 The workings of the UK HDR project near Penryn in Cornwall

13 *Biological sources*

Wood was one of the earliest sources of energy used by people. As recently as the early nineteenth century in Britain, wood was used to make charcoal for use in the iron industry. For hundreds of years it had been used for buildings, making ships and as a fuel for cooking and heating. Only remnants now remain of vast forests that once covered most of the British Isles.

With the advent of coal and more recently, oil, wood is no longer a major energy source, except in the developing countries. Here, dung is also used as a fuel (Figures 13.1 and 13.2).

CASE STUDY

Fuel wood in Sri Lanka

About thirty years ago most of Sri Lanka was covered by forest. Today most of it has gone (Figure 13.3). The demand for wood as a source of fuel is one of the major reasons. By the end of this century very few trees will be left unless there is a programme of replacement called **afforestation.** Table 13.1 shows how important firewood is as a fuel source in Sri Lanka. The articles in Figure 13.4 from people living and working in Sri Lanka, give some idea of the need for wood as a fuel and of the problems involved in collecting it.

If the wood runs out what will happen? Some of the better off households may use kerosene. However, an increasing use of petroleum products could add millions of dollars to Sri Lanka's oil import bill. Energy problems are already beginning to dominate Sri Lanka's economy.

Figure 13.1 Wood as an energy source

Figure 13.2 Dung as an energy source

	Electricity & petroleum %	Firewood %
Industry/commerce	37	25
Transport	36	–
Households	17	75
Other	10	–
Total	100	100

Table 13.1 Sri Lanka's energy consumption 1980

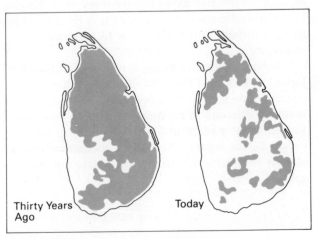

Thirty Years Ago

Today

Figure 13.3 Sri Lanka's shrinking forests

'The problem is immense for plantation women because they must depend a great deal on firewood for keeping the home fires burning. They have to ensure that they have enough firewood throughout the year to cook the meals, and also, because there is a rainy season, to have enough firewood to keep them warm… plantation women find that few, a very small percentage, of men are willing and able to go and bring the firewood for them.'

(Missionary)

'It's up to me to fetch the wood from the dealers and the jungle. This can take two or three hours each day. We use a lot of wood and there's only the children to help me.'
(Mother of five children)

If the women cannot afford to buy firewood, and there are no tea cuttings available on their estate, they are forced to steal wood from the forest.

'To go to the forest is not easy. We have to walk three miles. We carry what we can, but it's heavy, and there are guards there. If they catch us, they'll take our knives and the wood as well. Then they'll write to the estate and say we took it and we'll be fined or even sent to prison.'

(Plantation worker)

Figure 13.4 The problems of collecting wood in Sri Lanka

CASE STUDY

Biogas in Sri Lanka

Biogas is a useful alternative energy source to wood. Sri Lanka's three million head of livestock and six million poultry could provide enough dung to make biogas.

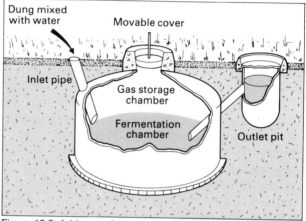

Figure 13.5 A biogas plant

Figure 13.5 shows how the Chinese system being introduced in Sri Lanka works. It is a form of **appropriate technology** (see Chapter 30). It uses locally available raw materials and relatively simple technology to meet an energy need. The main dome-shaped chamber is buried underground. Dung is fed in and heated by the sun and the warmth of the ground. This causes **fermentation** and the releases of **methane gas**. The waste from this process can be used to make fertiliser. Although these units are not expensive to build, they are unfortunately beyond the means of most people in Sri Lanka. Perhaps some form of assistance could be given by the Government.

ACTIVITIES

1 Study Table 13.1. What are the main uses of firewood in Sri Lanka?

2 a Study Figure 13.4.
   Whose job is it usually to find fuel?

   b Why is collecting wood sometimes a dangerous activity?

3 What is likely to happen if most trees have gone from Sri Lanka by the end of the century?

4 Suggest a possible solution to the problem of **deforestation** — the loss of trees.

ACTIVITIES

5 Explain how the use of biogas could slow rates of deforestation in Sri Lanka.

6 Summarise the arguments for and against the use of biogas in a table. Use two columns, one headed **Advantages** and the other **Disadvantages**.

It might seem that biological sources of energy are of little value in our modern world, but in recent years some have become of great importance.

CASE STUDY

Alcohol in Brazil

If you or your parents make home-made wine you will know that alcohol can be made or distilled from many biological sources such as potatoes, rhubarb, oranges and bilberries. It is the sugar in plants which turns to alcohol, so, not surprisingly, sugar cane is one of the best sources of alcohol.

Many developing countries, in particular, have been badly affected by rising oil prices. Valuable foreign exchange earnings have to be used to pay for imports of oil. Alternative energy sources are therefore very attractive. Brazil has developed one of the largest and best known alcohol programmes in the world. Around São Paulo in Brazil thousands of hectares of land are devoted to growing sugar cane for large distilleries. Nearly a quarter of vehicles in Brazil now run on alcohol fuel, called **ethanol**, made from sugar cane. The rest run on a mixture of alcohol and ordinary petrol. These fuels are readily available from pumps at filling stations just like petrol is in Britain (Figure 13.6).

Figure 13.6 Ethanol used as a fuel

Increasingly, the fibrous waste left over after sugar is made from cane, called **bagasse**, is also used to produce alcohol. This increases the potential of sugar cane as a source of energy even further. In Hawaii and South Africa, the sugar industry even provides electricity for the national grids.

From ethanol it is also possible to make **ethylene**, one of the most important products in the chemical industry. A huge list of manufactured goods can be linked to ethylene, including plastics, synthetic fibres, paints and even synthetic rubber. Perhaps the most important point to be made is that the use of biological sources as an energy source and in the chemical industry could represent a huge saving in oil, coal and natural gas. The potential is enormous.

ACTIVITIES

7 a What is the difference between **ethanol** and **ethylene**?

  b Which do you think is the most valuable?

8 Would you describe ethanol as a renewable or non renewable energy source? Briefly give reasons for your answer.

9 Why could alcohol based fuels become such an important source of energy in the future?

10 Study the cartoons in Figure 13.7 and list the uses of alcohol based fuels that are shown. Add others that you can think of.

11 What problems might arise if farmers in countries like Brazil are encouraged to grow sugar cane for alcohol refineries? They are often attracted by the steady market and good prices.

Figure 13.7 Some uses of fuel from sugar cane

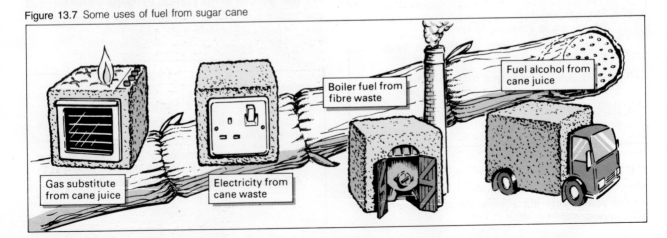

14 What is industry?

A dictionary would state that the word industry means 'diligence', or 'employment in useful work'. In its widest sense the word has the same meaning to a geographer. However, it is usually used to refer to mining and manufacturing. The type of work done by a coal miner, a car assembly line worker and a computer operator are all very different. Industry is only one of many different economic activities which include transport, communications and agriculture. To make things easier a **classification** or grouping of economic activities is often used. It has four divisions.

Figure 14.1 Figure 14.2 ▲ ▶
Primary occupations

Primary activities

This group includes activities which involve the extraction and/or early processing of natural resources, such as quarrying, mining, agriculture and fishing (Figures 14.1 and 14.2). The materials from this sector provide the basis for secondary activities.

Secondary activities

These are largely concerned with manufacturing industry, making things out of the **raw materials** from the primary sector (Figures 14.3 and 14.4). For example, aluminium for soft drink cans is made from a mineral called bauxite. Some manufacturing processes lead to products that are bulky and that use large amounts of raw materials, such as making ships or steel bridges. These industries are called **heavy industries. Light industries** are the opposite, for example, making calculators or television sets.

Figure 14.3 Figure 14.4 ▲ ▶
Secondary occupations

Tertiary activities

People who work in these activities do not actually make anything but they serve the primary and secondary sectors. These are usually called **services** and include transport, shops and teaching (Figures 14.5 and 14.6). Many people who are employed in this sector work in offices (see Chapter 25).

Figure 14.5 ▲
Figure 14.6 ▶
Services

Quaternary activities

This sector has only been identified in recent years and applies mainly to developed countries. It includes marketing, advertising and organisations specialising in research and development, and consulting work (Figures 14.7 and 14.8). Some industries are so large that they employ specialist firms to carry out research for them and to promote their products. Again, this sector supports or serves mainly the secondary sector.

The proportions of people employed in these different sectors, however, has not always been the same as Figure 14.9 shows. Several hundred years ago most working people in Britain were employed in primary activities, mostly farming. Since the Industrial Revolution growing numbers have worked in the manufacturing industry. Many left farming as a result of mechanisation. The growth of the tertiary sector in Britain has taken place in this century and the emergence of the quaternary sector only in the last few decades.

Figure 14.7 Figure 14.8
Quaternary occupations

Figure 14.9 Trends in employment sectors

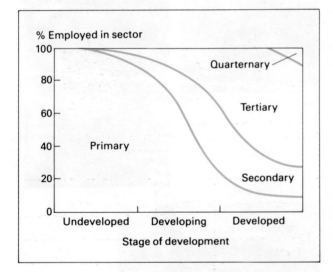

ACTIVITIES

1 Draw a table with four columns. Head the columns with the four different types of economic activities. Now try to list at least four different activities in each column.

2 Identify three examples of heavy industry and three examples of light industry other than those referred to in the paragraphs above.

3 Study Figure 14.9. For **both** developed and developing countries, approximately what percentages of the working population are employed in: i the primary sector; ii the secondary sector; iii the tertiary sector?

4 Briefly try and account for the differences in employment between developed and developing countries.

15 The needs of industry

Imagine that you wanted to set up a business and manufacture something – let's say a new electric car. One of the most important decisions you would have to make is *where* to build your factory. You need to understand the **locational needs** of your industry. **Extractive industries** have to be located where the raw materials are to be found. For example, coal mining takes place where there are coal seams. However, for manufacturing industries and services there are many more locational needs. The manufacture of an electric car is only an example.

Some industries, like iron and steel, use bulky raw materials that are expensive to transport. They tend therefore to locate close to their **raw material** sources. On the other hand many modern industries, like electronics, manufacture products that are fragile and more bulky than their components or raw materials. They are attracted to **market** locations to reduce the distance their products have to be transported. Usually, however, several locational influences are involved. For every factory the things shown in Figure 15.1 represent important needs.

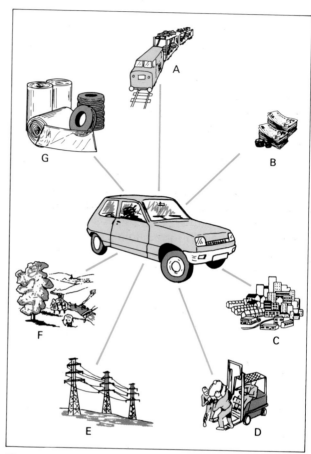

Figure 15.1 Factors which influence the location of industry

ACTIVITIES

1 a Name an industry, other than iron and steel, that tends to locate close to its raw materials. Briefly explain why.

  b Name an industry, other than electronics, that tends to locate close to its markets. Briefly explain why.

2 a Make a copy of the table in Figure 15.2 (you need to leave enough space for **2 d**).

  b In Figure 15.1 each locational need has been given a letter. Put the correct letter into the boxes in your table.

  c As a group, discuss what other locational needs might be important. After discussion choose *two* new needs to complete the last two rows in your table.

  d In the third column, write a short sentence explaining how each need might influence the location of a factory. Again, you should discuss this first.

Locational requirements	Letter	Explanation
Transport		
Market		
Raw materials or components		
Capital		
Land		
Labour		
Power supply		

Figure 15.2 Table for Activity **2 a**

16 Kaolin mining

Kaolin, or china clay as it is more popularly known, has many industrial uses in today's technical world. These include the manufacture of china, paper and some medicines. Kaolin has its origins in granite. In Britain it is found in the south west where granite was formed about three hundred million years ago. Kaolin developed around the edges of granite masses when they formed deep below the surface (Figure 16.1). The complete area of granite below Devon and Cornwall is called a **batholith** (see *People and the Physical Environment*). Gases and water vapour reacted in the intense heat with a mineral called **feldspar** in the granite. This mineral was chemically altered to form kaolin. Other minerals in the granite remained largely unaltered and have to be removed in the extraction process. The landscape has since been eroded and weathered so that the granite now appears at the surface forming moorland areas. The kaolin can easily be extracted from quarries (Figure 16.2).

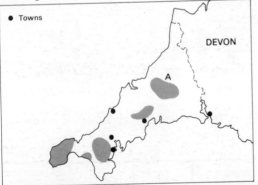

Figure 16.1 Map of Cornwall showing main granite areas

Figure 16.2 A kaolin quarry

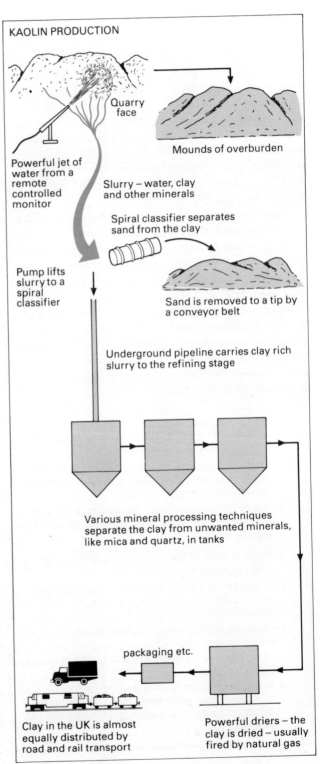

KAOLIN PRODUCTION

Quarry face

Powerful jet of water from a remote controlled monitor

Mounds of overburden

Slurry – water, clay and other minerals

Spiral classifier separates sand from the clay

Pump lifts slurry to a spiral classifier

Sand is removed to a tip by a conveyor belt

Underground pipeline carries clay rich slurry to the refining stage

Various mineral processing techniques separate the clay from unwanted minerals, like mica and quartz, in tanks

packaging etc.

Clay in the UK is almost equally distributed by road and rail transport

Powerful driers – the clay is dried – usually fired by natural gas

Figure 16.3 Simplified flow diagram of kaolin production

ACTIVITIES

1 Copy Figure 16.1. Use an atlas to identify the moor lettered A, and the towns.

2 Explain in a few sentences how kaolin is formed.

3 Study Figure 16.2.
   a What is shown in the top left of photograph?
   b What are the lines crossing the slope in the middle top of the photograph?
   c What are the white lines crossing the middle of the photograph?
   d Suggest what the trucks might be doing.

4 What type of mining is used to extract kaolin? (If you are unsure, turn to pages 35 and 36).

5 Describe and explain the process by which kaolin is extracted and produced in as much detail as you can. Relate what you can see in the flow diagram (Figure 16.3) to the photograph (Figure 16.2), and the landscape shown in Figure 16.5.

6 Study the map extract Figure 16.4 along with Figure 16.5.
   a The area in the photograph is contained within the map extract. Try to give an approximate six figure grid reference for the point above which the photographer was located.
   b In what general direction was the camera pointing when the photograph was taken?
   c Using the map and the photograph;
      i Name the settlement labelled A on the photograph.
      ii Identify the A road from X to Y on the photograph.
      iii Describe the ground surface and state what is at the following grid references − 000574, 010580, 980565. The photograph will help.
      iv Can you work out what has happened at B and C in the photograph?
   d Par (072535) is one of the small harbours from which the china clay is shipped. Suggest ways in which the china clay might be transported to Par from: a the area around Stenalees (010571); b from the area around High Street (966534) and Foxhole (965548).

The aerial photograph shows the environmental effects of kaolin extraction. The piles of overburden have been called the 'Pyramids of Cornwall'. About seven tonnes of waste is produced for every one tonne of clay. Old quarry pits can also be seen.
   e Pollution can take various forms − atmospheric, noise, water, pollution of the ground and soil, and aesthetic or visual. Write about the types of pollution that are most likely around St Austell.
   f Suggest some steps that could reduce the damage to the environment, particularly once extraction of the clay is finished.

Figure 16.5 Aerial view of china clay quarries near St Austell

Figure 16.6 China clay workings − before and after

Coursework ideas

Many minerals and rocks are mined or quarried in Britain. Find out if any resource is extracted near your home. What is it being used for? How is it extracted? Look at the environmental impact of the operation.

Figure 16.4

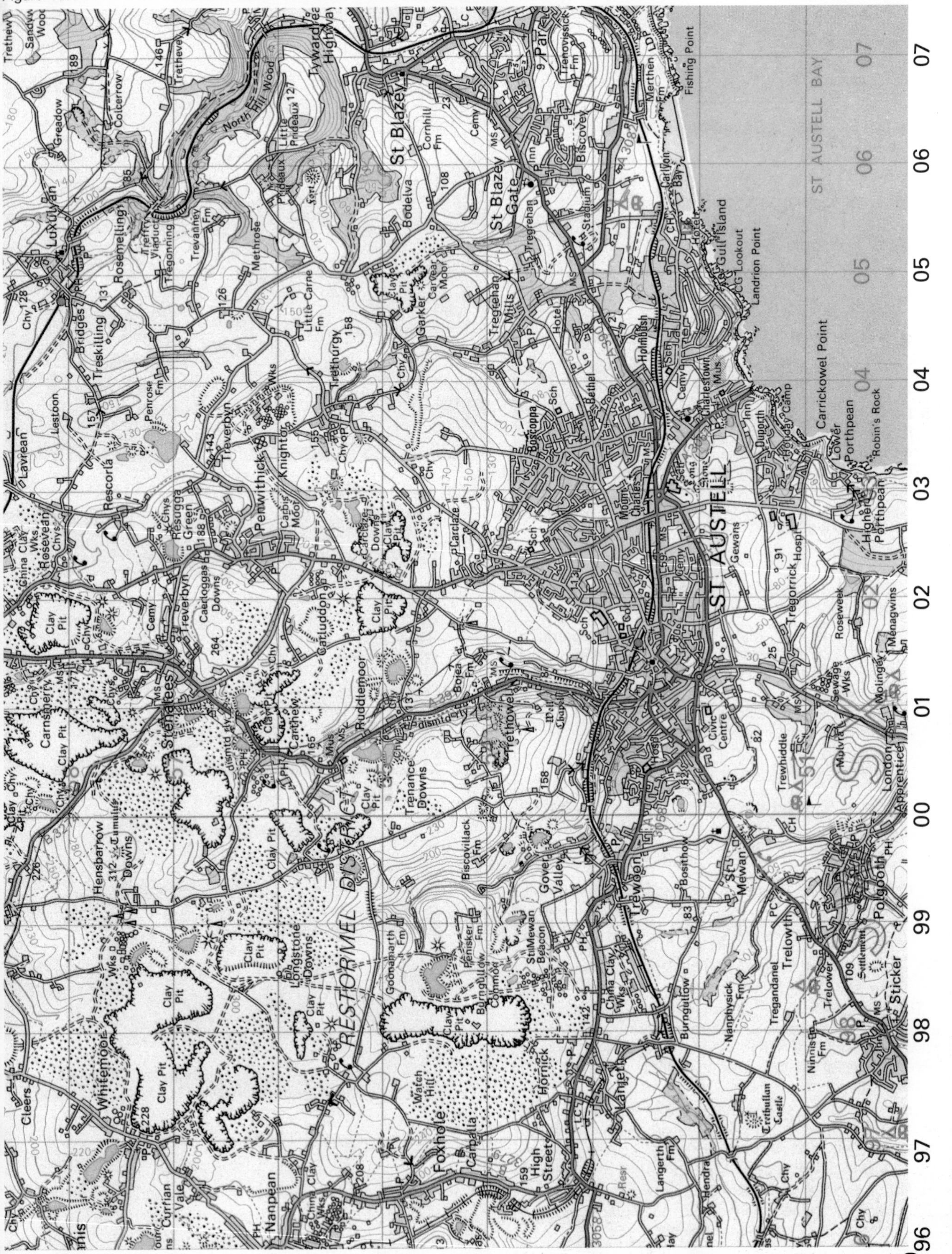

17 Coal mining

Mining in Britain probably started before the Romans arrived. However, it was not until the eighteenth century that large scale coal mining took place. Coal mining today makes a valuable contribution to the British economy. Coal provides power for industry, thermal power stations, transport and domestic purposes. In addition, the industry employed about 140 thousand people in 1988. It is difficult to measure the industry's value as its activities are so varied.

There are three basic methods of mining, **opencast, drift** (or **adit**) mining and **shaft** mining. Each has certain advantages and disadvantages.

Opencast mining

ACTIVITIES

1 Study Figure 17.1. Make a copy of the diagram. Write the correct labels from the list below into the spaces provided:
   walking dragline   excavator   spoil heap
   coal seam   truck   overburden

2 Explain in a few sentences how open cast mining works. Why do you think this was the earliest type of mining used?

3 Look at Figure 17.2. What do you think are the main disadvantages of this type of mining?

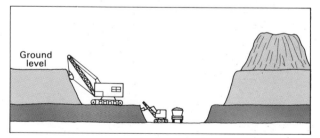

Figure 17.1 Open cast coal mine

Figure 17.2 Visually obtrusive open cast coal mining spoil heaps

Drift (or adit) mining

ACTIVITIES

4 Study Figure 17.3. Make a copy of the diagram and explain in a few sentences how this type of mining works.

5 Why might this type of mining only be found in a few locations?

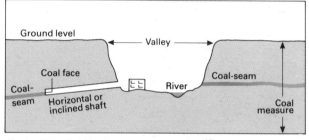

Figure 17.3 A drift mine

Shaft mining

This is the main method of mining coal in Britain and is often used to reach seams hundreds of metres below the surface.

ACTIVITIES

6 Study Figure 17.4. Draw a simplified copy of this diagram. Write the correct labels from the list below into the spaces provided:
   coal seams   coal face   mound of overburden
   coal measures   shaft

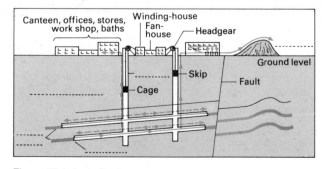

Figure 17.4 A shaft mine

ACTIVITIES

7 What is the difference between a **coal seam** and a **coal measure**?

8 Shaft mining is the most expensive of the three methods. Why do you think this is so?

9 In what ways is drift mining easier than shaft mining?

10 Many miners used to die in shaft mines from explosions and collapsing shafts. Explosions were caused when gases were ignited by lighting equipment. Study Table 17.1 and suggest why the safety record in British mines has improved since 1947.

11 When shaft mines are abandoned the surface often sags or may collapse; this is called **subsidence**. Look at Figure 17.5; the photograph shows evidence of subsidence. Describe and explain what the photograph shows.

Sometimes more than one type of mining is appropriate on a coalfield as the coal may be concealed in parts of the coalfield by overlying rocks. Elsewhere it may be exposed at the surface.

Table 17.1 Number of men killed in British mines

1947	1953	1968	1976	1986/7
618	401	130	55	15

Figure 17.5 Subsidence damage associated with coal mining

ACTIVITIES

12 a Study Figure 17.6 and make a copy of the diagram. Add the terms **concealed coalfield** and **exposed coalfield** in the spaces provided.

b Establish which type of mining is most suited to locations X, Y and Z in the diagram and for each briefly explain why.

Figure 17.6

Coal seam   Coal measure

CASE STUDY

The Selby coalfield

In the early 1980s a new coalfield project was completed at Selby in Yorkshire. Costing well over £1000 million, it is intended to exploit nearly three hundred square milometres of coal reserves between Selby and York (Figure 17.7a). It is now one of the largest and most modern coal mining projects in Europe.

The coal is brought to the surface through an old drift mine at Gascoigne Wood. The shafts provide access for men and materials. Figure 17.8 shows the main features of the coalfield and the surrounding area.

Figure 17.7a (top) A modern colliery — Kellingley, Yorks

Figure 17.7b (bottom) An old colliery — Cumberland

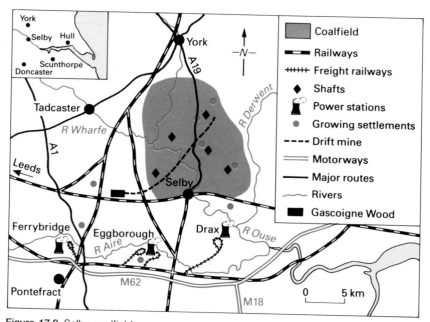

Figure 17.8 Selby coalfield

Figure 17.9 A merry-go-round train at a modern power station

Figure 17.10 Automatic loading of coal at a colliery

Figure 17.11 Britain's coalfields

Key 1mm = 1 000 000 tonnes output

ACTIVITIES

13 a Study Figure 17.7. Compare the view of the modern colliery at Kellingley with Cumberland colliery. List the main differences between the collieries at the surface.

   b Study Figure 17.8. What do you think most of the Selby coal is used for?

   c Use Figure 17.11 to discover of which British coalfield Selby is a part.

14 Study Figures 17.8, 17.9 and 17.10.

   a Explain any ways in which the coalfield and road and rail networks might be linked.

   b What do you think a 'merry-go-round' train is and how does this system of transporting coal work?

15 a Study Figure 17.11. On an outline map of the British Isles mark on the position of Britain's coalfields. Do not name them at this stage.

   b Using the key provided add proportional bars to the map based on annual production at each coalfield for which information is available in Table 17.2. The Scottish coalfield has been done for you. You can now add the names of all the coalfields to your map in the spaces provided; an altas will help you.

   c Now list the four coalfields with the highest output and the four coalfields with the lowest output. Put them in rank order.

Region	Output 1986/1987 (000 tonnes)
Scottish	3 442
North Eastern	10 229
North Yorkshire	14 797
South Yorkshire	12 549
North Derbyshire	6 064
Nottinghamshire	18 114
South Midlands	5 660
Kent	496
Lancashire	10 112
South Wales	6 479

Table 17.2 Production of coal by region 1986/1987

CASE STUDY

A new coalfield – Margam

Britain still has large reserves of coal which have not yet been exploited. A new mine is in the process of being constructed at Margam in South Wales. The colliery will be located between Tondu and Pyle not far from Port Talbot (Figure 17.12). It will be one of the largest collieries in the country when it is completed, designed to exploit about 19 million tonnes of high quality coking coal.

It is estimated that when complete the colliery will generate about fifteen years of work for about 800 men. This is particularly important in an area where unemployment is quite high. The local economy will gain about £9 million each year and local industry with links with the colliery will benefit too. The mine surface will be like Selby in appearance (see Figure 17.7 on page 36) with modern buildings and many steps will be taken to landscape the waste. The steepest slopes will be planted with trees and lower slopes restored for agriculture.

Figure 17.12 The location of Margam Colliery

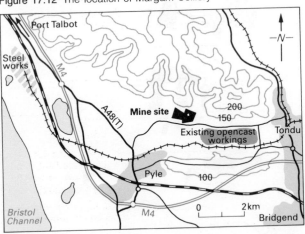

(see Figure 17.7 on page 36)

---

16 a Study Figure 17.12. What industrial activity in Port Talbot will benefit from the development of Margam Colliery?

   b Approximately how far is it from the colliery by rail?

   c Suggest how the coal might be transported from the colliery.

17 Some of the coal will be used locally, for example, in Pyle. Suggest how the coal might be transported. What do you think it will be used for?

18 List the various ways in which the local economy will benefit from the colliery both during its construction and when it is completed.

19 Study Figures 17.12 and 17.13. The Margam Mine will be in an attractive part of Wales; parts of the area are forested and used for agriculture. What conflicts might arise between farming and the coal mining operations?

20 A large mining project like Margam has many advantages and disadvantages. These often lead to arguments between different groups of people. A special meeting, a **public enquiry**, is usually held at which people can express their views.

   a Hold a 'mock' public enquiry in your class. Choose members of your class to represent the different groups of people who would have an interest in the project. The following is a list of suggestions:
      i a member of the National Coal Board;
      ii a local resident;
      iii a farmer and a landowner;
      iv an environmentalist;
      v a local MP.

   b As a class you might think of other interested parties.
      Each person playing a role would have to work out their arguments very carefully, either for or against the project, and then present them to the rest of the class. Your teacher could act as a chairperson. You could, if you wish, at the end, have a vote – it would be interesting after all the advantages and disadvantages had been considered, to see if your class thinks a project like this should go ahead.

Figure 17.13

INDUSTRY – MANUFACTURING

18 Iron and steel

The iron and steel industry has been changing rapidly throughout the world in the last few decades. Developed countries, like Britain, have largely cut back on production which has led to the closure of steelworks and the loss of many jobs. This is particularly due to a loss of overseas markets. A number of developing countries like India and Taiwan have been expanding their steel output from new steelworks.

Iron and steel is an example of a **heavy** industry. Several bulky raw materials are used in the manufacturing process.

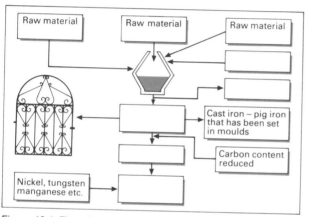

Figure 18.1 Flow diagram of the steel manufacturing process

Figure 18.2 Location of ironworks in the UK in 1967 and 1989

ACTIVITIES

1 a Draw an enlarged copy of Figure 18.1.

 b The following short statements cover all the stages in the manufacturing process, although they are not in the correct order. Fit the statements into the correct boxes. The diagram has been partly completed to help you, and there are also clues in the pictures and the statements:

 i Iron ore – such as hematite;

 ii Air – increases the temperature in the furnace;

 iii Blast furnace;

 iv Special alloy steels – qualities of hardness, resistance to corrosion, cutting capabilities;

 v Waste by-products after smelting called **slag**;

 vi Wrought iron – bends very easily;

 vii Steel – 0.5 to 1.5% carbon content;

 viii Coke – is one of the raw materials put into the blast furnace;

 ix Limestone – is used as a flux and added to the blast furnace;

 x Pig iron – very brittle, 3 to 5% carbon.

 Add the title **A diagram showing the stages in the manufacture of iron and steel**

2 Briefly explain why iron and steel is a **heavy** manufacturing industry.

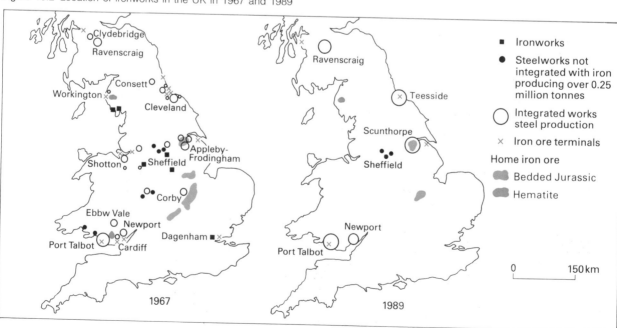

The iron and steel industry in Britain

The maps in Figure 18.2 show that the industry has changed dramatically in a very short period of time. The articles in Figure 18.3 appeared in British newspapers in the early 1980s and reflect the changes that were taking place in the iron and steel industry.

Figure 18.3 Change in the steel industry

BSC to shed 1,700 jobs in Sheffield area

BY IAN RODGER

THE British Steel Corporation is eliminating 1,709 jobs from its engineering steel works in the Sheffield area because of falling demand and the dismal outlook.

▲ Financial Times (26 Nov. 1982)

Teesside redundancies 'are necessary to break even'

2,200 jobs go as BSC begins more cutbacks

▲ The Guardian (3 Dec. 1982)

Union summoned to meeting

Axe poised over 1,000 steel jobs in Scunthorpe

By John Hatten
The British Steel Corporation is expected to announce next week at least 1,000 job losses at its Scunthorpe works, where 9,000 steelworkers have lost their jobs in the last three years as a result of contraction.

▲ The Guardian (3 Dec. 1982)

Foreign steel unloaded after picket at port is lifted

▲ The Guardian (27 Nov. 1982)

Why not open arms to cheap steel imports?

The Times (16 Dec. 1982) ▶

ACTIVITIES

3 a Study Figure 18.2. use an outline map of Great Britain to show the location and names of steelworks in Britain in 1988.

b Draw arrows to Teesside and Scunthorpe from the north east corner and label them **iron ore imports from Sweden.**

c Draw arrows to Port Talbot and Newport from the south west and label them **iron ore imports from South America.**

4 What were the main influences on the distribution of iron and steelworks in 1967?

5 Briefly explain the distribution of steelworks in 1988.

6 a Sheffield remains a relatively small iron and steel producing centre; it concentrates on alloy and high quality steel. Try to discover what is meant by **alloy steel.** Your chemistry textbook might help you!

b Name the steel product that Sheffield is particularly famous for. Clue − stainless steel is used in its manufacture.

c The steel industry in Sheffield was originally based on local supplies of iron ore, coal and limestone. Iron ore is now imported from abroad, the coal from other parts of the Yorkshire, Nottinghamshire, Derbyshire coalfield. The industry, however, has remained in Sheffield. What is this an example of? Explain your answer (see page 67).

7 Study the newspaper articles and Table 18.1.

a Draw a graph using the figures in Table 18.1. Label the vertical axis **numbers employed** (in thousands) and the horizontal axis **year.**

b Describe and try to explain the trend shown in the graph you have drawn by referring to the newspaper articles in Figure 18.3.

Table 18.1 Employment in the UK iron and steel industry (as defined by ECSC Treaty of Paris)

Year	Number employed (000)	Year	Number employed (000)
1972	200.5	1980	112.1
1973	198.5	1981	88.2
1974	196.9	1982	74.5
1975	185.3	1983	63.7
1976	182.3	1984	61.9
1977	178.9	1985	59.0
1978	165.4	1986	55.9
1979	156.6	1987	54.9

CASE STUDY

Steel in a developing country – the Tema steelworks in Ghana

The Tema steelworks was completed in 1964 as part of a growing industrial complex. It was developed with the assistance of several multinational corporations and foreign governments (see Figure 18.4). In 1962 a large artificial harbour had been built at Tema as Ghana lacked good natural harbours. Unfortunately, there is no iron ore mining in Ghana so the materials needed for the furnaces have to be imported from the UK and Sweden.

A steelworks is a useful industry for a developing country as so many other activities involve the use of steel; see how many you can think of. Most of the products from the steelworks are girders, bars and products for the building and engineering industries. Many people were employed in the construction of the steelworks and a great number are employed in its operation today.

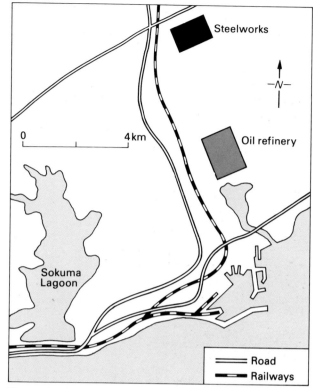

Figure 18.4 The port at Tema, Ghana

Figure 18.5 Industrial centres and infrastructure in Ghana

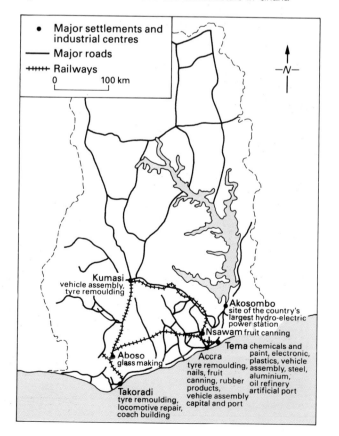

ACTIVITIES

8 Draw an outline map of Ghana showing Lake Volta, Akosombo and the HEP plant, the main settlements, the railways and label the steelworks at Tema. An atlas will help you.

9 a The steelworks has two electric arc furnaces. Where would the electricity for these come from?

   b Study Figure 18.5. What industries and other activities in southern Ghana would have benefited from the steelworks?

   c How would the steel be distributed to industries in southern Ghana?

   d Write a few sentences summarising why Tema was chosen as the location for the steelworks.

19 Shipbuilding

Figure 19.1 Shipyards at Tyneside at their peak

Shipbuilding is an excellent example of a traditional heavy industry which was once one of Britain's major industries. It was one of the principal sources of employment in areas like the north east and Clydeside in Scotland. Figure 19.1 shows part of Tyneside when it was at its peak as a shipbuilding centre. Shipbuilding also acted as a market for many other industries, notably steel. Yet, in this century Britain has moved from being one of the leading shipbuilding nations of the world to being one of the smallest. This is reflected in Tables 19.1, 19.2 and 19.3. It was even stated in the *Observer* newspaper in 1984 that 'in 10 years or less merchant shipbuilding in Britain may belong only in the history books'. The main reasons for this dramatic decline are as follows:

1 A world decline in trade by sea, partly due to the increasing cost of oil.
2 Over-capacity — so many ships had been built in the period just after the 2nd World War that few need building today.
3 Oil companies have reduced tanker fleets. Less Middle East oil is used by the main consuming countries. Britain now has oil from the North Sea and the USA from Alaska and Mexico, for example. Pipelines are now used instead of tankers in many places.
4 Increasing competition from foreign shipyards in countries like South Korea, Japan and Brazil.

Table 19.1 Shipbuilding — percentage of world output

	1938	1942	1946	1950	1954	1958	1962	1966	1970	1974	1978	1982	1983
Japan	--	--	1	6	8	23	27	46	49	49	50	48	47
UK	33	39	50	38	27	15	12	6	4	3	2	2	1

Table 19.2 Numbers employed in shipbuilding and marine engineering — UK

Year	Number employed
1977	182 000
1978	182 000
1979	175 000
1980	158 000
1981	148 000
1982	142 000
1983 (Sept.)	133 000

Table 19.3 Some trends in merchant shipbuilding — UK

Year	Completions	Orders placed
1978	96	70
1979	95	52
1980	88	61
1981	51	72
1982	67	57
1983	68	45

CASE STUDY

Shipbuilding in Japan

Since 1980 Japan's shipbuilding industry, despite fewer orders since 1983, has overshadowed those of other countries (see Figure 19.2). The industry was once largely concentrated around Tokyo near the main steel plants. However, space was very limited for expansion. The shipbuilding industry in Japan was severely damaged by World War II, although in one sense the war helped. New and larger shipyards were built after the war, particularly in southern Japan where plenty of land was available. The warm climate here also meant much of the building of ships could take place out of doors. This reduced the need for expensive buildings. Japan has produced ships at very competitive prices; some tankers have been the largest in the world.

Figure 19.2 however, shows that even Japan is facing a tough time at the top. Japan is facing competition from countries like Brazil and South Korea. Shipbuilding is a relatively easy industry to develop without a lot of advanced technology. Countries like South Korea can also provide the large labour force needed very cheaply (about 30 per cent cheaper than Japan in 1987). Ships from these newly industrialising countries are often cheaper than those from Japan although not always of the same quality.

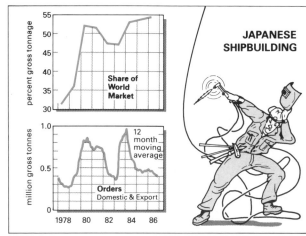

Figure 19.2 Japan's changing status as a shipbuilding nation

Figure 19.3 *The Guardian* 3.6.86

Tyneside loses 1,000 more shipyard jobs

By David Simpson,
Business Correspondent

The beleaguered North-east shipbuilding industry will reveal further casualties today when the recently privatised Swan Hunter yard on the Tyne confirms that it is to axe 1,000 workers by the summer.

The latest cuts will take the number of redundancies within the shipbuilding industry in Tyne and Wear to almost 2,500 over the past month alone, after cuts announced three weeks ago by the state-owned British Shipbuilders.

The area already boasts one of the highest unemployment levels in the UK, topping 20 per cent.

The first victims today are expected to be steel workers and boilermakers, reflecting the advanced state of most of the remaining ships at the yard.

Two Type 22 frigates for the Royal Navy have been launched and are now being fitted out, while the Sir Galahad, an auxiliary landing ship, is due to be completed early next year. The only material workload is another Type 22 frigate, HMS Chatham, whose keel was laid last month.

Further job losses could follow later this year unless Swan Hunter wins two other promised MoD orders. One of these is a Type 23 frigate contract due to be awarded within the next five or six weeks.

ACTIVITIES

1 a Draw a graph based on the data in Figure 19.1. Label the vertical axis **percentage of world output** and put the **years** on the horizontal axis. Use different coloured lines for Japan and the UK.

   b Describe and account for the changes shown by the graph. Use Figure 19.3 and Tables 19.2 and 19.3 to help you.

2 a What does the cartoon (Figure 19.4) suggest is one of the major reasons for the decline in British shipbuilding?

   b Make up a caption for the cartoon.

3 Read through the newspaper article (Figure 19.3). What types of large ships are still being built in British shipyards?

4 What was the main reason for the outstanding importance of the Japanese shipbuilding industry after World War II?

5 Why were new shipyards built in southern Japan after World War II?

6 Briefly explain why Japan is finding it difficult to keep its position as the leading shipbuilding nation of the world.

Figure 19.4

20 Oil refining and petrochemicals

Oil has become extremely important throughout the world as a raw material in the twentieth century. We use oil as a source of energy and to generate electricity. Oil products lubricate machinery and surface roads. Our transport systems are dependent on oil and a huge range of chemicals from oil help to clothe and feed us and improve our lives in many ways.

Crude oil is a complex substance, being not one but a mixture of several different liquids. For fertilisers, plastics, petrol and so on to be made, crude oil has to be separated and treated in many complex ways. This is done in a **refinery** (Figure 20.1). Figure 3.1 on page 5 shows the distribution of oil refineries in Britain.

The main refining process is called **distillation** which involves boiling the crude oil and turning it to vapour. In a refinery this process in done in a **fractionating column** (Figures 20.2 and 20.3). Each separate part of crude oil is called a **fraction**. Crude oil is heated and turned into vapour which then rises in the fractionating column. Some fractions condense more rapidly than others and are therefore released at various levels. This all sounds very simple, but in reality, the process is much more complex. The space-age appearance of a modern oil refinery is evidence of the sophisticated processes involved.

Figure 20.1 Aerial view of an oil refinery

ACTIVITIES

1 Study Figure 20.2 and explain briefly how oil is refined.
  a Study Figure 20.2 and describe why oil is so important today.
  b Make a copy of Figure 20.4 and complete the column headed **oil products**.

Figure 20.4

Fraction	Oil products
Lighter ↑ ↓ **Heavier**	

In addition many chemicals and petrochemicals are eventually produced – medicines, cosmetics, paints, insecticides, fertilisers, plastics etc. (**petro chemicals** are chemicals made from petrol).

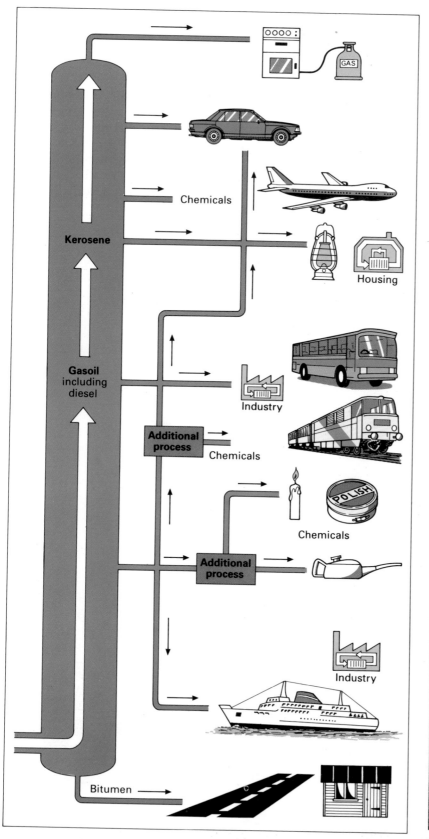

Kerosene

Gasoil including diesel

Chemicals

Housing

Industry

Additional process
Chemicals

Chemicals

Additional process

Industry

Bitumen

Figure 20.2 Diagram of fracionating column and oil products

Figure 20.3 Fractionating column at a modern oil refinery

CASE STUDY

Teesside, Britain

Look at the map extract Figure 20.5.

The Teesport area is dominated by three main industries, the Teesport oil refinery (530250, 550235, 525230), (Figure 20.6), Chemicals (570220, 535240), (Figure 20.7), and the Teesside steel works (565255). Industries have grouped together (**agglomerated**) because of the locational advantages of the area and because of the presence of other industries with which they are linked. One of the best examples is the attraction of the chemical industry to the area due to the presence of the oil refinery. Products from the oil refinery provide raw materials for the chemical industry. The River Tees and the port facilities largely explain this concentration of manufacturing industry (Figure 20.6).

Figure 20.5

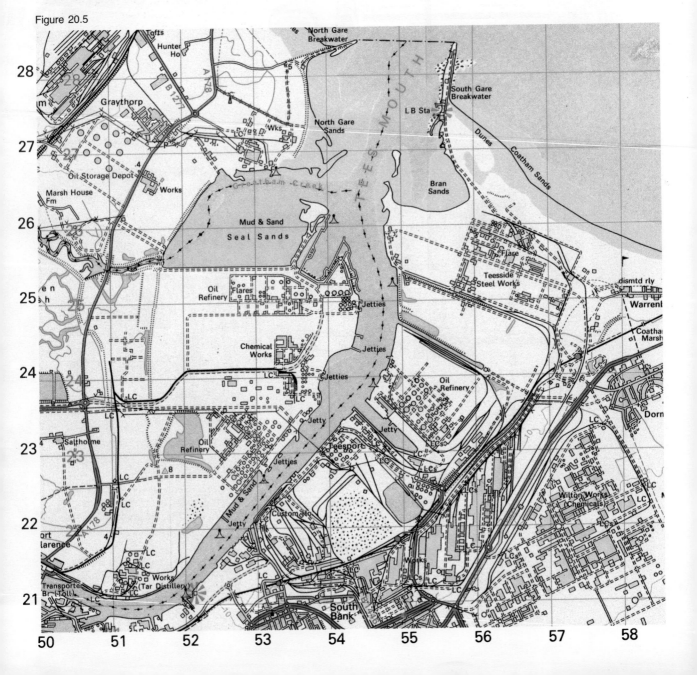

Figure 20.6 Teesside oil refinery

Figure 20.7 Wilton Chemical Works

Figure 20.8

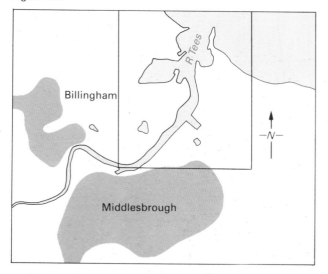

ACTIVITIES

Study Figures 20.6, 20.7, 20.8, and the Ordnance Survey map extract Figure 20.5.

2 Copy Figure 20.8 and then add the following to your map (your atlas will help)
A roads   railway   the chemical works
Teesside Steelworks   Teesport oil refinery
jetties

3 What evidence is there on the map extract that the Tees is heavily used by shipping?

4 Why have jetties been built, for example, at 544243 and 533228?

5 The following is a list of factors that influence the location of industries. Try to explain how each has influenced the location of the oil refining and petrochemical industries at Teesport (turn back to page 31 for help):

a raw materials (the chemical works produce petrochemicals, and use chemicals such as potash and sulphur from abroad and salt from the local area);

b land − why would an estuary environment be attractive? there are extensive areas of mud, marsh and sand;

c energy − give map evidence in support of your answer;

d water supply;

e labour;

f communications and access to market − give map evidence in support of your answer;

g Also think about safety; there are potential dangers with some of these industries.

As Teesport is in an Assisted Area (see Chapter 27) industries have also received Government help.

6 How might the development of North Sea oil have affected the industries in this area?

7 How has the chemical industry benefited from the presence of the oil industry?

8 How have the oil refineries benefited from the presence of the steelworks (the steelworks were built before the oil refineries)?

9 Describe and suggest reasons for the nature of the land at 543220.

10 The large scale industrial activity will in turn have encouraged smaller industries and activities providing equipment and services, for example. Try to list a few. See if you can find any on the map − remember to give a six figure grid reference.

11 In what ways might the Middlesbrough area have benefited from the presence of these industries? Note, they are largely capital intensive industries.

CASE STUDY

Rotterdam — Europoort, Netherlands

Figure 20.9 shows part of the industrial area of Rotterdam-Europoort, the largest port in western Europe (see *Settlement* book pages 78–79). The port and its industries have access to a huge market of about 120 million people within a 250 kilometre radius. Excellent communications — road, rail, canal and air serve this **hinterland** (the area served by a port).

The Rhine delta was once completely agricultural, but now, as Figure 20.10 shows, the area is a vast new docks system and industrial complex at the mouth of the Rhine. Oil imports are of great importance to Rotterdam-Europoort. The new waterway, dating from the late 1950s, allows huge tankers to reach Europoort. In addition to the oil storage tanks, and bulk cargo (coal, iron ore) terminals, the main industries are oil refining and petrochemicals. Many other supporting industries have been attracted to the region because of the rapid industrial growth providing another good example of agglomeration. The area has excellent locational advantages for these industrial activities.

Figure 20.9 The Rotterdam-Europoort industrial and port area

ACTIVITIES

12 Use what you have learnt about the oil refining and petrochemical industries at Teesport to explain the importance of Rotterdam-Europoort as a centre for these industries.

Figure 20.10 Industry at Rotterdam-Europoort

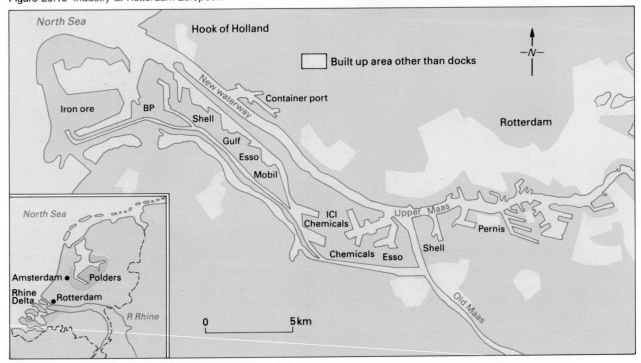

21 The motor industry

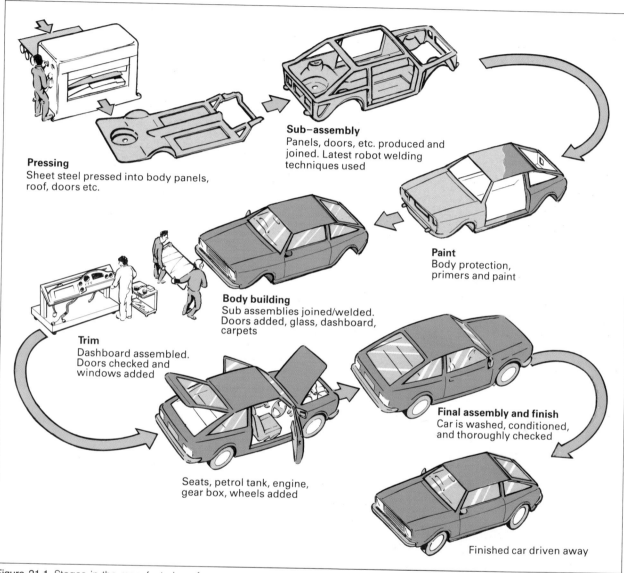

Pressing
Sheet steel pressed into body panels, roof, doors etc.

Sub–assembly
Panels, doors, etc. produced and joined. Latest robot welding techniques used

Paint
Body protection, primers and paint

Body building
Sub assemblies joined/welded. Doors added, glass, dashboard, carpets

Trim
Dashboard assembled. Doors checked and windows added

Seats, petrol tank, engine, gear box, wheels added

Final assembly and finish
Car is washed, conditioned, and thoroughly checked

Finished car driven away

Figure 21.1 Stages in the manufacturing of a car

The motor industry is one of the most important industries in Britain, both in terms of employment and the contribution it makes to the economy. The motor industry includes the manufacture of trucks and other road vehicles as well as cars. This chapter will concentrate on car production.

A modern car is a very sophisticated piece of engineering with as many as 14 000 different parts called **components**. Figure 21.1 shows the basic stages in the manufacturing process.

ACTIVITIES

1 Study Figure 21.1 and write a brief description of how a car is made.

In the early days of the car industry, vehicles were hand made. Not surprisingly, the industry was small scale and only a few cars were made each year. The industry was attracted to areas already specialising in engineering and metals with a skilled labour force. In Britain, this included Lancashire and the area around Birmingham in the West Midlands.

Since World War I, car manufacture has become highly mechanised and based on **mass production**, methods developed in the United States by Henry Ford (as in 'Ford' cars). The industry grew rapidly after World War II, as people had more money to spend on consumer goods, like cars.

Cars are now produced in vast numbers each year. The range of models is smaller, though, to streamline the manufacturing process and make mass production easier. A modern car plant does not employ as many people as a few decades ago, due to the use of machinery and, more recently, computers and robots (Figure 21.2).

It has changed from being a **labour intensive** to a **capital intensive** industry.

Figure 21.2 Robot-controlled operations in a car manufacturing plant

The location of the car industry

Three different types of location can be identified (Figure 21.3).

Figure 21.3 The location of the UK car industry

1 Some of the traditional manufacturing centres remain important, particularly in the West Midlands and Coventry. This can be explained by:
   a links with component producing industries in South Staffordshire;
   b the pool of skilled labour in the region;
   c the importance of the West Midlands market;
   d excellent communications with the rest of the country.
2 The components for a car including the steel, are transported quite easily and cheaply relative to a finished car which is bulky, fragile and expensive to transport. Consequently, the modern car industry, finds locations close to markets attractive. So the car industry spread to the South Midlands and to the South East; notably to Dagenham, east London and Luton, Bedfordshire (Figure 21.3). These locations are also close to ports for exports.
3 The final locations are in places like Merseyside, South Wales and Scotland. An important locational influence in recent years has been the Government (see Chapter 27). The growth of industry has been controlled in areas like the south east, while firms have been encouraged to move to less prosperous areas like South Wales.

CASE STUDY

Ellesmere Port

It was clear in the 1950s that General Motors car and truck plants at Luton, Dunstable could not cope with the demand for new vehicles. They needed to build a new factory. However, the Government would not let them build in the south east. Instead they encouraged General Motors to locate on Merseyside. GM invested £60 million in the Ellesmere Port site – a former RAF airfield (Figure 21.4). Today it plays a vital role in Merseyside's economy, more than 5000 men and women are employed in the plant. It uses the latest technology including robots and computers, to avoid repetitive tasks and to ensure efficiency. The Vauxhall Viva was one of the first cars to be made at the plant, and more recently, the Astra.

Figure 21.4 General Motors' car plant at Ellesmere Port

CASE STUDY

Ford in Spain

Car companies are good examples of **multinational corporations** (see Chapter 31). They often have branches in many countries to expand their markets and frequently to take advantage of cheap labour; Ford is a good example. In 1976 production started at a new Ford car plant at Valencia, one of the largest industrial complexes ever to be built in Spain (Figure 21.5). The Fiesta was the first car to be produced at this new plant. Ford wanted to expand in Europe and take advantage of the large market for small cars in southern Europe.

ACTIVITIES

2 a Using an outline map of Britain, make a copy of Figure 21.3. Use different colours for each of the major manufacturers.

  b Use an atlas to add motorways and major cities to your map.

  c Describe the pattern formed by the main concentrations of car plants in Britain.

  d What do you notice about the location of most of the car plants on the motorway network?

  e How would the car industry have benefited from the construction of the motorways?

3 Study Figure 21.4. The following locational factors influenced General Motors in locating their car plant at Ellesmere Port:

  a a suitable area of land;

  b good rail communications;

  c access to the M53;

  d Government support;

  e access to port facilities;

  f plentiful labour supply;

Explain the importance of each factor.

ACTIVITIES

4 Study Figure 21.5 and list the reasons why Valencia was chosen by Ford to be the location for a new car plant.

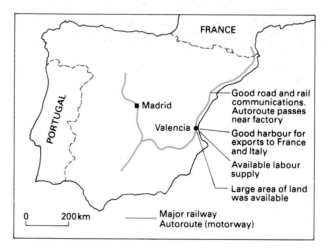

Figure 21.5 Locational advantages of Valencia for Ford car plant

CASE STUDY

The motor vehicle industry in Nigeria

Transport is very important to a developing country as mobility and economic progress are so closely linked. Communications are essential for the movement of goods, food and people.

As in many developing countries, the motor vehicle industry in Nigeria is located near to a port, Lagos (Figure 21.6). The emphasis is largely on commercial vehicles, such as trucks. Vehicles like land rovers can operate easily in the hot climate and cope with poor roads and open country. Cars are assembled in Lagos too.

Once a motor vehicle industry has been set up and skills have been learnt, it is also possible to manufacture similar products like tractors and agricultural machinery. This helps to reduce the need for expensive imported machines.

The motor vehicle industry is very attractive to a developing country as it has a large multiplier

effect (see page 59). Figure 21.6 shows a number of industries in Lagos that benefit from the motor vehicle industry. However, it is an expensive industry to establish and a great deal of capital is needed. Multinational corporations (see chapter 31) are often involved to provide investment and the practical and managerial skills.

ACTIVITIES

5 Study Figure 21.6.
   a Lagos is a major port and population centre. Why is it a good location for a car assembly works?
   b Which other industries in Lagos would you expect to have close links with the car factory?
   c Name any multinational corporations represented in the city and try to state what they make.
   d What problems might the site of Lagos have for an industry wanting to expand.
   e List any benefits the motor industry might have had in Nigeria that you have not already referred to.

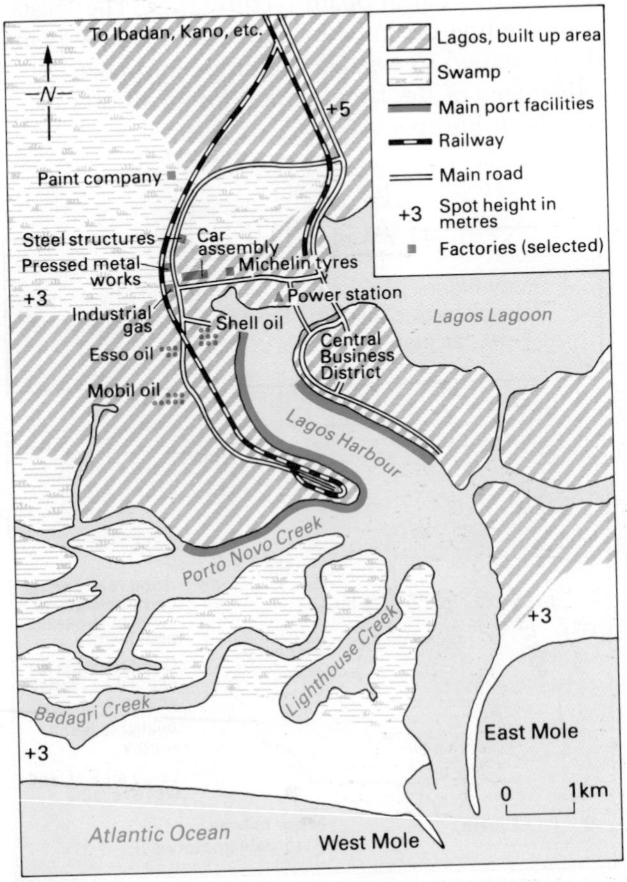

Figure 21.6 The distribution of the motor industry and related activities in Lagos, Nigeria

22 Hi-technology industry

Hi-technology industry, or 'hi-tech' as it is usually called, has largely developed in the last twenty years. Hi-tech involves the manufacture of computers, telecommunications and video equipment, and other micro-electronics. It is a modern industry and has very different locational factors to those industries which you have studied so far. The comparatively small factories use electricity for power and, as Figure 22.1 shows, can be located almost anywhere. Such industries are said to be **footloose**. Recently a large number of factories have been encouraged to set up in **Assisted Areas** (see Chapter 27). These are areas of high unemployment where a lot of old industries have been forced to close. Figure 22.3 is an advertisement trying to encourage hi-tech firms to locate in such an area. Figure 22.2 has a similar purpose for a **science park**, an estate of hi-tech firms in this case, near Bristol, a more prosperous part of the country.

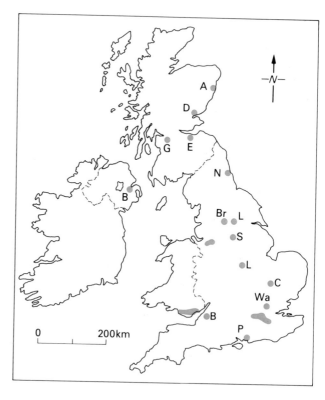

Figure 22.1 The main areas of hi-technology industry in Britain

Figure 22.2 Advertisement for a science park near Bristol

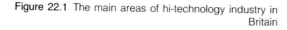

AZTEC WEST

Aztec West is Britain's newest and most dynamic concept of industrial development. It owes a lot to the outstanding achievements of California's world famous Silicon Valley, which has in the past decade provided the environment for some of the most advanced industries to develop and expand.

Aztec West is just a few miles north of Bristol at junction 16 on the M5 – that's just one mile from the interchange with the M4. So it's also one of Britain's best distribution locations.

Bristol is already noted for its high technology companies like Rolls Royce, Sperry Gyroscope, I.B.M., Marconi Avionics, British Aerospace and Westland Helicopters which means that Aztec West is building on an already well established concept that companies based in the right location attract good staff and maintain industrial relations.

When complete Aztec West will offer about 2 million sq.ft. of research, development and campus style office accommodation; plus shopping, recreation and local amenities like banks, cafes and tree lined walks.

You might say it's going to become Britain's technology hot shop in the '80s and '90s. What's more there'll be room for up to 10,000 new jobs.

The first phase of laying the infrastructure is now well advanced; so if you'd like to discuss setting up at Aztec West we'd be happy to show you around and discuss the possibilities.

BRADFORD ECONOMIC DEVELOPMENT UNIT

In recent years the City of Bradford has concentrated on modern high-technology industries and can now offer an impressive package of incentives to Micro-electronics companies wishing to expand into the city.

The local workforce in the area is dextrous, skilled, hard-working, cheerful and friendly, and they have one of the best strike-free records in the country – ideal virtues in Micro-electronics. The local council has an Economic Development Unit which can act quickly to offer companies finance, rent-free periods on council property (sometimes up to six months and over) and low cost-premises. Bradford has also retained Intermediate Area Status which means companies can benefit from Government and E.E.C. Grants.

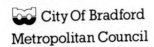

City Of Bradford
Metropolitan Council

ACTIVITIES

1 a On an outline map of Britain make a copy of Figure 22.1.

b Use an atlas to identify the main areas and cities associated with hi-tech industry. Write these on your map or put them in a key.

2 a Construct a pie diagram based on the employment figures in Table 22.1. The degrees for each sector have been given in column A.

b Does employment in hi-tech industry seem to be evenly distributed throughout the UK?

3 a Study Figure 22.2 and 22.3. Imagine you wanted to locate a new hi-tech factory somewhere in Britain. Why might you be attracted to these locations?

b Which of the two locations would you choose and why?

4 Figure 22.4 shows one of the main concentrations of hi-tech firms in the south east, the so called 'Thames Valley Corridor' (Figure 22.5).

a Make a copy of the map in Figure 22.4 and using an atlas locate the towns of Reading and Newbury.

b Add the position of Heathrow Airport and label it clearly.

c Name the motorway which passes east west across this area.

Figure 22.5 Hi-tech industry, Newbury

Figure 22.3 Advertisement intended to attract hi technology industries to Bradford

	Percentage of jobs	A	Percentage of firms
South East	56.0	202	66.2
Scotland	9.2	33	7.8
North east	7.9	28	8.2
South west	6.6	24	8.9
Others	20.3	73	8.9
	100%	360°	100%

Table 22.1 Regional variations in employment in hi-technology industry

Figure 22.4 Thames valley hi-tech corridor in Berkshire

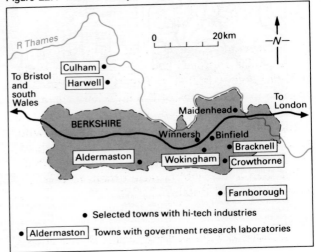

Technology in Wales

A look at the efforts being made in the Principality to attract and nourish 'sunrise' industries

Question *Why did the Parrot Competition choose Wales as the location for establishing Britain's first totally integrated floppy disc manufacturing unit?*

Answer *Because the hard headed American businessman behind the company discovered the Principality could offer the best deal.*

Frank Peters, the company's managing director, said it was the availability of the Welsh Development Agency-led investment package that clinched the decision.

The agency put up a £1m equity investment and was instrumental in negotiating the rest of the financial deal with City investors. Before embarking on the venture, the agency followed its usual practice of putting the company, the men behind it and the market under expert examination.

When it is ready, Parrot will move into a purpose-built 57,000 square feet production facility designed and built by the agency on a five acre sight on Cwmbran Development Corporation's high technology park. From that base it aims to capture eight per cent of the international market with a production of some 50 million disc units a year.

The company will join a remarkable concentration of advanced technology ventures in South East Wales, including Amersham International, Control Data, Ferranti, Inmos and Mitel. Further west along the M4 motorway Sony has just celebrated the production of the millionth television set at its Bridgend factory.

In Mid Wales, with its easy access to the midlands, smaller companies are thriving, backed and funded by Mid Wales Development which has initiated a whole range of industrial and social projects.

Companies are discovering it is possible to combine high technology operations within a beautiful rural area which offers a life style city dwellers can only envy. From the great and recent deindustrialization of its heavy base industrial backbone of steel and coal, the Principality has quickly emerged as a technology friendly location. Foreign based companies in Wales now employ far more people than the National Coal Board and British Steel combined.

One of the great benefits is the workforce. Contrary to ill-informed myth, the Welsh do not have a bad strike record. Sceptics can examine government statistics which support the statement.

Almost without exception, inward investment companies have been able to establish a one-union workforce which streamlines negotiating procedures at every level. The people are also dextrous and have proved to the immense satisfaction of such companies as Sony that they can readily embrace new skills.

Companies can rely also on an increasing pool of young people who have a basic mastery of new technology because of the work being done in schools and at the Information Technology Centres run by the Manpower Services Commission. These centres have now been established throughout Wales to provide school leavers with basic computer programming and practical work experience in the production of new technology items.

One asset the Principality provides that money cannot buy is an abundance of clean air which for many high tech companies can be a vital factor in determining where to locate their enterprises.

Another is first class communications providing easy access to markets. From the M4 corridor in South Wales, Heathrow airport is just two hours away and the problems and reports of delays on the Severn Bridge have been largely exaggerated for political purposes. Most of Mid Wales is under two hours from Birmingham and the north east has excellent road, rail, sea and air links.

In addition, a new understanding is being forged with the University of Wales to work closely with firms on Research and Development projects. In Clwyd, NEWTECH will be operating from the centre of the Deeside Industrial Park providing immediate and expert high tech services for companies.

Tim Jones
Welsh Correspondent

Figure 22.6 *The Times*, May 1984

ACTIVITIES

5 As a group discuss why this area has such a high concentration of hi-tech industry.

6 Figure 22.6, a newspaper article about hi-tech industry in South Wales identifies many of the locational factors which usually affect this industry. This area forms an extension to the west of the Thames Valley Corridor.

Identify the factors which have attracted hi-tech industry to South Wales. List the factors and write a few sentences explaining how they have been important in influencing the location of this industry.

7 Design a poster, perhaps like the one for Aztec West (Figure 22.2), intending to attract hi-tech firms to south Wales. You should use illustrations, a clear message or slogan and some written details.

8 a What is a **science park**?

b Locate the Science Parks in Figure 22.7. Identify and explain the factors on the map that have influenced their locations.

Coursework ideas

See coursework suggestions for offices on page 65. A similar approach could be used for a study of hi-technology industry. This will only be possible in those areas of Britain where there are concentrations of this type of industry.

You could also interview some employees who have moved to an area because of the growth of new hi-tech industries. You could find out what (other than a new job) attracted them to that particular part of Britain.

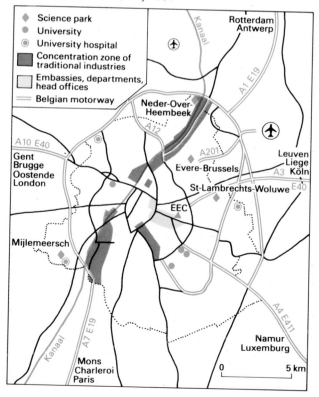

Figure 22.7 The location of science parks in Brussels, Belgium

23 *The frozen food industry*

In 1955 'Birds Eye' launched a new food product on to the market. Within weeks its success was certain and it was on sale throughout Britain. Today it is as popular as ever and as a product is worth over £900 million to Birds Eye each year. The mystery product is — fish fingers.

Frozen foods were not entirely a new idea though. Eskimos have been aware of the preserving properties of freezing food for centuries. However, it was not until the 1920s that a quick and commercially profitable method of freezing foods was developed by Clarence Birdseye in the USA. Soon the company that handled the new frozen foods had expanded throughout the world, including Britain.

After World War II the market for frozen foods grew rapidly. This trend seems likely to continue as Figure 23.1 shows. Fridges and freezers are accepted as basic pieces of equipment in most people's homes in Britain today. As the market for **convenience foods** has grown, more and more frozen food products have been developed. These range from frozen fish and meats to cakes and even frozen meals and exotic foreign foods. By 1986 a quarter of all households in Britain owned microwave ovens; many new products are aimed at this market.

CASE STUDY

Birds Eye/Walls

Birds Eye/Walls is one of the largest frozen food companies in Britain with hundreds of different frozen food products. The Company's head offices and one of its main factories are in Grimsby. This is a good location as it is near to important vegetable producing areas in Lincolnshire and South Humberside and on the coast for fish products. Many of the ingredients used in the factory are **perishable** and fragile so cannot easily be transported great distances. Food processing industries often locate close to their raw materials for this reason. The finished products, in contrast, are less perishable and have a widespread market — all of the United Kingdom and abroad.

Birds Eye/Walls has other factories in England at Gloucester, Lowestoft, Kirkby, Eastbourne and Hull. Each tends to concentrate on a particular group of products. For example, many frozen meals and frozen fish products are made at Grimsby whilst the factory at Gloucester concentrates on ice cream.

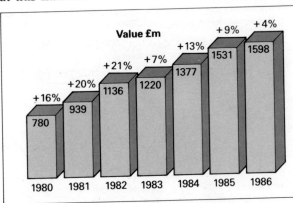
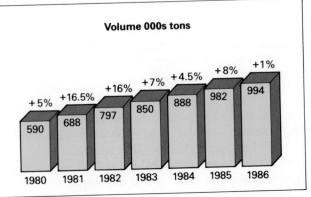

Figure 23.1 Changes in the size of the consumer market for frozen foods in Britain

---

ACTIVITIES

1 Suggest reasons for the rapid growth of the frozen food industry.

2 As a group, try to list ten very different frozen food products.

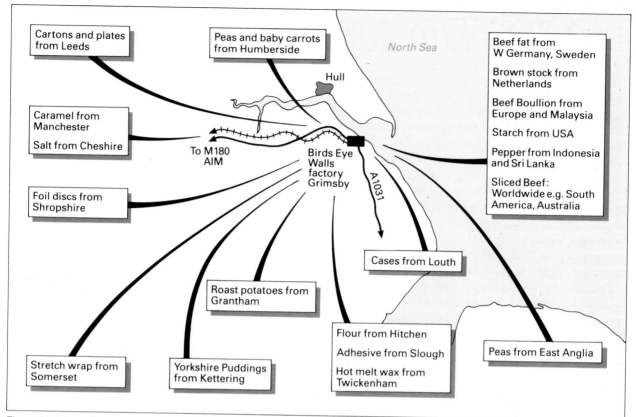

Figure 23.2 Ingredients and components used to produce Birds Eye's Roast Beef Platter

ACTIVITIES

3 Figure 23.2 shows the location of one of the Birds Eye/Walls factories at Grimsby and the main ingredients and components that are involved in producing their Roast Beef Platter. Use the list of ingredients and components below to produce a similar map for Cod Fish Fingers. Draw the map in such a way that you can tell which direction the ingredients have come from. An atlas will help you.

Ingredients/components From

Ingredients/components	From
Cod blocks	Denmark, Iceland, Norway, Baltic
Flour	Cambridge
Salt	Cheshire
Starch	Holland
Palm oil	Indonesia and African Ivory Coast
Breadcrumbs	Hull and Bristol
Carton and polywrap	Merseyside
Stretch wrap	Somerset
Shrink wrap	Cheshire
Corner pieces	Northampton

Sales/exports to:
UK, Republic of Ireland, Gibralter, Malta British Forces overseas
Special packs for Middle East and Far East

Main export ports: Felixstowe, Southampton, Harwich

4 Use the map you have drawn and Figure 23.2 to answer the following questions:

a How do ingredients and components reach the factory?

b How are finished products distributed in the United Kingdom and abroad?

c Explain why some ingredients have to be imported.

d Explain why the Birds Eye/Walls factory at Grimsby is located close to its raw materials and not its market.

e Frozen fish and vegetables were some of the earliest products made by the company at Grimsby. Think about the nature of the ingredients and the finished products carefully. Suggest why Grimsby is an ideal location for the frozen foods industry.

24 The tourist industry

The word tourism became part of the English language in the nineteenth century. It is a very recent development as travel was extremely limited before the growth of the railways and steamships. Following the Industrial Revolution, however, advances in transport soon made travel much more feasible both abroad (see Figure 24.1) and within Britain. Thomas Cook was one of the first tour operators to take advantage of this and was responsible for what was probably the first organised tour. In July 1841, 570 people rode on an excursion train from Leicester to Loughborough and back for the cost of one shilling (5p).

Tourism today is a major industry in many countries, including Britain and many developing countries. Tourism has been more widely recognised as an industry in Britain, and developed and marketed as such, since the 1969 Tourism Act. The British Tourist Authority (BTA) was then set up with overall responsibility for tourism in Britain, including overseas visitors. Also under the Act the English Tourist Board (ETB) was established which looks after the twelve regional tourist boards that promote tourism in their particular areas (Figure 24.2). Tourism is difficult to define as it can vary according to the length of stay, type of destination and purpose – holiday or business for example. We will concentrate on travel for holidays. It is also important to make a distinction between travel in Britain by British residents – called **domestic tourism**, and **foreign** or **overseas tourism**.

There has been a steady overall rise in the last decade in the number of British residents taking holidays. Increasing amounts of money have been spent helping the economies of the parts of the country they visit. Much can be learnt about domestic tourism from the Tourism **British Home Tourism Survey**. Tables 24.1 and 24.2 describe some aspects of holidays taken by British tourists.

Figure 24.1

Figure 24.2 Regional distribution of tourism 1987 – numbers of nights (millions) spent by British and overseas visitors

Table 24.1 Accommodation used by British tourists 1986

Accommodation	Trips %	Nights %	Estimated spending %
Serviced – hotels etc.	22	19	38
Self-serviced	35	43	40
Friends or relatives	43	36	21
Other	2	2	1

(Columns may add up to more than 100% because more than one type of accommodation was used on some trips.)

Table 24.2 Different locations visited by British tourists 1986

Location	Trips %	Nights %	Estimated spending %
Seaside	35	40	48
Small town	17	14	12
Large town	14	10	9
London	7	4	4
Countryside	25	25	22
Other	6	6	5

(Columns may add up to more than 100% because some trips were to more than one type of destination.)

Increasing numbers of Britains have also been taking holidays abroad, a trend encouraged by the development of **package holidays** largely since the 1960s. The tour operator, like Thompsons or Thomas Cook, sells a holiday that includes the air fare, hotel bill and possibly some meals, all at a fixed price. This is very convenient for the tourist and also works out more cheaply than arranging the holiday without the tour operator. Spain, Italy and France remain the most popular destination countries for British holidaymakers.

ACTIVITIES

1 Explain briefly what Table 24.1 tells you about the importance of different types of holiday accommodation.

2 a Study Table 24.2. Which types of destination were the most and the least popular for British holidaymakers in Britain in 1986?

   b London has always been one of the most popular parts of Britain for overseas visitors. Suggest a few reasons why it is less popular as a holiday destination for the British.

3 What is a **package holiday** and which parts of Europe are particularly popular destinations for British tourists?

There has also been a steady rise in the numbers of overseas visitors to Britain in the last decade with an increase, too, in the amount they spend. The **International Passenger Survey** is the main source of information on visitor movements in and out of the UK. Figure 24.2 shows which parts of Britain are the most attractive areas for tourists. (Note that it combines British and foreign tourists.)

ACTIVITIES

4 a Use an outline map of Britain to make a copy of Figure 24.2 leaving out the numbers.

   b Shade the different regions using the key below. Make sure you add the key to your map and give your map a title.
   0–19 Yellow   20–39 Orange   40–69 Red
   70–99 Brown   100+ Black

   c Describe and try to suggest reasons for the pattern of tourism on your map. An atlas may help you.

Tourism makes a major contribution to the economies of Britain and many other countries. Overseas tourist spending in Britain has risen steadily from over £1 million in 1976 to over £5 million in 1986.

One aspect of the economics of the tourist industry deserves particular mention and helps to explain why the tourist industry is so attractive to developing countries. Money spent by tourists can increase in value at a destination. Figure 24.3 shows how this **multiplier effect** works. It is particularly strong in the tourist industry.

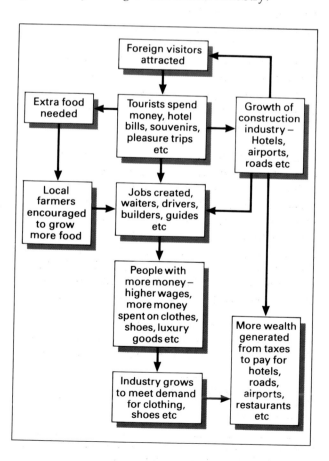

Figure 24.3 The multiplier effect

London theatres and clubs, concert halls and the Underground earn much of their annual income from tourists. It is certain that many of London's leisure facilities would not survive without tourists.

We have seen that the tourist industry is very diverse. It includes hotels, restaurants, airlines, travel agents, tour operators and amusement parks, to name but a few of its many parts. This helps to explain why it is also a **labour intensive industry** (see Table 24.3). It has been estimated that about one and a half million jobs in Britain are directly or indirectly related to the tourist industry.

Table 24.3 Number of employees in the hotel and catering industry in England 1976–1985

Year	Males	Females	Total
1976	300	540	840
1977	300	553	853
1978	301	572	873
1979–1981	no data available		
1982	308	620	928
1983	311	594	905
1984	345	655	1000
1985	362	679	1041

Source: Department of Employment
Note: Those employed in the hotels and catering industry represent only a small fraction of all those employed in the tourist industry.

ACTIVITIES

5 a Explain what is meant by the term **multiplier effect**.

 b In what other ways does tourism help the economy of an area or country? You could refer in your answer to attractions and businesses that benefit from tourism in a city like London.

6 A lot of employment in the tourist industry is **seasonal**. The ski industry in the Alps is a good example, being a winter industry (see *People and the Physical Environment* page 28).

 a What would you do to encourage people to visit the Alps in the summer?

 b What sort of things could visitors do?

 c Why is it important for ski resorts to try to cater for summer as well as winter demand?

CASE STUDY

Theme parks: Alton Towers

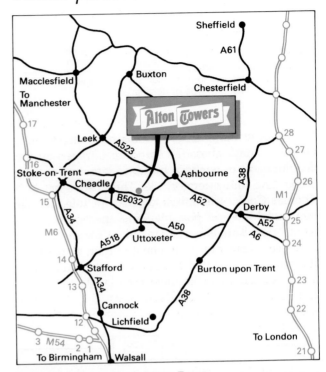

Figure 24.4 The location of Alton Towers

Alton Towers in Staffordshire (Figure 24.4) is one of the leading leisure parks in Europe and has been voted Britain's most outstanding tourist attraction by the BTA. There are over 100 attractions and rides set in beautiful grounds and gardens. Figure 24.4 shows how accessible Alton Towers is from various parts of the country. Most people who go to Alton Towers are day visitors but there is also a growing number of overseas visitors. Alton Towers is a fine example of a tourist attraction in the provinces outside London.

ACTIVITIES

7 a Which motorway would visitors from Birmingham and Manchester use as part of their journey to Alton Towers?

 b Which major population centres do you think most of the visitors to Alton Towers come from?

8 Why do you think the BTA has made great efforts in recent years to promote places like Alton Towers in the provinces? Figure 24.2 will help you.

Tourism in developing countries

Since about 1970 there has been a rapid development of tourism in many developing countries which previously had few visitors. Two major reasons for this trend are:

1 improvements in international transport, particularly airlines;
2 efforts made by governments to promote tourism.

We have already seen that tourism can generate employment, earn a lot of money, and have an important multiplier effect which benefits other activities like hotels, building and agriculture. Some developing countries have invested so much in tourism that the industry is one of their major sources of income. Examples include Mexico, Kenya, Tanzania, Tunisia and the Seychelles. However, hotels in particular are often owned by foreign multi-national companies (see Chapter 31). They can take a very large share of the profits. The government of a country has to decide how much multi-national involvement it will allow in its tourist industry.

CASE STUDY

Tourism in Kenya

Kenya is one of the most important countries for tourism in Africa. In the early 1980s about 300 000 visitors arrived each year, about five times more than twenty years previously. Most of the tourists came from Europe and North America and provided Kenya with about £25 million pounds in a year.

The Kenyan Government has encouraged the development of the industry. The Kenya Tourist Development Corporation is the equivalent of the BTA. It promotes and advertises tourism to Kenya and plans and supports the development of tourism in Kenya.
(See *Agriculture and Rural Issues* for more on Kenya's National Parks)

ACTIVITIES

9 Why are developing countries so keen to develop a tourist industry?

10 Use an atlas to find where Kenya is in Africa. Study the information about Kenya in Figure 24.5.

 a List all the different attractions referred to and try to think of others.

 b List some different jobs that you would expect to have been created in Kenya as the result of tourism.

 c What special facilities would have to be built in a developing country like Kenya before large numbers of tourists could visit it?

11 a Explain briefly what attractions developing countries have to offer tourists that are different or not available in developed areas like Spain or Italy.

 b List some other developing countries that you know have tourist industries and state their main attractions. Obtain some travel brochures to help you.

 c It is very important to note that tourism brings with it certain problems as well as all the advantages. Think of Kenya — a relatively poor country in comparison with Britain, now receiving several thousand tourists each year.
  As a group or in pairs, try to list some of the problems that could arise and explain how each could be solved (Figure 24.5 will give you some ideas). Here is one problem — without a solution, to help you get started:
 Many tourists enjoy exploring the coral reefs and fishing in the Indian Ocean. Pollution from hotels and other tourist developments has been disposed of in the sea. This is destroying some of the marine life that some tourists come to see.

Figure 24.5 Travel brochure advertisement for Kenya

One element which makes Kenya so different from other exotic destinations is its wildlife. And where better to witness nature's pageant than in the Amboseli Reserve – a truly magnificent setting beneath the snowy summit of Mt. Kilimanjaro – a photographer's dream. But if your budget allows, do take the optional tour to Treetops, the world famous tree hotel, deep in the forested Aberdare mountains – it's a unique and memorable experience…

Mombasa is Kenya's premier resort offering wide stretches of soft white sandy beaches backed by feathery palms and casuarinas, its hotels set well apart from each other in their own secluded grounds. 80 miles to the north, Malindi is a smaller, more compact resort with a 'village style' atmosphere. Whichever you choose, swimming in the deliciously warm Indian Ocean is a delight. In season, hotels offer a variety of local entertainment and each is ideal for those seeking to relax, unwind and forget the pressures of the world…

25 Offices

Figure 25.1 Offices in the centre of London

ACTIVITIES

1 As a group, try to list a number of examples of each of the four types of office. Try to include some local firms or perhaps firms your parents work for.

The location of offices

Traditionally, the major location for offices has been in city centres. The aerial view of the city of London in Figure 25.1 clearly shows how offices dominate the scene. Many third world cities also have a concentration of offices at their centres (Figure 25.2).

A hundred years ago in Britain the pattern of employment was very different from today. The largest group of workers was domestic servants (1.25 million) followed by agricultural labourers (1 million) and labourers working mainly in factories (about 0.5 million). Today, increasing numbers of people will work in the services, largely in offices (see Figure 14.9). This is a trend which has been growing in Britain in recent years. In 1960 about 45 per cent of the working population worked in offices and by 1985 the proportion had risen to about 60 per cent. There are several different types of offices:

1 *Government offices.* These include Government departments such as the Ministry of Defence and the Department of Trade and Industry. Many are concentrated in London, being the centre of government for Britain. These deal with national, rather than local, or regional needs.
2 *Head offices.* These are concerned with the administration and control of large organisations or companies. Smaller offices, shops and industrial activities are controlled from head offices. They are often located in major cities like London and New York. For large organisations, like British Airways, the image and prestige of a location in a city like London is very important.
3 *Commercial offices.* These provide services for other offices and include activities such as banking, advertising and marketing.
4 *Services.* Job centres, estate agencies, solicitors and betting offices are examples in this category, they provide services to the public.

Figure 25.2 Centre of Mexico city

Offices tend to have rather different locational needs compared with industries. City centres are traditionally very accessible by public transport and car. This is important for office employees and clients. Offices also need links with other offices, for example the head office of an airline would benefit from being close to banks, insurance firms and advertising agencies. Shops are also available in city centres for office staff. Land rents are very high in city centres. Offices are one of the few land uses that can afford to pay such high rents. It is clear from the photograph of the centre of London (Figure 25.1) that many office blocks are tall — they are saving on land rents by building upwards rather than outwards.

In recent years a new type of office location has emerged, sited on the fringes of cities (Figure 25.3). These are often near motorway junctions and in areas where the surroundings are quieter and more attractive. This is important for office employees who may live nearby rather than having to commute long distances. Land will also be cheaper than in city centres and more space could be available for car parking and expansion.

Figure 25.3 Modern offices on the outskirts of Brentford, Essex

The pattern of office employment and the distribution of offices throughout England and Wales is not even. Although there has been a gradual growth in office development in England and Wales, one part of the country still remains dominant whether assessed in terms of employment, offices or amount of office space (Figure 25.4).

Figure 25.4 Employment in offices in England and Wales 1971

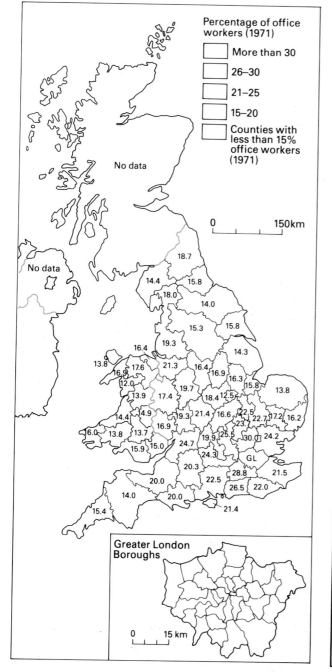

Percentage of office workers (1971)

- More than 30
- 26–30
- 21–25
- 15–20
- Counties with less than 15% office workers (1971)

No data

No data

0 ____ 150km

Greater London Boroughs

0 ____ 15 km

ACTIVITIES

2 a Briefly describe why most offices have traditionally located in city centres.

  b Why are many modern offices choosing to locate away from city centres?

3 a Study Figure 25.4. Which parts of the country have over 20 per cent of the workforce employed in offices? Name some of the counties with the highest percentages.

  b Name the parts of the country where less than 15 per cent of the workforce is employed in offices.

  c Try to explain the pattern of office employment you have described.

4 The following exercise is based on a copy of Figure 25.4.

  Complete a choropleth map using the data in Figure 25.4. Use the following key to shade in the map.

  less than 15% Yellow   15–20% Orange
  21–25% Red   26–30% Dark brown
  more than 30% Black

(Questions **3 a**, **b** and **c** could be answered after completing this map.)

Office location in the south east and London

The south east and London remains the main area for offices in England and Wales. London, however, has many problems as far as offices are concerned. For example, there are very high rents, a lack of space for expansion and car parking, many employees have to commute long distances and the surrounding urban environment is often noisy and unattractive. It is not surprising that many offices have moved away from the city. However, London retains many offices because of its importance as the capital city in Britain and one of the world's major cities. It is a centre of government, finance and commerce.

Since the completion of the M25 in 1986, many new office developments are taking place on the fringe of the capital (see Figure 25.5). It is, however, not always easy to build on the fringes of major cities as conflicts often arise with Green Belt restrictions. Another recent trend has involved offices being built in newly-developed inner city areas such as London's docklands.

Figure 25.5 The motorway network around London and areas of development pressure

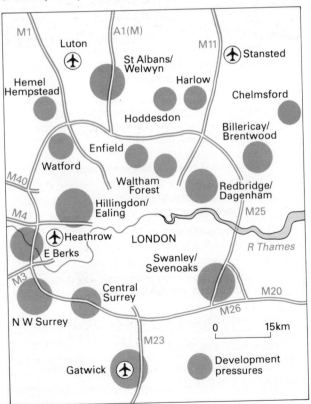

CASE STUDY

Office location outside London

There are a number of centres outside London where office development is expanding. These areas include Manchester, Liverpool, Birmingham, Chester, Ipswich, Swansea and Peterborough. Definite efforts have been made to move some Government offices away from London to the provinces. A good example is the Driver and Vehicle Licensing Centre in Swansea. This process is called **decentralisation**. Many major firms are also aware of the problems of having offices in London and of the advantages of making a move to other parts of England and Wales.

The East Midlands is one area where office development has been expanding. Peterborough Development Corporation has advertised extensively in the Press and on television, the advantages to be obtained by locating in Peterborough (Figure 25.7).

Figure 25.6 Newspaper advertisement for Thurrock

THURROCK
THE EUROPEAN LINK

. . . The right location for a European headquarters in the UK.

If you're looking for a base from which to reach all of the UK and Europe your answer must be Thurrock. Situated on the north bank of the Thames and bordering the eastern edge of London, Thurrock combines miles of pleasant English countryside with modern centres for industry and commerce. Yet London's financial centres in the City are less than 18 miles away.

London's new orbital motorway, the M.25, links Thurrock into the whole of the UK motorway system giving easy access to the whole country.

The same motorway means that London Heathrow Airport is less than one hour's drive while London Stansted and London Gatwick can be reached in less 40 minutes.

Even closer is the new City Airport in London's docklands only 10 miles along a main highway.

The frequent railway service means that London's financial centres are only 20 minutes away while an interchange with the new docklands railway will bring London's new business centres within half an hour of Thurrock by train.

The giant Tilbury Docks, the UK's largest and best equipped container port, is at the centre of Thurrock's busy 17 miles Thames waterfront. From here some of Europe's major ports are less than a day's sailing away.

Thurrock has a large workforce of nearly 100,000 people, comprising around 15% management and professional, with the remainder divided equally between intermediate and non-manual, skilled manual and semi/unskilled manual.

Figure 25.7 Newspaper advertisement for Peterborough

ACTIVITIES

5 Compile a table showing the advantages and disadvantages for office location in London.

6 Thurrock in Essex is one area on the fringe of London where Green Belt restrictions are less severe. Study the advertisement (Figure 25.6). Imagine you are an overseas investor. Explain why Thurrock might be a good place in which to work and live and build new offices.

7 Explain what you understand by the term **decentralisation**.

8 Add to the map you may have drawn as part of Activity 4 on page 63 the major centres outside London where office development is expanding.

9 Study Figure 25.7 and list the advantages for office location to be obtained by locating in Peterborough.

Coursework ideas

There are a number of ways in which offices in a town or city could be studied.

1 Find out why the offices are located in a particular part of a town or city. This would involve using a questionnaire.

2 The distribution of different types of offices could be mapped and then explained.

3 Many types of offices such as travel agencies, have increased in number in town and city centres in the last few years. It would be interesting to use maps to see how the numbers, types and location of offices had changed over a particular period of, say, 20–30 years. GOAD town and city maps would be helpful. Your teacher should be able to tell you how these can be obtained.

It is important to choose a town or part of a city that is not too large.

26 Industrial regions

Before the Industrial Revolution in Britain, industry was scattered and small-scale. Many industrial processes, like weaving were carried out in people's homes; this was called **cottage industry**. Today, industry is largely concentrated in industrial regions. The stages of this gradual change are shown in Figure 26.1.

A Replica of Hargreaves' "Spinning Jenny"
Development of Machinery

1740–1900 – factory interior
Purpose built to house Machinery

Railway, canal and industry
Transport helped to concentrate industry

City in industrial revolution
Growth of cities like Manchester

Certain industries came to dominate certain parts of Britain; as is shown in Figure 26.2. However, in the twentieth century, and particularly in the post war period, these traditional industries, like textiles, declined dramatically. This decline has largely resulted from the recent industrialisation of the Newly Industrialised Countries (NICs). Countries like Singapore and Taiwan are using modern technology and need fewer British exports. As a result, the British regions that depended on these industries have experienced serious social and economic problems. Perhaps the most serious is high levels of unemployment.

Modern industries are also less reliant on coal as a power source. Instead, electricity, oil, or gas are largely used which are more widely available. Industry, no longer tied to the coalfields, has been free to locate elsewhere in the country.

ACTIVITIES

1 a Make a copy of Figure 26.2. Mark on the position of Britain's coalfields using Figure 17.11 on page 37. Then use different colours or types of shading and add the other information.

 b Where was the major industrial region in the mid-nineteenth century?

 c Which parts of Britain had very little industrial development in the mid-nineteenth century?

 d What do you notice about the distribution of the industrial regions and Britain's coalfields? Briefly explain why.

Figure 26.1

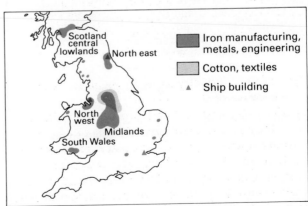

Figure 26.2 The regional distribution of major industries in the 19th century

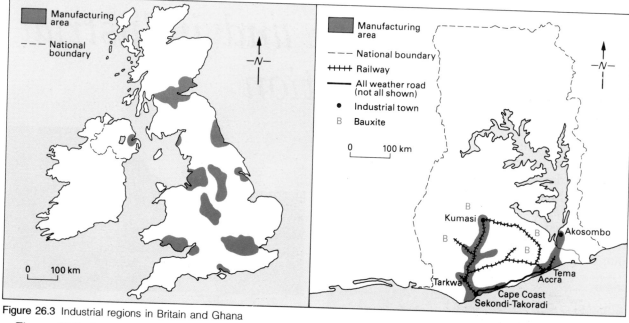

Figure 26.3 Industrial regions in Britain and Ghana

Figure 26.3 shows the main industrial regions in Britain today. Many of the old industrial regions are still important. However, as you can see, new ones have emerged. These are based on twentieth century growth industries like consumer goods, such as washing machines, fridges, electronics and vehicle manufacturing, particularly in south east England. The finished product from these industries is often more fragile and expensive than the components or raw materials. These industries, as a result, have tended to locate close to markets, which can also provide a source of labour. The older regions are now trying to attract some of these growth industries too.

Despite the fact that conditions have changed so much for the old industrial regions, some industries find it is uneconomic to move. The pottery industry around Stoke on Trent in Staffordshire is a good example; it has a long history dating from the seventeenth century, the name of Wedgewood particularly, has an association with fine china manufactured in the area.

The pottery industry grew in the area around Stoke because of the availability of the raw materials, clay and coal and good communications in the form of the Trent and Mersey canal. However, increasingly, the industry had to bring in raw materials from a long way away, for example, china clay from Devon. It would have seemed sensible to move to reduce transport costs, yet the industry remained in Staffordshire – this is called **industrial inertia**.

The main reasons for industrial inertia are as follows:

1 large amounts of money have often been invested in buildings and factories;
2 a pool of skilled labour close to the industry exists. It would be difficult to replace;
3 money has often been invested in infrastructure – roads, railways, waste disposal, for example.

Industrial inertia is very common in developed countries like Britain and the USA where there are many areas of long established industries.

ACTIVITIES

2 What do you understand by the term 'industrial inertia'?

3 Study Figure 26.3.

a How many industrial zones are there in Britain and how many in Ghana?

b In Britain and in Ghana, are the industrial regions concentrated in one particular part of the country, or are they fairly evenly spread?

c In which country do lines of communication seem to have the greatest influence on the shape of the industrial regions?

d How have mineral and energy resources affected the distribution of industrial regions in both countries?

e In both countries, some of the industrial regions tend to be near the coast. Why do you think this is so?

27 Government and industrial location

We have seen in Chapter 26 how industrial regions grew in Britain largely based on one or two particular industries and near major coalfields. For various reasons, many of these traditional industries, like shipbuilding and iron and steel, have declined in the twentieth century. The regions where they were once important then went into decline and have experienced much

Figure 27.2 Old industrial buildings in Rochdale

Figure 27.1 Inner city Manchester – Moss Side

unemployment. Figures 27.1 and 27.2 give some idea of other problems such regions face. Other parts of Britain, particularly the south east, have at the same time grown, often at the expense of the declining regions. Such growth is the result of labour and industry moving to these more prosperous and dynamic areas.

Since the 1930s high levels of unemployment in the problem regions, forced the Government to intervene. A number of Government Acts have followed with the aim of controlling growth in parts of the country like the south east and encouraging growth in areas like the north east and south Wales. The measures are often called 'carrot and stick' measures; the cartoon in Figure 27.3 will help you to understand why.

The Government has divided the problem regions in Britain into two types of **Assisted Areas**. **Development Areas** obtain the greatest amount of assistance from the Government and the **Intermediate Areas** the least. Figure 27.4 shows where these areas are in Britain. and some of the different forms of assistance that are available to firms that move there.

'Carrot' measures Incentives	'Stick' measures Disincentives

Some 'carrot and stick' measures

New factories available – rent free for two years	Loans available at low rates of interest
Higher levels of taxation	Grants towards moving costs
Grants towards the cost of new buildings and machinery	No loans or grants available for expansion
Difficult to obtain permission to expand existing factories	Tax allowance on the cost of new machinery

Figure 27.3 Cartoon illustrating the idea of 'carrot and stick' measures

Figure 27.4 a) Assisted Areas in Britain (DTI – Department of Trade and Industry)

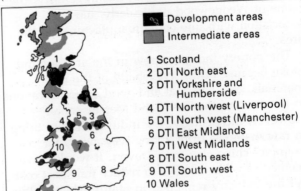

- Development areas
- Intermediate areas

1 Scotland
2 DTI North east
3 DTI Yorkshire and Humberside
4 DTI North west (Liverpool)
5 DTI North west (Manchester)
6 DTI East Midlands
7 DTI West Midlands
8 DTI South east
9 DTI South west
10 Wales

	Intermediate areas	Development areas
Government built factories	Various	
Factory buildings (RDGs no longer available after 31.3.88)	25–35%	up to 45%
Government loans available	No	No
Grant to clear derelict land	100%	100%
Training grants	Various	
Grants for moving & housing key workers	Yes	Yes
Other grants for training centres	Yes	Yes

RDG = Rural Development Grant

Figure 27.4 b) Some of the incentives available in Assisted Areas

---

ACTIVITIES

1 Study Figures 27.1 and 27.2. List the problems that a declining industrial region experiences. You might be able to think of more than those shown in the photographs.

2 a Make a copy of Figure 27.3a illustrating the idea of 'carrot and stick' measures.

   b Decide from the list of measures in Figure 27.3 b which ones are intended to attract industry to an area and which are intended to discourage. Write each measure in the correct place on your copy of Figure 27.3a.

   c What are 'carrot and stick' measures, and why are they needed?

3 a On an outline map of Britain make a copy of Figure 27.4. Use different colours to show the different types of Assisted Area.

   b Add the major cities or towns in each Assisted Area to your map.

4 How does the distribution of the Assisted Areas relate to Britain's coalfields and traditional manufacturing areas? (see Figure 17.11 on page 37).

5 Why do you think different types of Assisted Areas are needed?

6 With your partner, or in larger groups, try to list the likely problems and advantages faced by industry in inner city areas. What problems might be caused by the presence of industry in the inner city?

7 How are Enterprise Zones different from Assisted Areas?

(Inner Cities are dealt with in detail in the *Settlement* book.)

In 1980 the Government set up **Enterprise Zones**. These are concerned with inner city areas only. Industry has been moving away from inner city areas for some time, being attracted to the Assisted Areas and new towns. This has led to high levels of unemployment and decaying urban/industrial landscapes in many inner cities. Similar measures are available in Enterprise Zones as in the Assisted Areas, designed to attract industry. They have been very successful in starting new industrial growth in many urban areas in Britain.

CASE STUDY

Corby – a success story

In the 1970s, Corby in Northamptonshire was very much a steel town. Many of its residents and much of its economy relied on the fortunes of the steelworks located there. Then came what could have been a disaster – in 1980 the main steelworks in Corby closed. Thousands lost their jobs. Corby was immediately made an Assisted Area with all the special grants and financial incentives that this involves. Work quickly began to clear away old industrial buildings (Figure 27.5) and to create new sites for modern growth industries. Many of the new areas were **green field** sites; thousands of square metres of open countryside for industrial development (Figure 27.6).

Figure 27.5 Steelworks being demolished (Corby)

Figure 27.6 Green field site on the outskirts of Corby, Northamptonshire

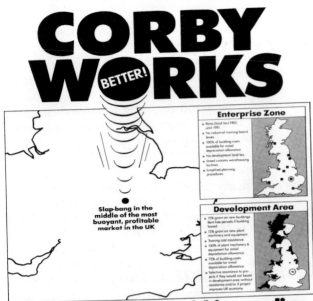

Figure 27.7 A newspaper advertisement for Corby

In June 1981 Corby made history in becoming England's first Enterprise Zone. The Enterprise Zone has attracted so many new companies and used the available land so quickly that in 1984 it was voted by a Government survey as '... the most successful Enterprise Zone'. More than 250 companies have made Corby their home. They include a Weetabix plant, the Oxford University Press national distribution centre, Avon Cosmetics and many hi-technology and electronics companies. Five thousand jobs have also been created.

Corby is also an example of a **new town** (see *Settlement* book) but one of the older and well established examples. It has an attractive and modern centre, a wide range of housing types, new schools and many recreational facilities. Corby is also surrounded by some of the finest countryside in England. Apart from its attractions as a town and the incentives available due to its Assisted Area and Enterprise Zone status, Corby has a number of other important locational advantages for industry as the advertisement in Figure 27.7 shows.

Coursework ideas

If you live in or near an area where industries have been attracted by Government assistance, you could base a study on this. Try to discover the types of industry attracted to the area. Have any large retail outlets been established? (This might have had an effect on a nearby town centre). You could visit the different companies and find out why they located in the area and in what ways they were attracted by assistance from the Government or local agencies. The area might be an Enterprise Zone or an industrial estate. You might contrast several areas, perhaps a modern Enterprise Zone with an area of older industry.

28 Unemployment

We all hope when we leave school to find a job. Sadly, though, unemployment is one of the most serious social and economic problems in Britain today. In 1988 about three million people in Britain were unemployed (Figure 28.1). However, as Figure 28.2 shows, the problem is more severe in some parts of the country than others.

Figure 28.1 A job centre

Figure 28.2 UK unemployment January 1988

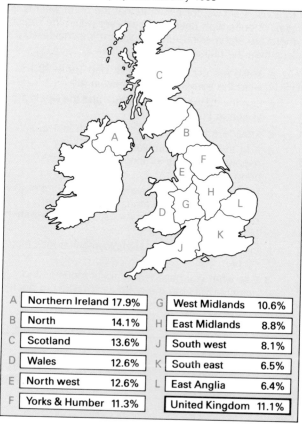

A	Northern Ireland 17.9%	G	West Midlands	10.6%
B	North 14.1%	H	East Midlands	8.8%
C	Scotland 13.6%	J	South west	8.1%
D	Wales 12.6%	K	South east	6.5%
E	North west 12.6%	L	East Anglia	6.4%
F	Yorks & Humber 11.3%		United Kingdom	11.1%

CASE STUDY

Unemployment in the north west

Since the Industrial Revolution, the north west has been an important area of manufacturing industry. In the twentieth century, these industries included chemicals, metal manufacture, engineering, vehicles, textiles and associated industries. They were particularly concentrated in areas like Liverpool, Widnes, Runcorn, the Wirral, Bolton, Bury, Oldham and parts of Manchester (see Figure 28.3). To varying degrees, these industries have all contracted during the recession of the late 1970s and early 1980s. The demand for some of the goods manufactured by these industries has also fallen in the latter half of the twentieth century. Increasing foreign competition has also had a damaging effect. One of the results has been very large numbers of unemployed left with few alternatives for employment.

Figure 28.3 The North West

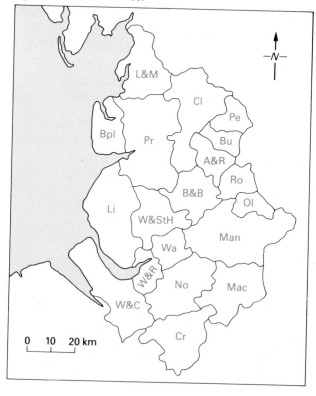

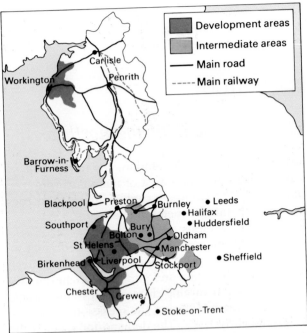

Figure 28.4 Development and Intermediate Areas in the north west

Produce a choropleth map based on the statistics in Figure 28.2

1 a Draw an enlarged copy of the map of the UK and its regions, with the letters in each region.

b Use an atlas to locate the major cities or towns in each region, e.g. Birmingham in the West Midlands.

c Use the following key to shade in the map:
0–9% Yellow   9.1–12% Red
12.1%–14% Brown   14.1%+Black

d Add the UK average figure to your map as shown in Figure 28.2

e Provide a key to identify the regions lettered A to L, and a key for the employment percentages using the information under c.

f Add the title Regional variations in unemployment – UK 1988.

g Write a few paragraphs to describe the pattern of unemployment in the UK. In a group, discuss some possible reasons for the pattern you have described.

The map you have produced is in some ways misleading as it hides variations in unemployment within regions. It shows, for example that the north west is an area of quite high unemployment, yet within the region there are areas where unemployment is considerably less than the national average.

2 a Make a copy of Figure 28.3. Use Table 28.1 to write the name of each region in full.

b Use the statistics in Table 28.1 and the key in 1 c to shade in the map.

c Provide a key using the information in 1 c.

d Add the title A choropleth map showing regional variations in unemployment in the north west 1988.

e List the areas where unemployment is higher than 14 per cent and the areas that have been established as Assisted Areas by the Government (Figure 28.4).

f List the areas where unemployment is less than 10 per cent.

g Use what you have learnt about industry and unemployment in the north west region to briefly account for the answers you have given to 2 e and f.

There has, however, been a notable growth in services, such as banking, administration and education. Modern industries have also grown, such as electronics, consumer goods and hi-tech industries. This has particularly benefited Manchester and the area to the south of the City. They have been partly attracted by female labour released from the contracting textile industry. Government assistance has also been available in several parts of the north west where economic problems have been severe (see Figure 28.4 and Chapter 27).

Travel to work area	Unemployment rate* %
Accrington and Rossendale	10.0
Blackpool	14.7
Bolton and Bury	13.4
Burnley	11.0
Clitheroe	5.0
Crewe	9.3
Lancaster and Morecambe	13.6
Liverpool	18.7
Macclesfield	6.2
Manchester	12.3
Northwich	10.5
Oldham	12.4
Pendle	10.2
Preston	9.3
Rochdale	12.6
Warrington	10.3
Widnes and Runcorn	17.0
Wigan and St Helens	16.2
Wirral and Chester	16.3

*The number of unemployed as a percentage of the estimated total working population.

Table 28.1 Unemployment in the north west, January 1988

29 Industry in the Third World

Most developing countries in one sense share a common background; they were once colonies. Africa and South America particularly were divided up as colonies of powerful European nations such as Britain, Spain, France and the Netherlands. Many countries were exploited as sources of slaves, cheap labour, minerals and agricultural produce.

The majority of developing countries have now achieved independence. In Africa, for example, many became independent in the 1950s and 1960s. Some have found it difficult to manage their own affairs through having been administered by overseas nations for so long. They face many other problems too. It is particularly hard to bridge the gap that has developed between the rich and poor nations. Industrialisation is often seen by newly independent countries as the answer to quick development and wealth. However, developing countries lack so many of the basic requirements for industrial and economic growth.

Problems of industrialisation in the Third World

1 *Money*. A shortage of money or **capital** is perhaps the most serious problem. One of the main ways for a country to earn money is through its exports and investments overseas. Yet many developing countries do not have enough money to invest and to build up manufacturing industry to create exports. This **poverty trap** can be shown as a simple diagram (Figure 29.1). One way out of the trap is to borrow money from developed countries or international financial organisations like the World Bank. Some countries, like Brazil and Mexico, have industrialised in this way. However, much income is taken by foreign multi-national corporations. Serious difficulties have also been created by rising energy costs and foreign competition for their goods. Many countries are now in debt often amounting to many millions of pounds (Figure 29.2).

2 *Markets*. Both domestic and overseas markets have limitations for developing countries. The

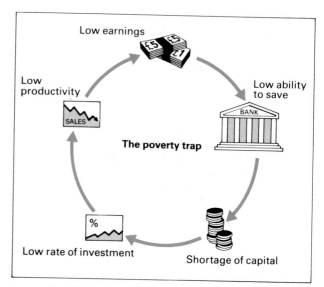

Figure 29.1 The poverty trap

Figure 29.2 Countries with the largest national debts

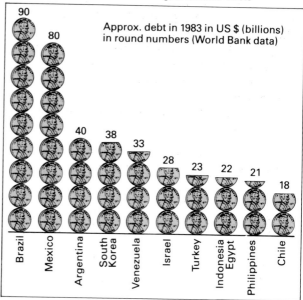

Approx. debt in 1983 in US $ (billions) in round numbers (World Bank data)

domestic market, as in Brazil, may be large but, as so many people are poor, they have little money with which to buy goods. Developing countries also find it very hard to break into overseas markets and compete with countries like the USA and Japan. Developed countries often have **trade barriers** which are designed to make imports expensive and encourage the purchase of their own goods.

3 *Labour.* Labour in many developing countries is often cheap, however, it is largely unskilled. Managerial and administrative expertise is also lacking. There are shortages of training colleges and universities. Young people often feel duty bound to work for money to help their families, rather than to take time to learn new skills. Factory work is often difficult for people brought up in open fields or small scale industry. Imagine how people in Britain would feel if they had to suddenly work in fields rather than air-conditioned, clean offices.

4 *Infrastructure.* Industrial growth is dependent on good communications and the availability of energy. Both are poorly developed in most developing countries. Roads and railways often only link ports to inland mining centres or agricultural regions. This is often a reflection of a colonial past (see Figure 18.5 on page 41). Port facilities are usually inadequate and insufficient energy is available for modern industrial growth.

Developing countries that have overcome some of these limitations have been able to break out of the poverty trap. For example, since oil was discovered in Nigeria, it has been used to encourage industrial growth. Exports of Nigerian goods and oil too helps to generate capital which can then be further invested. It should be clear by now that money answers many of the problems faced by developing countries. Yet, in the form of debt it can also create some of the most serious problems.

Figure 29.3

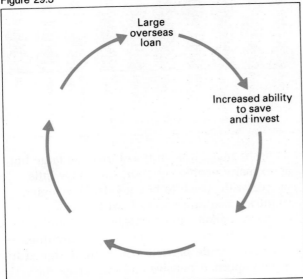

Some countries like Tanzania have deliberately chosen a different approach. Small scale rural industries have been encouraged based on traditional skills and employing local people. Encouragement is also given to agriculture. Although the pace of development may be a little slower, the final outcome is perhaps more healthy, in that employment possibilities are created for more people and a huge debt is not built up.

ACTIVITIES

1 Explain why developing countries often build up large debts.

2 Make a copy of Figure 29.3. It shows how a large sum of money might help a country break out of the poverty trap. Using Figure 29.1 to help you, complete your diagram.

3 In groups of about three or four try to suggest some ways in which developed countries could help developing countries to industrialise. Be careful not to suggest ways that might be harmful in the long term. Write down the three main ideas decided by your group. Now, as a class, try to decide by discussion and a vote which are the three best ideas.

CASE STUDY

Ghana

Ghana is situated in West Africa and has a population of about 14 million (1987). It gained its independence from Britain in 1957.

Ghana made a number of serious mistakes in the early years of its independence. To try to achieve rapid growth money was spent on prestigious projects and large scale industries. For example, the aluminium smelter and oil refinery at Tema. These are capital intensive industries that created very few employment opportunities for Ghanaians. The jobs that were created were largely of the wrong type for an unskilled, poorly educated labour force. The domestic market was also not growing as fast as it could because few people were earning high wages.

Ghana also lacked adequate supplies of energy and a sound infrastructure. So money was spent on the Volta Dam HEP scheme, a new artificial harbour at Tema an a motorway between Accra and Tema; all at enormous expense. The motorway today is hardly used. The other two projects have proved to be more successful. Not surprisingly though, Ghana soon amassed a huge debt.

Virtually all of the development that occurred since independence has been concentrated in the south. The country has, in effect, been divided into two. Strong rural to urban migration has taken place as people move to where they believe the jobs are and where development is taking place.

Only through very strict Government controls, on wage levels and inflation, for example, has Ghana been able to develop a more stable economy.

ACTIVITIES

4 Read through the newspaper article in Figure 29.4 and, with the help of the text and Figure 18.5 on page 41, list the problems and mistakes that Ghana has suffered.

5 Study the map of Ghana on page 41 (Figure 18.5).
   a Describe the pattern of roads and railways in Ghana.
   b Try to suggest reasons for the pattern you have described. An atlas will help to give you some ideas.

Figure 29.4 Sombre mood as Ghana celebrates 30 years

Sombre mood as Ghana celebrates 30 years

THIRTY YEARS ago Ghana was an African beacon. The first country south of the Sahara to win independence, it symbolised the hopes of millions of Africans across the continent who were seeking an end to the colonial rule of Britain and other European powers.

Today Ghana marks its independence anniversary in a sombre mood which reflects the state of a continent which has successfully severed the colonial ties, but now battles against drought and famine, Aids and civil wars, corruption and mismanagement, deteriorating terms of trade and a crippling external debt.

It is ironic that Ghana today is a very different example on the continent: an example of how a government, with the advice and financial support of the International Monetary Fund and the World Bank, might reverse years of decline.

Despite some signs of revival, the centre of Accra still resembles a waste land, while sea front monuments such as the huge Black Star Square stand as a souvenir of misallocated resources and grandiose ambitions of the early 1960s.

When Ghana gained independence on March 6, 1975, under President Kwame Nkrumah, it was the richest and most developed of Britain's African colonies. But it quickly slipped down a long spiral of economic and political decline marked by corruption, inefficiency, misguided policies and military coups.

President Nkrumah was overthrown by a military coup in 1967 and his policy of Pan-African socialism abandoned, but the situation continued to deteriorate during the 1970s under a succession of military rulers.

A bloody purge and a return to civilian rule following Mr Rawlings' "first coming" in June 1979 failed to stop the rot.

As a result his return two-and-a half years later was termed a "revolution" intended to "change the character and face of the nation."

The main aims were to rescue the economy, curb the abuse of power and corruption, redistribute wealth and encourage popular participation in government.

But the new radical regime was forced by the country's effective bankruptcy to adopt a policy of economic pragmatism and turn to the IMF and World Bank for assistance.

Over the past three-and-a-half years it has implemented one of the most radical economic reform programmes in Africa. In a continent where IMF/World Bank successes are rare it is seen as an important test case.

After three successive years of economic growth the country has moved from rehabilitation and crisis management towards structural adjustment and sustained growth.

Three years ago the economy was in deep crisis. Visitors to Accra stayed in candlelit hotels as there was electricity only every other day. Bathtubs were filled with stagnant water as a precaution against frequent water cuts. There was virtually nothing to eat in the dining room and visitors often brought their own provisions of tinned sardines and Nescafe. Few cars were on the streets due to petrol shortages and most telephones did not work.

Visitors now have an easier time, though lampbulbs are still scarce in bedrooms and butter and marmalade are carefully rationed at the breakfast table of Accra's main state-owned Ambassador Hotel.

A lot remains to be done before Ghanaians recover their real incomes of the 1960s. But for the time being Ghana continues to enjoy unaccustomed stability and prospects of economic improvement.

Financial Times 26 November 1982

30 Appropriate technology

In 1986 a severe famine affected much of the Horn of Africa. Thousands of nomads were transported by air to fertile land in Somalia. They were provided with modern new houses and some of the best farming equipment available. However, today the area is deserted. Many of the nomads returned to the bush to continue their old way of life. Tractors, combine harvesters and bulldozers lie rusting and the buildings are abandoned, some hardly used. What went wrong?

These people were suddenly expected to make the change from a very simple way of life to one based on modern 'western' technology. They lacked the skills and knowledge to make such a change. Not surprisingly the millions of pounds invested were wasted – the technology was **inappropriate**.

In developing countries **appropriate technology** often has to be used. It is of little use, for example, to provide tractors for people who have not been shown how to use them and, more importantly, do not know how to repair them when they break down. Spare parts are often expensive too, and may take a long time to obtain from abroad. Here are two examples of technology appropriate to solving a particular problem and meeting people's needs.

CASE STUDY

Wind pumps in Kenya

When the wind pump shown in Figure 30.1 was put up near Lake Turkana in Kenya, local police chiefs had to be called in to control the crowds. Thousands of tribes people in an arid part of Kenya suddenly had a supply of fresh water. Now it serves about 4000 people each day. Designed for heavy duty, it has a life span about 20 times that of a diesel pump.

Trained Kenyans are now making wind pumps like the one in the photograph. Similar wind pumps and workshops where they can be made have been set up in other parts of Africa and elsewhere in the world.

CASE STUDY

Fishing boats in India

Fishing boats in the village of Quinlan in southern India have traditionally been made of wood from the giant mango tree. However, when it became illegal to cut these trees down, the 5000 Quinlan fishermen and their families who depend on the sea for their living became desperate. Modern, new boats with engines would have been too expensive and difficult to maintain. The old boats were being patched up and becoming increasingly unseaworthy.

A British company called 'Intermediate Technology' studied the problem and as a result produced some new boats made of plywood which is available locally. The simple, low cost design was much like the traditional boats but safer and easier to handle. Local villagers have been trained to repair and maintain the new boats which have enabled many families to continue with the only way of life they know.

Clearly governments have a key role to play. Their decisions will influence the pace of development and the sort of technology used. Mistakes could be very costly, vast sums of money might be wasted adding perhaps to an already large national debt. This in turn affects the whole

Figure 30.1 Windpump near Lake Turkana, Kenya

country, the reputation of the Government and the wealth of its people.

Appropriate technology is particularly important for small scale activities. We have seen in earlier chapters that large scale industrial activities, such as iron and steel works, car manufacturing and tourism, are possible in developing countries. The pace of development does not have to be slow.

Figure 30.2 a) A set of carpenter's tools designed in Africa and made from local materials. Cost £38 (imported cost £240)

Figure 30.2 b) Bikes an intermediate form of transport often used to carry goods in the third world

31 Multinationals

A **multinational corporation** (MNC) is one which is usually based in a developed country such as Britain or the USA. It is often made up of many separate firms. Many of its operations or factories are located overseas, often in developing countries. A lot of industry in developing countries is owned by MNCs (Figure 31.1). A few examples include large oil companies like Shell or BP with refineries around the world and car firms like Ford and Volkswagen with factories in countries like Brazil.

Figure 31.1 Some multinational companies operating in Nigeria

Peugeot	Beechams	Colgate/Palmolive
Seven-Up	Longman	ICI
Mobil	Tate and Lyle	Cadbury
ICL	Ovaltine	GEC
Taylor Woodrow	Michelin	EMI

Virtually every major industry has overseas multinational involvement. The presence of MNCs in developing countries, however, has been controversial. There are clearly two points of view; that of the multinational and that of the developing country.

The multinational's point of view

There are several major attractions for multinationals in developing countries.

1 *Cheap labour*. While wage levels have soared in countries like Britain and West Germany, they remain relatively low in the developing countries. MNCs frequently locate the most labour intensive part of their operations overseas to take advantage of cheap labour (Figure 31.2). Even in Western Europe, firms use Spain and Portugal in a similar way (see Chapter 21).

Figure 31.2 Manufacturing in Singapore

2 *Markets*. There are often large markets for goods produced by MNCs. Advertisements for products that we are familiar with in Britain can often be seen in Third World cities (Figure 31.3).
3 *Raw materials*. Many developing countries have huge reserves of minerals and timber that are highly valued in the developed countries.

Figure 31.3 Advertisements in the Third World

The developing country's point of view

Developing countries frequently see industrialisation as the path to progress and rapid development. However, they lack many of the needs of modern industry that are present in developed countries (see Chapter 29). Therefore, industrialisation is often slow. Many newly industrialising countries recognise that MNCs can provide the capital, skilled labour and many other things that they lack. They can therefore develop more rapidly, and benefit from the products being manufactured and the employment opportunities created.

One further advantage is called **technology transfer**. The developing country can benefit from the technology and skills brought by the multinational. Its own people can acquire new expertise and thus a developing country slowly builds up a skilled labour force. When the Volta Dam was built in Ghana (see Chapter 8) the American firm involved trained Ghanaians to understand the scheme and its operations.

However, many developing countries have come to mistrust MNCs. Their advertising is often seen to be harmful. Poor people are often easily influenced by glossy advertisements, longing for the things rich people can afford. In many countries men have been tempted to spend part of their wages on junk food, drink and tobacco while their families go hungry. Alcoholism is a major problem in some South American countries.

A further problem is the way in which some MNCs take advantage of the lack of proper pollution control and safety measures in developing countries. This, of course, makes their operations cheaper than in a developed country. Some of the world's most polluted areas are industrial regions in the developing countries; places like Mexico City and the Rio–São Paulo industrial axis in Brazil (Figure 31.4).

Not surprisingly, perhaps, some developing countries have restricted MNC involvement or at least laid down strict controls.

ACTIVITIES

1 Write a short definition of what a multinational corporation is.

2 Ford has major car plants in the following locations around the world. Use an atlas to find these places and plot them on an outline map of the world. Give your map a suitable title.

Auckland	Jaboatao
Buenos Aires	Lavilla
Brisbane	Melbourne
Cologne	Saarlouis (nr Saarbrucken)
Dagenham	
(east London)	São Paulo
Genk	St Thomas (nr London, Canada)
Halewood	
(Liverpool)	Valencia
Hermosillo	Wellington

3 Copy Figure 31.1 and underneath each company's name write the type of products with which they are linked. Work in pairs.

4 List the products that are being advertised by multinationals in Figure 31.3. Where possible, try to identify the company as well.

5 Explain briefly why MNCs locate part of their operations overseas.

6 Study the section on Tema steelworks in Chapter 18. Assume that a company wants to build a large steelworks, like the one at Tema, in a developing country. List some of the problems it would face by locating in a developing country.

7 Draw up a table with two columns. Head one column **Advantages** and the other **Disadvantages**. Then in note form summarise the argument for and against MNC involvement in developing countries. Give the table an appropriate heading.

Figure 31.4 Heavy smog and pollution in the Rio de Janeiro/São Paulo industrial axis in Brazil

Index